Raising Cane in the 'Glades

Raising Cane in the 'Glades

THE GLOBAL SUGAR TRADE AND THE TRANSFORMATION OF FLORIDA

Gail M. Hollander

The University of Chicago Press   Chicago and London

GAIL M. HOLLANDER is an associate professor in the Department of International Relations at Florida International University.

The University of Chicago Press, Chicago 60637
The University of Chicago Press, Ltd., London
© 2008 by The University of Chicago
All rights reserved. Published 2008
Printed in the United States of America

17 16 15 14 13 12 11 10 09 08   1 2 3 4 5

ISBN-13: 978-0-226-34950-3 (cloth)
ISBN-10: 0-226-34950-0 (cloth)

Library of Congress Cataloging-in-Publication Data

Hollander, Gail M.
   Raising cane in the 'glades : the global sugar trade and the transformation of Florida /
Gail M. Hollander.
       p.   cm.
   Includes bibliographical references.
   ISBN-13: 978-0-226-34950-3 (cloth : alk. paper)
   ISBN-10: 0-226-34950-0 (cloth : alk. paper) 1. Sugar trade—Florida—Everglades.
2. Drainage—Florida—Everglades. 3. Rural development—Florida—Everglades.
4. Sugar—Manufacture and refining—Florida—Everglades. I. Title.
   HD9107 .F6H65 2008
   338.1′73610975939—dc22
                                                                        2007035931

*To my parents, Julianne and Robert,
and to Rod, the light of my life.*

Contents

Illustrations

Tables

Acknowledgments

The germ of this book dates to my graduate studies in geography, but the roots go further back to the experience of being a teen-age transplant to south Florida from the Midwest. The discipline of geography in its myriad forms — political, feminist, developmental, economic, urban — has helped me in many ways to understand that landscape as a political-economic-cultural construction. But what initially compelled me was that my experience of south Florida did not fit well within the rural, environmental, or agrarian studies of my graduate years. I carried vivid memories of an agro-industrial landscape that did not accord with the "family farming" model presumed to characterize first-world agriculture. My first glimpses of plantation agriculture date to a nighttime bus trip across the southern portion of Florida during a drought. The Everglades, from the peaty-muck soil upwards, were burning. Flames shot up along both sides of the road, a tunnel of fire. Once through the flaming and smoke-filled swamp, the bus stopped in a handful of small towns, South Bay, Belle Glade, Clewiston, and Moore Haven. In the rural spaces between these towns were the prisonlike housing compounds, surrounded by chain link and barbed-wire fences, where the agricultural workers lived.

I am grateful to the many people I have met in the years since that bus trip who have contributed immeasurably to the development of this book. At The University of Iowa, Rebecca Roberts, David Reynolds, George Malanson, and Kathleen Newman each provided a key piece of the intellectual puzzle. However, no one has been more helpful than Marc Linder, who quite literally showed me the way to the next archive drawer. At Florida International University, I have been fortunate to have as colleagues a diverse group of scholars who have encouraged me in numerous ways: Mahadev Bhat, Ralph Clem, Francois Debrix, Damian Fernandez, Antonio Jorge, Ken Lipartito, Laura Ogden, Patricia Price, Lisa Prugl, and John Stack. I have been doubly fortunate to work with a wonderful group of graduate students: Cheeku Bhasin, Serena Cruz, Astrid Ellie, Monalisa Gangopadhyay Roy, Donna Goda, Rachel Martinez, Cristian Melo, Jan Solomon, Frances Spiegel, and Jason Weidner.

A geographically dispersed group of colleagues and friends have inspired me with their work: Altha Cravey, Mona Domosh, Margaret Fitz-Simmons, Susanne Freidberg, Julie Guthman, James McCarthy, Frank Magilligan, Brian Page, Nathan Sayre, Richard Schroeder, Dick Walker, and Michael Watts. Outside these academic circles are friends whose support was essential: in Iowa City, Jan Weissmiller, John Dilg, Hutha Sayre, and Bob Sayre; and in Albuquerque, David Weisberg, Natalie Marshall, and Julius Marshall-Weisberg. And then there are the friends to whom I am related, Julianne Hollander, Robert Hollander, Laura Rebol and Matthew Rebol.

My research in the environs of Clewiston and Belle Glade, Florida, was made possible by the kindness of strangers, especially Charlotte and Henry Drake, who not only let me live in a camping trailer in their yard but also introduced me to many farmers and ranchers in the area. The staff of the Clewiston Library could not have been more helpful. Judy Sanchez, the late Pete Rosendahl, and Barbara Meidema, of the United States Sugar Corporation, Florida Crystals, and the Sugar Cane Growers Cooperative of Florida, respectively, each gave generously of their time. At Florida International University, I have received invaluable help from three talented librarians: Stephanie Brenenson, Sherry Mosely, and Gail Clement. Carl Van Ness, curator of manuscripts at the University of Florida, provided his expertise in navigating the Department of Special and Area Studies Collections, especially the Braga Brothers Collection. At the National Archives in College Park, Maryland, Joseph Schwartz and Michael Hussey, civilian records archivists, were especially helpful. I would also like to thank Joan Bernhardt, who kindly provided permission to use as an illustration one of the political cartoons drawn by her mother, the late Anne Mergen. A timely and generous grant from the Florida International University Foundation provided support for a portion of the research. I am very grateful for the suggestions of three anonymous reviewers, which improved the manuscript substantially.

At the University of Chicago Press, I would like to thank Katherine M. Frentzel, Stephanie Hlywak, Dmitri Sandbeck, Matt Avery, Mara Naselli, and Tisse Takagi. I would also like to express my gratitude to Nick Murray, whose diligent copyediting of the manuscript improved it immeasurably. I am also grateful to Tom Willcockson for his cartographic expertise.

All along the way, Christie Henry's encouragement and her enthusiasm for this project have been critical to its success. Her patience and understanding made all the difference in the world.

As always, I've saved the best almost for last. I can begin to thank Rod Neumann, but it seems unlikely that I will ever finish, given the magnitude of his contribution to this book from its inception to completion. Other authors should be so lucky. He did get to choose the last word. Finally, I would like to express the opposite of thanks to Charley, Katrina and Wilma, hurricanes from which we have only just recovered.

Abbreviations

AAA	Agricultural Adjustment Act of 1933
ASCS	Agricultural Conservation and Stabilization Service
BES	Bureau of Employment Security
BMPs	best management practices
BWI	British West Indies
BWICLO	British West Indies Central Labour Organisation
C&SF Project	Central and Southern Florida Flood Control Project
CAFTA	Central American Free Trade Agreement
CERP	Comprehensive Everglades Restoration Plan
CIA	Central Intelligence Agency
CIEP	Council on International Economic Policy
DER	[Florida] Department of Environmental Regulation
EAA	Everglades Agricultural Area
EPA	U.S. Environmental Protection Agency
FFCA	Florida Flood Control Association
FRLS	Florida Rural Legal Services
FSA	Farm Security Administration
FSCL	Florida Sugar Cane League
FSES	Florida State Employment Service
FWRA	Florida Water Resources Act
GAO	General Accounting Office
GM	General Motors
HFCS	high fructose corn syrup
ISA	International Sugar Agreement
ISCGA	Interstate Sugar Cane Growers Association

IIF	Internal Improvement Fund
K-O-E	Kissimmee-Okeechobee-Everglades
LOTAC	Lake Okeechobee Technical and Advisory Council
NAFTA	North American Free Trade Agreement
NP	[Everglades] National Park
NSC	National Security Council
NSRB	National Security Resource Board
SFWMD	South Florida Water Management District
SOE	Save Our Everglades
STA	stormwater treatment area
SWIM	Surface Water Improvement and Management
UFWU	United Farm Workers Union
USDA	U.S. Department of Agriculture
USES	U.S. Employment Services
USSC	United States Sugar Corporation
USSEB	United States Sugar Equalization Board, Inc.
WRDA	Water Resources and Development Act
WTO	World Trade Organization

CHAPTER ONE

From Everglades to Sugar Bowl and Back Again?

Beginning in the 1980s, environmental activists identified south Florida's sugarcane agro-industry as the primary culprit in the deterioration of the remnant Everglades ecosystem, including Everglades National Park (NP). Within the context of the broad historical sweep of political discourse in the United States, environmentalists were not especially original in pointing an accusatory finger at the sugar industry. Populist political discourse has a long tradition of demonizing the sugar industry's "unfair" trade advantages and exploitation of the "little guy," going back at least to the 1880s and the formation of the American Sugar Trust. The movers and shakers of the industry behind the trust came to be known as "sugar barons," a label intended to classify them as a subclass of the infamous "robber barons" of American capitalism. Sugar barons operated in a complex network of growers, refineries, and financial institutions that stretched into the "marble antechambers and the smoke-filled back rooms of the politicians of several nations" (McAvoy 2003, 6). In south Florida, the industry and its owners have come to be known simply as "big sugar," represented by the two largest producers of sugarcane in the continental United States, Flo-Sun and the United States Sugar Corporation (USSC). Together with several other companies and a handful of independent farmers, USSC and Flo-Sun harvest 403,000 acres of cane and produce from seven mills 13,621 tons of sugar, or approximately 20 percent of the domestic sugar supply (USDA 2005).

Environmentalists were, however, significantly original in pinning on big sugar an environmental disaster of international importance. The Everglades ecosystem is said to be on the verge of collapse, as evidenced by the decline in species diversity, diminished numbers of native species, and invasion of non-native species in Everglades NP. Most of what Marjory Stoneman Douglas famously labeled the "river of grass" actually lies north of the park (Douglas 1988). Scientists and park managers became increasingly aware that land and water uses upstream were threatening the viability of Everglades NP and the 'Glades as a whole. The region is internationally recognized as a World Heritage Site, an International Biosphere Reserve, and a Ramsar Convention Wetland of International Importance,[1] and in 1980 a broad-based, statewide preservation movement emerged with the slogan "Save our Everglades." By the early 1990s the Florida sugar agro-industry was at the center of a contentious political struggle over the ecological "restoration" of the Everglades. The criticism directed at big sugar has focused not only on the negative environmental impact of cane cultivation, but also on the politics and economics of plantation production in south Florida. Price supports guaranteed by the U.S. Sugar Program insure stable, often above-market prices to growers, leading critics to argue that consumers and taxpayers are paying big sugar to destroy the Everglades. Any effort to understand the historical role of sugar in the diminishment and restoration of the Everglades ecosystem, therefore, must place the issue squarely within the political economy of food and agriculture in the United States.

Upstream and Downstream of Florida's Sugarcane Production Region

Just after New Year's Day, 1994, the local newspaper in Clewiston, Florida, alerted its readers that their community was going to be inundated. The warning came in the form of a letter, signed by more than forty local government officials, which explained that the U.S. Army Corps of Engineers had devised a plan to flood "the entire region south of the lake." The lake in question was Lake Okeechobee, and the plan to which they referred was part of a multiagency effort to reenvision the south Florida landscape with an eye toward restoring the Everglades. That week, galvanized by the news, hundreds of the town's residents turned out for a prayer vigil on the high school football field.

As it turned out, neither Clewiston—"America's Sweetest Town"—nor the fields of sugarcane surrounding it were flooded in the ensuing decade, nor are there plans to do so. But the event points to key elements of the once-contentious relationship between the community that owes its livelihood to the sugar agro-industry and a welter of initiatives concerning "Everglades restoration." As the impetus toward restoration took hold, it was conceptualized in terms of returning the Everglades to its historic state.

Thus the scientists on the advisory committee, who were asked to develop a range of alternative methods to further the goal of restoration, suggested as one possibility the return of wetland function to all former wetlands in current agricultural use. In doing so, they echoed a chorus of environmental activists and policymakers who were questioning whether the sugar industry belonged at all in the landscape of what had once been a portion of the Everglades.

But how, exactly, do the activities of sugarcane producers so significantly affect the downstream ecology? The answer requires an understanding of the fundamentals of water movement through the Everglades and the effects of a decades-long effort to control and redirect that flow. In considering the movement of water in the Everglades, systems ecologists have provided different perspectives on the geographical extent of the ecosystem (Davis and Ogden 1994). On the largest scale, the system comprises a drainage basin that extends from the Kissimmee River in the north through Lake Okeechobee and the freshwater marshes to the south, the wetlands of Everglades NP and Big Cypress Swamp, and finally the mangrove and salt marsh estuaries that empty into Florida Bay (fig. 1.1). The South Florida Water Management District (SFWMD), which was established to manage the distribution of the water for flood control, agriculture, and urban development, refers to this as the "greater Everglades system" or the K-O-E (Kissimmee-Okeechobee-Everglades) watershed. It is an enormous drainage basin, covering 10,890 square miles and parts of sixteen counties (fig. 1.2). Although the complexities of water flow in the system are the subject of ongoing research, there is an overriding north to south movement of water from Kissimmee to Florida Bay. Thus the quality and quantity of water reaching the protected areas of the Everglades NP, Big Cypress Preserve, and Florida Bay in the southern portion of the system are determined by "upstream"activities.

Historically, water moved through what we might call the "Everglades proper"—the wetlands stretching from the south shore of Lake Okeechobee to Florida Bay (fig. 1.3)—in an expansive sheet flow at a rate equivalent to "one-hundredth the speed of a leisurely walk" (Holling, Gunderson, and Walters 1994, 745). Geology and climate largely determined the system's hydrology. The geology consists of a gently southward sloping, flat, limestone plate overlain with peat and marl soils. A depression in the bedrock and the limestone ridges that restrict drainage to the east together create the geological conditions for the sheet flow and the periodic inundation of the Everglades' soils. These soils are subject to two counteracting

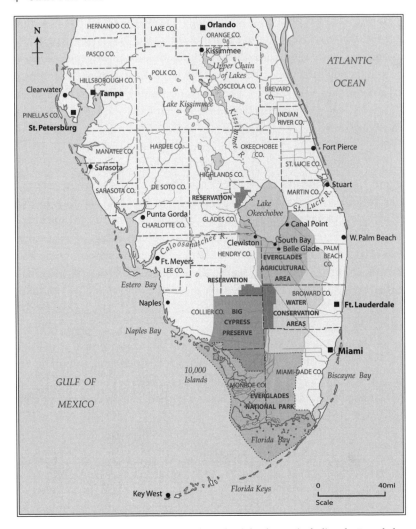

Figure 1.1. Key features of the south Florida wetlands landscape, including the Everglades Agricultural Area. Cartography by Mapcraft.com.

processes: deposition and oxidation. Oxidation, which occurs when soils are exposed under dry conditions, results in soil reduction at much greater rates than deposition, making "the Everglades ecosystem highly unstable with respect to disturbances in the hydrologic cycle" (DeAngelis 1994, 309). The hydrologic cycle is in turn the product of a regional climate marked by distinct differences in seasonal precipitation: most rainfall occurs during the summer and early fall, and there is an extended dry season from late fall through the winter. Thus, historically, the movement of water through

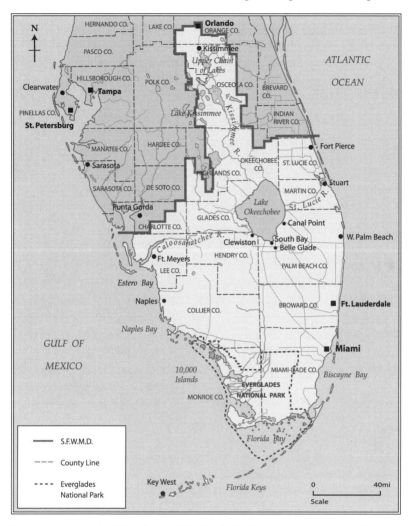

Figure 1.2. Geographical jurisdiction of the South Florida Water Management District. Cartography by Mapcraft.com.

the Everglades proper involved a great deal of temporal and spatial dynamism, resulting in a heterogeneous landscape and relatively high biodiversity (DeAngelis 1994).

A century of efforts to drain portions of the watershed by constructing canals, levees, and dikes has greatly altered this historic pattern. Beginning in the 1880s a parade of maverick real estate speculators, sugar agro-industrialists, progressive populists, conservative Republicans, and, ultimately, one very large federal agency, the U.S. Army Corps of Engineers,

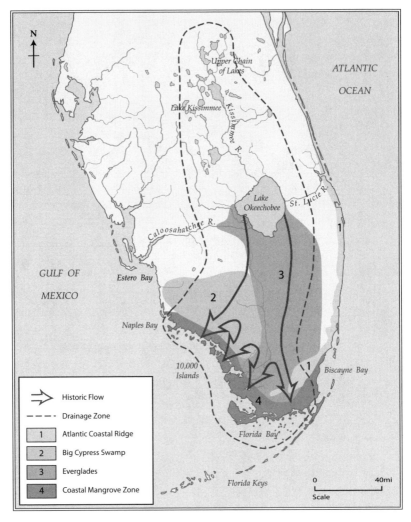

Figure 1.3. Historic, pre-drainage pattern of water flow in the Kissimmee-Okeechobee-Everglades. Cartography by Mapcraft.com.

effectively diverted most of the water from the downstream wetlands. Drainage, begun haphazardly in the nineteenth century and organized successfully in the twentieth, resulted in the construction of a "second nature" (Smith 1990; Cronon 1991), creating the premier sugar-producing region of the U.S. mainland. This environment of canals, pumps, and drained muck soils is demarcated as the Everglades Agricultural Area (EAA), comprising approximately 700,000 acres, where big sugar, as well as smaller, independent growers, now cultivate most of the 403,000 acres of cane grown in south Florida. The EAA lies adjacent to the south-southeastern shore of Lake Okeechobee, occupying a significant portion of the historic head-waters of the river of grass.

The "high modernist" (Scott 1998) solution for the problems caused by earlier piecemeal drainage took specific form in the Central and Southern Florida Flood Control Project (C&SF Project), which the U.S. Army Corps of Engineers began in 1948. After the Corps completed the perimeter levee and canal from Palm Beach to Dade County in 1952, the next stage, begin-ning in 1955, was to partition the agricultural and water-storage areas. When completed in 1962, the C&SF Project, which had installed enormous pump-ing stations to regulate water flows, gave the Everglades landscape a dis-tinctly industrial cast (fig. 1.4). The canals, levees, and pumps eliminated the historic hydrologic cycle and circumvented the underlying geologic basin structure (fig. 1.5). The C&SF Project was virtually synonymous with the es-tablishment and maintenance of the EAA, which encompassed most of the sugarcane agro-industrial region. The U.S. government, in collaboration with sugar capitalists, had fused environment, place, and commodity.

No sooner had the C&SF Project been implemented than critical anal-yses began to question the environmental and economic costs involved in the final transformation of the upper Everglades into the EAA. There were two central questions. The first concerned the validity of the cost-benefit analysis used to justify the project: Were the full costs of EAA development taken into account, and had the benefits been exaggerated? The second question concerned the ecological attributes of the region, specifically the uncertainty regarding the longevity of agriculture on drained muck soils: Was the entire production system sustainable for more than a few decades (Allison 1956; Ford 1956)?

Adding up all of the public investments in the EAA, including not only previous water-control infrastructure, but also publicly funded agricultural research stations, the Federal Farm Labor Housing Centers, and the Ever-glades fire-control district, revealed a tremendous, if not unprecedented, degree of state subsidization of private accumulation in the sugar industry (Ford 1956). The magnitude of the environmental transformation required and "the short term usability of the organic soils that support the sugar cane industry" (Ford 1956, 84) made this subsidization seem all the more mis-guided and irrational. The boundaries of the EAA are defined by the suit-ability of the soils therein for agricultural development. They encompass organic soils highly susceptible to shrinkage, compaction, and oxidation, and predictions at the time regarding subsidence—sound but somewhat overstated—suggested that existing EAA agriculture would be effectively

Figure 1.4. One of many pumping stations in the U.S. Army Corps of Engineers reengineered Everglades drainage. Courtesy of SFWMD.

Figure 1.5. Current pattern of water flow in the engineered Kissimmee-Okeechobee-Everglades system. Cartography by Mapcraft.com.

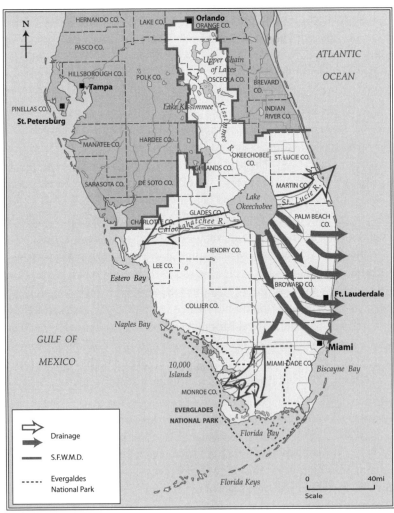

nonexistent by the year 2000. Producing a commodity already in surplus was economically irrational; instead, when productivity declined, "Restoration of natural conditions in the undeveloped units would then be undertaken insofar as possible" (Ford 1956: 120). Other observers at the time echoed this vision, suggesting that as the EAA became less productive, "transition into a wildlife area of ultimate world fame [would] follow" (Allison 1956). Thus, even as the Army Corps of Engineers finalized the C&SF Project to suit the needs of Florida's sugar industry, critics raised the possibility of ecological "restoration," a concept that would eventually come to dominate public discourse on the fate of the Everglades. Big sugar was and is still at the heart, geographically and politically, of the debates over the ecological condition of the Everglades and what to do about it.

The Sugar Question: Framing the Everglades Transformation

Many scholars and environmentalists tend to read the history of the Everglades backwards, to look at sugar production and the EAA from a contemporary perspective and think of them as "mistakes." In this reading, people were not previously aware of the ecological value of the Everglades, which they deemed "suitable only for the habitat of noxious vermin, or the resort of pestilent reptiles . . . now worse than worthless" (U.S. Senate 1911: 54). Only later, as environmental awareness developed — at a point in time usually unspecified — did the Everglades make the transition from worthless swamp to cherished wetland. Even the most recent historical treatment of the Everglades follows this tendency to identify a shift in the cultural valuation of wetlands and grant it causal power. The explanatory power of this view is limited and unsatisfactory on several counts. First, peoples' ideas about the Everglades have long been clouded by ambivalence and contradiction. Buckingham Smith, Esq., quoted above on the worthlessness of the Everglades, also commented on the "wild solitude of the place" and the "feelings bordering on awe" that it produced in him during his 1848 survey of the region (U.S. Senate 1911, 54). Second, the timing of the cultural shift from the notion of "swamp" to that of "wetlands" that allowed people to see the folly of drainage is, by implication, quite recent. Yet in his 1920 geography of south Florida, the naturalist Charles Torrey Simpson observed that although "only the preliminary work of drainage has been done yet it has had a marked effect on vegetation" (1920, 126). For Simpson, whose aim was to record the natural abundance "that is so rapidly disappearing — and forever," this was not a desirable outcome (1920, vi). In 1943, even the vice president of USSC testified on the importance of removing unoccupied

Everglades land from the real estate market in order "rewater" it and "restore the natural and unique beauty of such country" (Bitting 1943, 21). Finally, the focus on a cultural shift takes the "swamp" talk at face value, rather than as part of a discursive strategy through which interested parties of investors, speculators, and politicians sought to frame the question of what to do about the Everglades. In the end, little is explained by shifts in cultural values, especially when we consider the current rate at which wetlands are being lost across the United States and the around the globe.

In actuality, the development of Florida's agro-industrial region for sugar production did not result from a lack of appreciation of swamps, though prevailing cultural values regarding the environment were important; rather, it was the outcome of battles that reached the highest political offices in the United States and in countries around the world. These battles set productive regions against one another, pitted domestic against foreign political interests, and created many surprising alliances and rivalries among economically and politically powerful sectors of U.S. capital. Beginning in the late nineteenth century, such struggles were captured in the phrase, the "sugar question," coined by contemporary economists and journalists to refer to the intense international debates on the political economy of global sugar production and trade. On one level, we can understand the sugar question as a series of international and domestic political disputes and compromises over the use of various instruments employed to control global trade in the commodity. These instruments include bounties, tariffs, duties, and quotas. Briefly, bounties are state inducements, typically an income tax rebate, designed to encourage sugar producers to produce for export. Tariffs and duties, though technically different, are often conflated in writings on the sugar question. For the purposes at hand, both can be considered as taxes imposed on imported commodities that make them relatively more expensive than those that are domestically produced. Quotas allow states to restrict the supply of a commodity by creating an allocation system among producing regions, thereby keeping domestic prices higher than the world market price. Political struggles over which of these instruments could be used, where, under what circumstances, and for how long defined the sugar question.

On another level, the sugar question involved an interrelated series of discursive constructions of regions and commodities that worked to frame the range of potential political outcomes. In the case of sugar, these discursive constructions are important for two reasons. First, modern sugar production has depended on state support provided through the various instruments of trade regulation. Discursive regional constructions aimed at promoting state support and attracting investment were thus imbued with moralistic and nationalistic rationales for encouraging domestic production. Florida sugar boosters, for example, constructed local regional identities in opposition to those of regions "othered" on the basis of the putative moral inferiority of their production practices. Second, the material properties of sugar make regional discursive constructions critically important in the competition among producing regions for state support. Sugar is a fungible commodity—refined sugar is the nearest thing to a chemical that is consumed as a food—that is obtained from both temperate (beets) and subtropical (cane) crops. The significance of this lies not in a naive, geographical, determinist explanation of investment and production, but in the way sugar's material properties both constrain and provide opportunities for discursive practices. Because there is little difference between sugar from one source and that from another, commodity and regional discursive constructions play critical roles in the competition among sugar-producing regions for favorable state regulatory interventions.

The sugar question, then, is the common theme uniting the material and discursive practices that transformed the wetlands of south Florida and constructed it as a sugar-producing region of international importance. Using this approach, I examine how the region came into being, the role it played in shaping global production and trade, and how shifts in broader political-economic and geopolitical contexts have affected its political and economic potency. In the course of answering these questions, the role of Cuba—as a model, as a competitor, and as the regional "other" to Florida's "self"—looms large. The sugar question, in a sense, provided a way to conceptualize regional competition, and the competition between south Florida and Cuba has been fundamental to the fortunes of each. Though the sugar question long ago fell out of fashion in the economists' lexicon, the debates and struggles that it generated continued through the twentieth century and into the twenty-first. The recent political battles over the Free Trade of the Americas Agreement (FTAA) and the Central American Free Trade Agreement (CAFTA)—replete with moralistic and nationalistic discursive framings—centered on the terms of trade of agricultural commodities, especially sugar. The latest permutation puts the sugar question at the center of the push to develop crop-based ethanol for fuel. As they were in the late nineteenth century, the outcomes of such battles are crucial to the development and existence of the south Florida sugar agro-industrial complex.

Constructing Commodities and Regions

Three streams of literature inform this work, and—to push the metaphor a bit—each has several sources, and the streams sometimes converge. The first stream comprises a welter of approaches grouped within agrarian studies and global food-system theory. Rather than undertaking a comprehensive review, I note several themes of importance here. First, the book builds on the approach of global food-systems theorists to the internationalization of food production and consumption. Specifically, their study of agricultural commodity production suggested a way to develop a relational theorization of place and regional transformation that would overcome the limitations of a binary global/local framework. From rural sociology, Harriet Friedmann, a scholar of rural sociology, is credited with initiating a world systems approach to the study of rural society (Buttel and Goodman 1989). Friedmann demonstrated the need to understand particular rural production systems in the context of the international division of labor in agro-industry through a case study linking the overproduction crisis of U.S. wheat farmers to processes of proletarianization in the Third World (Friedmann 1982). A second key influence is the work of Sidney Mintz, who analyzed the global food system historically, linking relations of production with patterns of consumption. Mintz argued that both the Caribbean sugar plantations and their product played key roles in European industrialization: the plantations, because they "were the closest thing to industry that was typical of the seventeenth century" (Mintz 1985, 48), and sugar, because it provided cheap and easily available calories for the nascent European working class. Food-system theory has developed during the last quarter century, revealing causality and connections in numerous ways. Crucial concerns are the unequal power relationship between the Global North and South manifested in diverging food, trade, and agricultural policies, and the increasing globalization of food production and consumption (Goodman and Watts 1997).

While food-system theory made a significant advance by linking local food production to international political economy, these analyses tend to focus on commodity chains or networks, ignoring the embeddedness of production complexes in particular places. Investigating the "sugar question" required understanding how the region of the Everglades became the EAA and how those invested in that territorial production complex wielded power at the national and global scales. Linking global food-system theory, with its rich conceptualization of commodity networks, to geographic theorizations of regions and places, proved particularly fruitful. Holding both in focus—the dynamic networks of collaborative and competitive relationships that characterize commodity production *and* the embeddedness of an agro-industrial territorial production complex—provided a way to overcome the limitations of each, localizing the commodity chain and globalizing the place of production.

In the following analysis, I conceive of regions as socially constructed and historically contingent, understood to be processes rather than fixed objects (Pred 1984; Paasi 1991). Places and regions are not mere containers or contexts for social action but are "produced and reproduced as part of the broader social production of space" (Paasi 2002, 802). Treating regions or places as constructed and contingent raises the questions of how and why regions emerge as distinct geographic entities, which leads in turn to explorations of historical struggles over systems of production and consumption, territorial identity, and political ideology. Steering clear of investing power in place per se, I am concerned instead with the sort of question posed by David Harvey: "By what means, and in what sense do social beings individually and, more importantly, collectively invest places . . . with sufficient permanence to become a locus of institutionalized social power and how and for what purposes is that power then used?" (Harvey 1996, 320). Moreover, a central premise of this study is that we must understand the social construction of places relationally. Regions and places are discursively constructed through practices that both constitute and reflect relations of power, including the relative power among them (Harvey 1996; Peet 1996; Trigger 1997; Allen 1999). A relational theory of place-construction, therefore, is necessary to explain fully how place-formation is directly linked to social processes in various geographic locations and at various geographic scales.

The third stream from which I draw is that of political ecology. Political ecologists have articulated the concept of "regional discursive formations," which, as defined by Peet and Watts, are "certain modes of thought, logics, themes, styles of expression, and typical metaphors" that "run through the discursive history of a region" (Peet and Watts 1996, 16). Questions regarding relations among discourse, representation, knowledge, power, and ecological transformation are central to this inquiry, which shares with other political ecological studies an "approach to politics as a contested and negotiated domain in continual dialectical relationships with biophysical environments" (Paulson, Gezon, and Watts 2003, 205). In assessing the "virtues of political ecology as a theory of complexity" Taylor (1997, 122) highlights its emphasis on differentiating among unequal agents and on historical process, and its attention to processes operating at different spatial and temporal scales, which allows for analysis that is at once locally centered and translocal. In similar fashion, Batterbury characterizes political ecology as "ecumenical" in its explanation of landscape change as "the result of a suite of processes all operating at different scales and with different underlying forces" (Batterbury 2001, 439). However, for political ecology "the politics that matter most" are at the local scale, "the politics of everyday life," while at other scales "the work fares less well" (Robbins 2003, 643). Here I expand the repertoire of political ecology to include formal institutional politics, including the politics of U.S. presidential campaigns, the positioning of national political parties, and the often contentious relationship between domestic political-economic interests and foreign policy.

By what discursive and material processes are regions constructed? And how are they maintained over time? Specifically, through what historically and geographically contingent processes did the Florida sugar region come into being? In answering these questions, I narrate the history of the social relations of the commodity (sugar) and of the region/place (an intermittently flooded limestone basin in a subtropical climate) that has been socially and materially constructed as the EAA. In the process, I consider how this geographic location became a resource frontier, with all that entailed: certainly the Florida experience resonates with that recounted by Anna Tsing, "Frontiers aren't just discovered at the edge; they are projects in making geographical and temporal experience. Frontiers energize old fantasies, even as they embody their impossibilities" (Tsing 2005, 29). In the case of the region now known as the Everglades, its definition as a frontier depended on which state or empire claimed the space.

At the global scale, the case of sugar production on resource frontiers clearly illustrated a central dilemma of modernity: "The geography of capital produced a landscape of obscured connections. The more concentrated the city markets became and the more extensive its hinterland, the easier it was to forget the origins of things bought and sold" (Cronon 1991, 340). This was especially important for that precociously global good, sugar, which had a unique role in historical processes of commodification and changing cultures of consumption as the first nonluxury item widely consumed that was not locally produced (Mintz 1985). Modern European sugar production was initiated with state support, first for sugarcane in colonial territories and then for sugar beets in Europe, when commodities such as wheat from the settler colonies displaced local food systems (Galloway 1989; Mintz 1985). Thus, for more than a century, the global production of sugar has invoked a nationalist rhetoric that pretends to ask what "space . . . hides," that is, to inquire into "the mysteries of things unknown because done by others or misunderstood because known only by others" (Sayer and Walker

1992, 5). These moralistic geographic arguments, used to promote domestic production in various locations, quickly found their way to the space of south Florida to aid in constructing its regional identity in opposition to "othered" regions, which were said to be more exploitative.

The discursive aspects of place-construction have been central to the social process of creating the "nation's sugar bowl" in the historic Everglades and, later, to the attempts to "restore" it. Examining discursive practices allows us to explore the spatial and temporal nature of commodity production without reducing it to the combined effects of geographical, material, and ecological exigencies (Harvey 1996; Peet and Watts 1996; O'Tuathail 2002). I pay particular attention to the way that shifts in the broader political-economic and geopolitical contexts of sugar production provided discursive opportunities for industry investors and supporters to construct the U.S. sugar supply as critical to national security or regional economic development, depending on the situation. These discursive practices were effective, in turn, in shaping the geography of capital investments in sugarcane production in Florida and producing effects that rippled through the world food system.

My analysis has two methodological components. The first documents the role and form of discursive practices in the establishment, expansion, and defense of Florida's sugar-producing region. The primary sources are printed texts, recovered from a variety of locations supplemented with on-site interviews.[2] An underlying assumption in searching for relevant written texts is that powerful social actors, including politicians, business people, government bureaucrats, and the owners and editors of various mass media outlets play crucial roles in shaping discourse (Peet 1996; Paasi 2001). Therefore, I highlight throughout the biographies of such actors, particularly their political ideologies and economic interests. Key textual sources include corporate promotional and public relations materials; newspaper and magazine articles and editorials; the publications of industry associations and booster organizations; U.S. government documents, including Senate and Congressional testimonies and the memoranda of executive government departments; and various Florida state government documents. The second component addresses the broader and larger-scale political-economic contexts of the discursive practices of Florida sugar interests—specifically, a century of shifting global geopolitics and the role of the United States in international affairs. I trace the discourse of one commodity, sugar, and one region, the south Florida sugarcane plantations, as global geopolitical conditions shifted and U.S. influence on the structure of the world food system grew. The power and resonance of a particular discourse depend largely on historical events of national and global importance, widely accepted notions of national interests, and political beliefs and ideologies about environment, development, and progress.

While I pick up the sugar story in the nineteenth century, the globalization of sugar production and consumption began much earlier (Fogel 1989; Galloway 1989; Mahler 1986; Mintz 1985). Two long-standing characteristics of the sugarcane industry evolved during the period from approximately 700 to 1600, when it centered on the Mediterranean: the use of forced labor and the geographic separation of milling and refining (Galloway 1989). Rising demand for sugar in Europe led Spain and Portugal to encourage the cultivation of sugarcane on island colonies established on both sides of the Atlantic; Columbus took sugarcane to be planted in Hispaniola in 1493. By the sixteenth century, the heavily capitalized New World sugar industry "was not an easy activity for colonists to enter" (65). The English, French, and Dutch developed their own sugar colonies, so that by the end of the eighteenth century sugarcane was the most important crop in tropical America, from Louisiana to Brazil. From 1600 and into the nineteenth century sugar "was the single most important of the internationally traded commodities, dwarfing in value the trade in grain, meat, fish, tobacco, cattle, spices, cloth, or metal (Fogel 1989, 21). Thus, "it was Europe's sweet tooth, rather than its addiction to tobacco or its infatuation with cotton cloth, that determined the extent of the Atlantic slave trade" (18). The spectacular increase in sugar production and consumption had profound impacts from the micro-level of individual metabolism to the macro-level of global restructuring of capital, labor, and trade. Then, during the nineteenth century, European nations enacted policies of import substitution, using bounties to support domestic sugar beet production and tariffs to deter cane-sugar imports. The story of the Florida sugar industry starts at the close of the nineteenth century, when nationally organized sectors such as sugar were subject to attempts to develop international agreements, and continues in the twenty-first century with the emergence of the global ethanol assemblage.

Scope and Organization of the Book

My narrative follows an historical arc, from the first visions of a "reclaimed swamp" in late-nineteenth century south Florida to the imagined ecology of a "restored wetland" in the twenty-first century. I begin by examining the relationship among the international, domestic, and local politics of sugar during a critical period in the early development of U.S. sugar policy and the Florida industry. Thus chapter 2 focuses on the half century preceding

World War I when the international political economy of sugar production and trade had captured the world's attention. During this time, published debates, international conferences on the "sugar question," and attempts by national governments to forge international sugar agreements proliferated. The development of the European beet-sugar industry and accompanying global surpluses posed a challenge to promoters of U.S. agricultural interests, who saw in domestic sugar production a regional development strategy. Florida boosters, especially, saw in the modernizing Cuban industry both an exemplary model and a formidable competitor. I show how changing ideas of sugar, of Florida, and of the U.S. role in the Caribbean shaped the context in which southern agricultural boosters promoted the establishment of a sugar industry. In doing so they articulated what I call an "imagined economic geography," a necessary but not sufficient precursor to the development of a regional cane belt. By this I mean the thinking of interested economic parties—boosters—who envisioned and described an elaborate economic landscape that would transform a previously undeveloped region. These "imagined economic geographies" were quite detailed, including land measurements and speculation on potential labor sources, profits, and economic multipliers.

Chapter 3 focuses on explaining an apparent economic contradiction: the transformation of the Everglades into an agro-industrial complex for sugarcane at great ecological and monetary cost during a period of oversupply and depressed prices in the global sugar market. I begin by discussing the impact of World War I on U.S. sugar politics. The prospect of sugar shortages and the need to share the Cuban crop with allies provided Florida boosters with a powerful discourse in their struggle to construct a Florida sugar bowl; politically and economically interested parties developed a nationalistic and moral-geographic discourse concerning the transformation of the Everglades for sugar production. I show that in the postwar period, under markedly different political-economic conditions for the global sugar market, protectionist policies sparked development, bringing long-promoted ideas about Florida's agro-industrial potential to fruition. The chapter demonstrates how, once established, the "Nation's Sugar Bowl" in Florida became a locus of power in the political struggles over domestic and international sugar policy.

In Chapter 4, I argue that the newly established Florida sugar industry now faced three problems in expanding its production and profits: competition for quota share, labor supply, and water control. I begin by offering a fresh interpretation of the domestic political maneuverings behind the 1930s Sugar Acts, arguing that Florida's sugar interests played a hitherto unrecognized role. The Sugar Acts were the main tool of New Deal sugar policies, which were intended to balance the competing interests of sugar-producing regions—mainland, offshore and foreign—through a system of quota allocation. With the legislative establishment of quotas, political competition among producing regions intensified, and new discursive strategies of place-based comparisons emerged. Because they were based on historical levels of output, quotas seemed especially restrictive to the new and expanding Florida industry. Florida sugar capitalists based their discursive strategies largely on their representation of the place and social relations of production as morally superior to those of the foreign locations. Here I demonstrate how ideas about sugar and national security were used again to great effect, this time to restructure the geography of the regional labor market with the help of the federal government. Finally, I examine how the federal government addressed the third problem facing the industry when the Army Corps of Engineers undertook the C&SF Project, economically justified in part by the potential for increased sugar revenue.

In the context of the Cold War, the 1959 Cuban Revolution produced a profound historical shift in the economic geography of U.S. sugar sourcing. Chapter 5 concentrates on the years immediately preceding and following the revolution. Prior to the revolution, Cuba was both Florida's principal rival and its production model. When, in 1960, President Eisenhower suspended the Cuban quota, a terrific scramble to increase allotments ensued among producing regions, both domestic and foreign. Florida emerged as one of the significant "winners" in the fight to secure a larger share of the U.S. market. I show why that was so, how the industry was able to expand five-fold within five years, and how the relationship between the Florida and Cuban sugar industries was transformed. The explosive growth of south Florida sugarcane production, historically unprecedented in U.S. agro-industry, made the region the single most powerful player in the competition over quotas during the Cold War.

Chapter 6 turns to the regional impacts of industrial expansion to consider the challenges posed by the rapid growth of plantation production in rural Florida. The geographic expansion of the industry raised new questions about the treatment of labor and the downstream environmental impacts of agriculture in the EAA. From 1965 to 1985, the sugar question gained prominence during successive administrations, as presidents from Lyndon Johnson through Jimmy Carter sought to achieve a balance in U.S. sugar policy between foreign policy initiatives and domestic political realities. Moreover, each administration sought not only to balance U.S. sugar policy, but to use sugar quotas as the means to build and maintain circles of influence in foreign affairs that extended beyond the realm of commodity interests per se. In the context of the Cold War, sugar was still seen as a tool of regional agro-industrial development; however, the emphasis was on foreign rather than domestic regional development.

By the 1990s, the Everglades had completed a symbolic transition from swamp to wetlands in the public imagination, marked by the 1996 Everglades Forever Act. In political terms this meant a shift in emphasis from drainage and canalization to preservation and restoration. During the same period, the popular image of the sugar industry also shifted from that of an engine of regional development and producer of essential food to that of "big sugar," a corporate giant growing fat off of government subsidies and despoiling the environment. Chapter 7 thus brings the historic arc of the sugar question to the present, illustrating how changing political-economic and geopolitical contexts along with shifts in the scientific and symbolic meanings of the Everglades transformed the Florida sugar industry from a political powerhouse to a pariah. Elements of the industry promoters' discursive strategies have persisted, but they have been reworked to respond to these new challenges, manifested in the likes of NAFTA, CAFTA, and Everglades restoration plans.

CHAPTER TWO

The Sugar Question in Frontier Florida

For three centuries following Juan Ponce de León's disappointing 1513 expedition to the Florida peninsula, the territory served as a borderland for three empires, first claimed by Spain, then Britain, Spain again, and finally the U.S. in 1821. Recognizing south Florida's frontier history is crucial to understanding its symbolic and material roles in later U.S. policies concerning foreign relations, agriculture, and trade. The process of transforming the Everglades from swamp to agro-industrial site, of its "becoming" a sugarcane region, was contingent on the complex interactions of geography, shifting international political-economic circumstances, and domestic agricultural politics. Florida—a latent U.S. frontier in the subtropics—played a unique role in the geographic imagination of agricultural promoters as the state that would free the nation from its dependence on imports of foreign tropical commodities. After the Second Seminole War (1835–42), when south Florida became an active resource frontier, wetlands drainage and the establishment of sugar plantations were fundamental to its transformation.

One of my principal aims in this chapter is to establish the importance of this single commodity, sugar, in the early ideas of Everglades transformation. During the second half of the nineteenth century, the political economy of sugar production and trade captured international attention. Sparked by the development of the European beet-sugar industry and accompanying global surpluses, widely published debates and international conferences on what became known as the "sugar question" proliferated. As one contemporary observer noted, "No other food product enters so largely into the domain of state and international politics" (Crampton 1901, 283). Countries producing beet or cane sugar made numerous attempts to forge international agreements on the production and trade of the commodity. In the United States, agricultural interests saw in sugar production a powerful engine for regional development and sought federal support for increasing domestic production and inhibiting competition from foreign producers.

As the sugar question suggests, analysis of the transformation of the Everglades into a region of agro-industrial production must be relational. Toward the turn of the nineteenth century, where this chapter concludes, relations between Cuba and Florida took on an importance that for decades remained central to the transformation of the Everglades. Initially, southern boosters hoped that new import duties on Cuban sugar would provide a window of opportunity for regional development based on sugarcane. During this period, changing ideas of sugar, of Florida, and of the U.S. role in the Caribbean shaped the context in which boosters promoted the establishment of a southern "Cane Belt," analogous to the Midwestern Corn Belt. In doing so, they articulated what I call an "imagined economic geography," a necessary but not sufficient precursor to the development of a regional sugarcane agro-industry. Following the 1898 Spanish-American War, Florida's significance as a subtropical frontier was temporarily diminished. Agrarian boosters recognized that their primary rivalry was now with the new sugar-producing protectorates. Florida boosters, in particular, saw in the modernizing Cuban industry both an exemplary model and a formidable competitor.

Before analyzing the relationships among the international, domestic, and local politics of sugar during this critical period, it is helpful to examine the spectacular and ultimately tragic failure of the first attempt to drain the state's wetlands for sugar. A key turning point in the relation between place and commodity—that is, between the Everglades and sugar—resulted from what came to be known as the Disston land sale of 1881. Hamilton Disston, a northern industrialist and founder of the Disston Enterprise, led the first commercial effort to drain the wetlands for sugarcane cultivation. Although the cultivation site was a hundred miles from the historic Everglades, the Disston sugar plantation was instructive for both future agro-industrial investors and scholars interested in Everglades transformation. The endeavor proved that, once drained, the swamp and overflowed lands, which the state of Florida held in abundance, could be planted in sugar. Furthermore, the Disston land sale and the initial success and ultimate failure of the Disston Enterprise illustrate themes fundamental to the development of the Florida sugar industry. Disston, a northern industrialist backed by a group of northern financiers and investors, proved to be something of an archetype of sugar producers in Florida. The following section examines the lessons his successors learned from his efforts to raise cane in the swamps of postbellum Florida.

An Instructive Failure

From the perspective of state officials and boosters, postbellum Florida faced numerous interrelated and mutually reinforcing problems with regard to rural development: a lack of population, of infrastructure, and of capital. Development was stalled because investors were reluctant to invest in infrastructure without customers, and, conversely, potential customers needed infrastructure to develop commercially viable farms. The state did have one critical ingredient: land. Twenty-two million of Florida's thirty-five million acres of land had been transferred from the federal government to the state (fig. 2.1) under the Swamp and Overflowed Lands Act of 1850 (Dovell 1952, 263). However, the state's Internal Improvement Fund (IIF), which included both the land and the money received from sales, was tied up in litigation due to the loss of property value as a result

Figure 2.1. Political cartoon, circa 1916, depicting the federal transfer of wetlands to the state of Florida. Courtesy of the Historical Museum of Southern Florida.

of the destruction of railroad infrastructure during the Civil War. As a consequence, the IIF was in receivership, which meant that the state of Florida could not conduct business as it had in the past, by advancing land sales through credit, without first paying bondholders.

At that time, and for nearly a century thereafter, Hamilton Disston was widely regarded as the savior of Florida (or at least of the state's fiscal credibility), whereas today his intervention is seen as the beginning of the end for the historic Everglades. Disston's role in Florida, his endeavors and his fate, exemplify the forces shaping Everglades transformation and state development. President of the Disston Saw Company, which his immigrant English father had started in Philadelphia, Hamilton Disston had some of the capital and many of the political connections that were necessary to advance development in Florida. Disston interested "a group of moneyed friends" (Harner 1973, 16) in his vision of an extensive Florida undertaking and, on behalf of Disston and Associates, entered into an agreement with the state to drain nine million acres of land in exchange for half the reclaimed acreage as payment for the work of drainage. However, because the IIF was in receivership, it had to operate on a cash basis until freed from debt. Under an alternative agreement, the associates would buy four million acres of swamp and overflowed lands at twenty-five cents per acre; this million dollar sale allowed the IIF to settle its debt and to operate unencumbered by judgment creditors. Disston gained national notice and celebrity when he was reported to be the largest landholder in the country. The deal was not without controversy, however: critics noted that not all the land being claimed was swamp, nor was it all unsettled. The land claimed by Disston and Associates comprised parts of twenty-five counties, with the largest contiguous portion extending from Ocala to just south of Lake Okeechobee.

Disston oversaw a diverse operation that assembled dredges, undertook drainage, and established commercial peach orchards and sugar and rice plantations and mills. He attempted to finance these undertakings with land sales. To attract population, he set up emigration offices in the United States and Europe, established model towns, and published promotional brochures. Disston's venture into sugarcane cultivation and milling built on an undertaking of Rufus E. Rose, engineer for the Disston Drainage Co., who later became the Florida State chemist. In 1885 Rose, who came to Florida from Louisiana, had purchased 420 acres of marshland on which he planted 20 acres of cane, as well as rice and corn. In 1886–87, after Rose had planted 90 acres in cane, Disston bought a half interest in the plantation, St. Cloud. He initially backed expansion to 400 acres and provided capital to construct a sugar mill with grinding capacity of two hundred tons per day. According to reports of the time, both yields and sucrose content were high, reputedly the best American record to date.

Encouraged by this success as well as the passage in 1890 of the McKinley Tariff—which provided a two-cent per pound sugar bounty—and advised by "sugar experts," Disston sought to expand the sugar enterprise as rapidly as possible. Disston and Rose differed in their assessment of the future of the bounty; whereas Disston expected it to last at least until 1905, as stipulated by Congress, Rose had less faith in its duration. Thus, Rose "counselled against reorganization and sold his stock in the enterprise" (Manuel 1942b, 12956); he also resigned as superintendent, whereupon Disston replaced him with another Louisiana sugar man and set out to increase production. Disston reorganized St. Cloud Plantation as the Florida Sugar Manufacturing Company, which was capitalized at one million dollars with a bond issue of another million dollars floated to buy thirty-six thousand acres of land. He spent approximately three hundred and fifty thousand dollars to erect a factory capable of handling the yield of more than triple the cane acreage then being cultivated.

Disston designated a forty-acre tract on the plantation for the United States Department of Agriculture (USDA) to conduct research on sugarcane grown on drained muck soil.[1] Harvey W. Wiley, USDA chief chemist, oversaw the experiments, which involved thirty varieties of cane, half from the United States and half from Cuba and Java (Hanna and Hanna 1948). In his 1891 report, "The Muck Lands of the Florida Peninsula," Wiley described findings regarding soil samples at the station and the nearby plantation of the Florida Sugar Manufacturing Company, both adjacent to Lake East Tohopekaliga, in the environs of the town of Kissimmee. However, for reasons of soil and climate, he identified the south shore of Lake Okeechobee as the best region in the state for sugar. He imagined a network of canals that would provide transport and access for steam-powered cultivation, creating efficiencies which, coupled with his expectation of high yields, meant that "even the island of Cuba could not compete with Florida in the production of sugar." Four years before Disston's company began dredging southward from Lake Okeechobee, Wiley extolled the potentialities of the land they sought to drain: "There is practically no other body of land in the world which presents such remarkable possibilities of development as the muck lands bordering the southern shores of Lake Okeechobee" (Wiley 1891, 170).

Despite Wiley's glowing prediction regarding the future of sugar in Florida—or, perhaps, in part because of it—Disston's enterprise failed. Encouraged by sugar experts and by the 1890 bounty to expand, Disston was financially and managerially overextended. Retrospective assessments differ as to the relative importance of the various reasons for his failure, but the basic recipe is consistent. According to Rose, Disston was led to fail by "extravagance, by ignorance of proper methods of culture, and neglect of drainage, by want of proper business methods on the part of the company and its managers, and most important—by speculation" (quoted in Dodson 1971, 369). Samuel L. Lupfer, who had also worked as superintendent for Disston, agreed that "stock-jobbing" and mismanagement had led to failure, but noted that when Disston resumed active management in 1895, the outlook had improved. Those less intimately involved with the company emphasized the broader economic context, specifically the panic of 1893, and then asserted, even more strongly, that "the final coup was dealt by the repeal of the sugar bounty in 1894, contrary to the intent of the Congress which had passed it" (Manuel 1942b, 12956). In 1896, unable to pay his bills or continue operations, Disston committed suicide. His heirs were either apathetic or hostile regarding his Florida ventures. In 1901, Disston Land Company properties, previously valued at more than two million dollars, were sold for a mere seventy thousand; other landholdings near Kissimmee were forfeited for taxes; and the sugar estate was sold to Cuban planters who moved the mill to Mexico (Hanna and Hanna 1948; Sitterson 1953).

Though he failed spectacularly, Disston's intervention in Florida accomplished several things. His partnership with Rose in the St. Cloud plantation demonstrated the feasibility of growing sugarcane and rice on drained land. The USDA research station brought national scientific attention to Florida's muck soil. And, finally, the initial Disston land purchase had rescued the IIF. That influx of capital freed the state from obligations to creditors, which allowed the IIF to pursue the objective of development as defined by the state, through such strategies as land grants to railroads (Blake 1980).

Beyond these economic impacts and the personal tragedy, the tale of Disston in Florida introduces three themes that are critical to the story of the development of the Florida sugar industry. The first concerns the status of Florida development in 1880; as Hanna and Hanna said of Disston, "Union of the man and the place resulted from the time" (1948, 91). That is, from the perspective of boosters and bureaucrats, south Florida in the late nineteenth century was a frontier tabula rasa awaiting northern capitalist investment and energy to transform it into real estate for the production of agricultural commodities. This raises the question of why, so late in the political evolution of the United States, Florida had the geography of a frontier, where a single individual could still lay claim to four million acres of land. The second theme, raised by Wiley's encomium, concerns the national context of agricultural boosterism and, specifically, sugar boosterism, in which Disston's venture took place. Boosters in the South, particularly in Florida, were to play a central role in the geographic imaginaries of Everglades transformation. The third theme emerges from the way the repeal of the bounty highlighted the importance of the political economy of national and international markets to the dynamics of that transformation. But the story of how sugar bounties—and quotas and tariffs—became so central to the environmental history of south Florida begins even earlier, when Florida's fortunes turned on the fate of empires.

"America's Oldest Frontier and at the Same Time Her Newest!"

From the perspective of the mid-twentieth century, numerous writers noted with irony that, while Florida was home to the oldest European city in North America (St. Augustine), the state remained relatively undeveloped. Two characteristics set Florida apart from other southern states: its settlement history and the climate and ecology of the peninsula. Together, these conspired to make Florida a latent and subtropical frontier, situated— geographically, politically, and economically—between the U.S. South and the Caribbean. The peninsula's peculiar history led twentieth-century historians to proclaim it "America's oldest frontier and at the same time her newest" (Quaife 1948, i). The apparent emptiness, or frontier quality, of late nineteenth- and early twentieth-century Florida had been brought about by centuries of conflict, sovereignty transfers, and evacuations, as well as the displacement and marginalization of the Seminole and black populations. One of the strands in the historical narrative of Everglades transformation concerns the abrupt shifts in political geography that changed the meaning and identity of the region. Over the last three centuries, Florida's relative significance as a frontier territory depended greatly on which empire claimed ownership. Another strand addresses the historical geographic relationship between Florida and Cuba. The intertwining of their political-economic and environmental histories—first as proximate locations in the colonial network and then through the international politics of production and trade—has been influential in the development of Florida's sugar industry.

During the first half of the nineteenth century, the United States fought three wars in Florida in an effort to claim and control the territory. As a result of the First Seminole War, which began when Andrew Jackson invaded Spanish Florida in 1818, Spain ceded Florida to the United States in 1821. But the hostility that gave rise to Jackson's invasion of Florida had much deeper roots in the international disputes of the seventeenth and eighteenth centuries, during which "the southeastern portion of the continent had been a focal point for the triangular contest of the major European empire-builders" (Tanner 1989, 7). The battles, skirmishes, and treaties of European disputes were connected in complex and sometimes obscure ways with events occurring around the world when "the southern frontier was but one of many borders, in India, Africa, the West Indies, and North America, where Englishmen of the late seventeenth and eighteenth centuries vied with their rivals . . . for commercial and colonial supremacy" (Crane 1956, 4). In general, what was at stake throughout the southeast was control of territory, indigenous trade networks, and sea trade routes.

What made Florida's border disputes especially vehement were differences in the social relations of production. Spain's policy in Florida was to resettle Native Americans in mission villages that produced enough grain to supply St. Augustine and Havana. In the seventeenth century, large cattle ranches were established in central Florida, again to supply Cuba. However, agriculture in Spanish Florida was not as labor-intensive as in English Carolina, where displaced West Indian planters had established plantations. The hottest point of contention between the two empires was that Spanish Florida provided refuge for runaway slaves from English plantations in the Carolinas. As geopolitical pressures increased, Spain's ability to resist British encroachment depended increasingly on black homesteaders and the black militia, so in 1738 the Spanish governor established a free black settlement north of St. Augustine (Landers 1995).

Decades of international rivalry culminated in the resolution of the Seven Years War, when British forces captured Havana and Pensacola. To save the former, Spain agreed to the terms of the 1763 Treaty of Paris, which ceded "the Floridas" to Britain. In January 1764, three thousand residents of St. Augustine were evacuated: "Spaniards, slaves, free blacks and allied Indians all boarded ships for Cuba" (Landers 1995, 25). Britain had numerous reasons for taking the relatively undeveloped Florida in trade for Cuba. Perhaps foremost was the chance to gain control of the eastern seaboard, but the geopolitics of sugar also played a role through the lobbying of West Indian, especially Jamaican, planters, who did not want Cuban sugar to enter the British market duty-free (Kuethe 1988; Williams 1984). Also, the prospect of holding populous Havana by force was daunting, whereas the challenge posed by the acquisition of Florida, which was to attract population to its evacuated lands, was more manageable (Kuethe 1988).

The British government undertook a massive publicity campaign to promote Florida and offered generous land grants to "influential Englishmen, Georgia and Carolina planters, and former British soldiers" who would pledge to settle white Protestants (Sitterson 1953, 8; Proctor 1977; Sturgill 1977). One such undertaking involved transporting Greek and Minorcan indentured laborers to fulfill the colonial aspiration of producing Mediterranean crops on British soil. The coastal colony of New Smyrna became the first site of sugarcane cultivation in Florida. It was undercapitalized and unable to prevent workers from leaving, and therefore failed within a few years. Meanwhile, James Grant, Governor of British Florida, using South Carolina rather than the Mediterranean as his model, promoted plantation development using black slaves, first obtained from the Carolinas and then increasingly from the Caribbean and Africa. Thus, as British control brought Florida into the triangular trade, Anglo planters established large plantations to produce indigo, rice, cotton, and sugar, and "black freedom in Florida became only a remote possibility" (Landers 1995, 25).

Settlement prospects aside, Britain's primary reason for wanting Florida was strategic. For the same reason, Spain felt its loss (Fabel 1988; Sturgill 1977). While it remained peripheral in terms of the economic geography of the Caribbean and the world beyond, Florida was strategically important to Spain: its loss deprived Spain of any friendly East Coast ports between Havana and Spain and enabled Britain to interfere in Spanish trade routes. Charles III of Spain undertook colonial reorganization to strengthen Spain's position in the Caribbean, with an eye to regaining Florida. This involved military reform and build-up in Cuba, and therefore higher taxes, which in turn led to policies meant to stimulate the Cuban sugar industry so as to encourage the generally supportive aristocracy. Havana had so far been permitted to trade solely with Cádiz; Charles and his ministers enacted a series of free trade regulations, beginning in 1765 with an act that allowed Havana to trade with nine Spanish ports, which became part of the more comprehensive Regulation of Free Trade of 1778. These free trade acts were crucial to the rapid growth of the Cuban economy in general and the sugar industry in particular. Sugar production tripled between 1760 and 1791 as Cuba moved from eleventh to fourth place in world sugar production (Kuethe 1988; Tucker 2000). By the onset of the American Revolutionary War, Spain was in a position to provide support to patriot forces, and, by 1779, to enter the war for the purpose of regaining Florida. Under the terms of the 1783 Treaty of Paris, Florida once again became Spanish and "resumed its role as a military appendage of the Cuban colony" (Kuethe 1988, 74).

Much of Florida's population was again evacuated. Some slaves were sent to Dominica, the Bahamas, and South Carolina; others took advantage

of the chaos of transition to escape to Seminole villages or gain other refuge from the returning Spaniards (Landers 1995; Schafer 1995). The second
Spanish period posed difficult problems for governance for a variety of reasons, including confusion over land rights and the status of the black population, British control over critical Native American trade networks, and
the shift in political geographic context as the nascent United States gained
power. Even without this new neighbor, the project of developing Florida
would have been difficult. Being flanked by the slave-holding states of the
expansionist United States, however, Spanish Florida was forced to fortify
a hostile frontier. Under these conditions, Spain and Britain felt the need
to cooperate to maintain their economic and geographic presence in the
southeast.

Spain's hold on Florida weakened after the United States acquired the
Louisiana territory in 1803, and especially after the elections of 1810, which
brought more southern "War Hawks" into Congress (Tebeau 1971). Their as-
pirations to acquire Florida for the United States helped fuel at least three
organized attempts between 1811 and 1814. The Spanish governor turned
to the two groups who had most at stake in preventing U.S. acquisition—
Native Americans and blacks—to defend the frontier (Dovell 1952; Porter
1971). What transpired was a three-way struggle, with England ostensibly
helping to contain the United States by arming and training Native American and black troops along the Florida border, and by encouraging slaves to
flee southern plantations. These activities made relations between the Florida colony and the United States even more volatile. In 1815, Andrew Jackson entered Pensacola with the purpose of driving out the British. In 1818 he
again invaded Florida, this time intending to gain Florida from Spain, thus
initiating what came to be known as the First Seminole War, "primarily a
struggle against black maroons and their Indian allies" (Genovese 1979, 73).
This marked the end of Spanish occupation of Florida.

"Feelings Bordering on Awe": The United States Claims the Everglades

Under the terms of the Adams-Onis Treaty, Florida was ceded to the United
States in 1821, and in 1822 Jackson accepted the governorship of the territory. Annexation by the United States marked the most critical juncture in
the history of Florida's shifting geographic identity. To that point, Florida's
most salient role resulted from its geopolitical position. That is, it served as a
buffer state between competing imperial powers and, after 1783, the nascent
United States. Development during those three centuries of European occupation had been desultory. Now Florida was joined to an expansionist

nation intent on agrarian settlement from coast to coast. The buffer zone became a resource frontier. Florida, whose climate had not been unique or even particularly desirable in the context of Spain's New World Empire, became the only U.S. subtropical territory, valuable for its distinctive agricultural potential. In the ensuing century the project of developing this subtropical frontier was driven by a nationalistic economic ideology, an autarkic vision in which Florida, and specifically the Everglades, would free the country from dependence on foreign trade by producing tropical commodities. This was a role only south Florida could play, according to agrarian boosters and sugar promoters, who returned to this theme again and again throughout the nineteenth century and thereafter.

In the immediate aftermath of annexation, three interrelated issues faced the territorial government: land disposal, internal improvement, and "Indian affairs." The two groups—often closely associated—whose status would be most changed by Florida's inclusion in the U.S. South were the Seminoles and the free black population. In part this reflected a change in the scale of conflict. The trouble between whites and allied Seminoles and blacks, which had long produced international disputes, was now solely a U.S. domestic affair (Klos 1995). Native American "removal" in Florida was motivated by disputes over possession of black slaves as much as by conflict over land. As white settlers moved into Florida and initiated a slave-based plantation economy, they displaced resident Seminoles southward. In 1823, the Treaty of Moultrie Creek set aside four million acres for the Seminoles. The treaty satisfied neither the Seminoles, who were being displaced, nor white settlers, who favored Indian removal. After Jackson assumed the presidency, the government enacted policies authorizing Indian removal, and pressured the Seminoles to move to western reservations in Oklahoma. Tribal leaders held to the Treaty of Moultrie Creek, which was binding for twenty years, while whites ratified the Treaty of Fort Gibson, which set the date of January 1, 1836, to begin the removal (Tebeau 1971).

Late in 1835, the first skirmishes occurred in what would become known as the Second Seminole War. At the outset, Seminoles attacked sugar plantations on the east coast of Florida, located along the Matanzas, Tomoka, and Halifax rivers, to the south of St. Augustine, in the environs of New Smyrna and present-day Daytona. Many of these dated to the British or Spanish colonial periods. Though well-established and nationally prominent before the Second Seminole War, sugar plantations in this area were never restored (Dovell 1952). The onset of the war ended sugar production here, but policies enacted to end the war stimulated the development of sugarcane plantations elsewhere. In 1839 the "ardent expansionist Thomas

Hart Benton of Missouri" introduced a measure that passed to become The Armed Occupation Act of 1842, "the country's first homestead act" (Schene 1981, 69). The purpose of the act was to place white settlers on the contested resource frontier southeast of Gainesville by providing 160 acres of land to any whites who settled there. Among those who moved into the region were Major Robert Gamble and Dr. Joseph Braden, who, with 180 slaves, established a dozen plantations that produced sugar commercially from 1845 to 1857.

By the end of the war, approximately four thousand Seminoles and blacks had been "removed" west of the Mississippi, while about three hundred took refuge in the Big Cypress Swamp or the adjacent Everglades. The war had involved the army, navy, and marines in an effort to rout the Seminoles from the Everglades. Considered "the longest, costliest, and bloodiest Indian war in United States history" (Dovell 1952, 237), the seven-year war inflicted several thousand casualties, depopulated the frontier, and cost "thirty to forty million dollars" on top of regular army expenses (Genovese 1979, 73). Florida development stalled: "The war ended immigration and cut short economic development. The organization of counties continued, but the counties did not grow" (Rohrbough 1990, 216).

On March 3, 1845, Florida was admitted into the union as a slave-holding state. One of the issues that had faced the territorial government seemed to have been resolved—most of the Seminoles had been displaced to the west. The projects of "internal improvements" and "land disposal" remained, and would become tightly linked during the course of the century. Almost immediately, the Florida legislature submitted a resolution to Congress asking for federal help in studying the Everglades for the purpose of reclamation, stating that "at a relatively small expense, the aforesaid region can be entirely reclaimed, thus opening to the habitation of man an immense and hitherto unexplored domain, perhaps not surpassed in fertility and every natural ability by any other on the globe" (U.S. Senate 1911, 34).

In 1847 funds were secured for reconnaissance, and the U.S. Senate appointed Thomas Buckingham Smith, a St. Augustine lawyer, scholar, historian, philanthropist, legislator and unionist, to conduct the survey. Buckingham Smith's report is considered to be the first official publication on the Everglades. His writing style reflects his scholarship as an historian who had translated manuscripts relating to the exploration of Florida, including the narrative of Cabeza de Vaca, the memoir of Fontaneda, and the writings of Hernando de Soto (Dovell 1952). Ranging from the poetic to the prosaic, Buckingham Smith's expressions of wonder ("The profound and wild solitude of the place . . . add to awakened and excited curiosity feelings bordering on awe") gave way to his pragmatic conclusion that it would be entirely possible to lower the water table and expose fertile land so that the region could supply various tropical crops, especially sugar. He suggested that the nation would thus become less dependent on the West Indies and that the southern and eastern portions of Florida could form a new state, separate from middle and western Florida. In Buckingham Smith's report, the issues of drainage, reclamation, agricultural production and state-building coalesce, and it is for that sentiment that he is usually remembered.

> The Everglades are now suitable only for the haunt of noxious vermin, or the resort of pestilent reptiles. The statesman whose exertions shall cause the millions of acres they contain . . . to teem with the products of agricultural industry; to be changed into a garden in which can be reared many and various exotics, introduced for the first time for cultivation into the United States, whether necessaries of life, or conveniences, or luxuries merely . . . will have created a State. (U.S. Senate 1911, 54)

But Buckingham Smith also voiced prescient concerns regarding Everglades soils, which were understated in subsequent booster accounts but remain critical issues today:

> This deposit is exceedingly light and when dry and broken to pieces becomes an impalpable powder. If it should be found to be a good compost, its speedy exhaustion and its liability when dry and exposed to the surface to be removed by the winds are obstacles to its extensive successful use in the cultivation of sugar, rice, tobacco, cotton, or corn that should be anticipated. (U.S. Senate 1911, 52)

Of the dozen or so letters that Buckingham Smith appended to his report, most were sanguine about the prospects of Everglades drainage and the potential for cultivating tropical commodities. Typical of these were the testimonies of officers who had served in the Everglades during the Second Seminole War, such as General S. W. Harney, who wrote, "It is my opinion that it would be the *best sugar land* in the South, and also excellent for rice and corn. It affords the Union the best kind of cultivated land that is wanted to render us, to a great extent, independent of the West Indies" (U.S. Senate 1911, 57; emphasis added). General Thomas S. Jesup concurred: "The practicability of draining I take for granted. Were the surface of the Lake and the Glades lowered, these fine lands would be reclaimed and soon be converted into valuable sugar plantations as rich as any in the world" (56). Both men noted that development in the Everglades would be desirable from "a military point of view." Jesup suggested that a "numerous white population" be "interposed between the sugar plantations, cultivated by slaves and the free blacks of the West Indies." This, he thought, would "add greatly to the strength and security of the South" (56). Harney noted that south Florida was "extremely exposed in time of war," and predicted that populating the region would enhance "the security of the entire southern portion of the Union" (57). Only one correspondent, S. R. Mallory, refrained from joining the chorus that prophesied tropical abundance, claiming that, though some tracts could be drained, "it will be found wholly out of the question to drain all the Everglades" (63). His depiction of the Everglades is also exceptional for mentioning the smell "of its snakes and alligators" and his "acquaintance with its mosquitoes and horse flies," details left out of most booster accounts. Mallory's skepticism aside, what stands out from Buckingham Smith's report and accompanying documents is that, from their first reconnaissance, the Florida Everglades elicited visions of sugar.

Florida received two federal land grants during the early years of statehood, which, with subsequent modifications, allowed the state to claim more than twenty-two million acres, representing 59 percent of all land in the state (Vileisis 1997). In 1851 the state legislature passed an act to establish a Board of Internal Improvements comprised of state officials and to consolidate various land grants within the IIF. Though initially state officials promoted the construction of canals to both drain land and provide transport, their interest shifted quickly to railroads, with the result that land grants were transformed into railroad stocks. While hundreds of miles of railroads were being built, drainage languished, in violation of the spirit of the Swamp Lands Act. Nelson Blake notes that "if their policy had been successful, this flaunting of the laws might have been defended on pragmatic grounds" (Blake 1980, 41). However, development in south Florida stalled during the Third Seminole War (1849–59), as "fear haunted the prospective settlers and dozens of land patents were canceled in consequence of failure to establish claims" (Dovell 1952, 260). Then, in January 1861, Florida became the third state to secede from the Union. Because of its sparse population, minimal development, and extensive, nearly indefensible coastline, the state, though peripheral, was vulnerable during the Civil War, and most of the antebellum railroad infrastructure was destroyed. State boosters and officials had gambled on railroads and lost. At the close of the war, the railroads, no longer solvent, defaulted on their bonds, which the IIF then attempted to sell at 20 percent of their original value.

What gave the IIF a chance to recover after the war was the land—swamp and otherwise—owned by the state. Whereas elsewhere the federal government retained title to most public land, Florida's twenty-two million acres were several million more than even the largest western states could claim. The result was intense, contested politics concerning land development, and, at times, a close association in public perception between state government and land swindles. During the immediate postwar period and Reconstruction, various would-be developers petitioned the state for drainage contracts. These corporations sought land from the state on the basis of what they claimed they would achieve, either a certain number of drained acres or successful settlers: "Carpet-bagger trustees were transferring land to various northern financiers and corporations on conditions of settlement and reclamation which were rarely observed" (Manuel 1942a, 12867). Though state officials were willing to secure such contracts, they were hampered by a lawsuit brought against the trustees and several companies by Francis Vose, "a New York iron manufacturer who had taken construction bonds in payment for rails sold to the Florida Railroad" (Blake 1980, 50). Vose "charged that the value of bonds he had purchased at par was being jeopardized by the trustees, who had failed to pay past-due coupons, had mishandled funds, and had illegally transferred land to corporations" (Manuel 1942a, 12876). He secured an injunction against the trustees, "which prohibited them from selling trust land for script, State warrant, or anything but the current money of the United States" (12876). When the trustees ignored the restrictions of the Vose settlement, the IIF was placed in receivership. Thus Hamilton Disston's one-million-dollar bid was an offer Governor William D. Bloxham and the trustees could not refuse.

"The Grandest Political Machine": Sugar and Agrarian Populism

The emphasis that Disston placed on agriculture in his vision of Florida development was not unusual for the time. During the nineteenth century, as the institutional framework of U.S. agriculture evolved, the promotion of agriculture through private and public channels achieved national momentum. Agricultural boosterism—a popular theorization of agrarian-led regional development that was "inclined toward enthusiastic exaggeration and self-interested promotion" (Cronon 1991, 34)—flourished. The nineteenth-century phenomenon of agricultural journals reflected the popular interest in agrarian development and provided a primary outlet for boosters. Following the publication of the first U.S. agricultural journal in 1810, their numbers increased steadily until, on the eve of the Civil War, there were approximately fifty. In the antebellum period the number rose to an estimated four hundred. Likewise, during the 1840s agricultural societies also became popular, and by 1856 there were more than nine hundred state and local organizations across the country. At the national level, a standing committee on agriculture was established in the U.S. House of Representatives in 1820 and in the Senate in 1825. In the 1830s, farming advocates called on Congress to fund an agricultural survey modeled after English surveys, and in 1839 the Patent Office appropriated funds to gather statistics as well as to distribute free seeds. Regarding nineteenth century agriculturalists, historian Paul Gates observed that "no other economic group was the recipient of so much free advice nor was any other group so well represented by state and national societies, by fairs, by weekly and monthly journals, and by numerous 'experts'" (Gates 1960, 338).

However, in the antebellum period, national momentum did not override sectional interests. Most of the agricultural journals served regional constituencies, catering either to southern planters, New England farmers, Midwestern pioneers, or Western settlers, whose crops, climate, soils, methods of production and, most important, relations of production differed greatly. In 1852 agricultural leaders organized the United States Agricultural Society, a nongovernmental organization that combined technological, educational, and political missions. At their annual conferences, held in Washington, D.C., one of the policy questions they could not resolve was whether the group supported the creation of an agricultural department at the federal level, which southerners generally opposed. Sectional interests were especially keen in the sugar politics of the day. Southern planters were disgruntled by the interest of the Patent Office and the United States Agricultural Society in developing sorghum as an alternative sugar source. Meanwhile, the Patent Office came under severe criticism in 1856 and 1857 when seventy-five thousand dollars intended for seed distribution were allegedly used instead to acquire and distribute sugarcane slips for the southern sugar-growing states (Gates 1960). The differences between southern planters and northern farmers were voiced in 1858 by the legislature of Iowa in opposition to sugar duties, which Louisiana planters favored: "The hope to successfully compete with other countries in the growth of the cane is proved a fallacy by the experience of more than half a century; and an annual premium of ten millions of dollars to one class of industry seems disproportionate and oppressive to those who are not rich, but struggling with the stern realities of rural and pioneer life" (U.S. Senate 1858, 1–2).

In 1862, no longer divided along sectional lines, Congress was able to establish the United States Department of Agriculture (USDA). One of the critical commodities of interest to the newly formed USDA was sugar, which, like other southern supplies, had been cut off during the Civil War. The initial Everglades reconnaissance reports had made it clear that sugar was a topic of national interest, but now concern for sources took on new urgency. Searching for an alternative to tropical sugarcane, the USDA began experiments as early as 1863 with sugar beets, and the July 1866 report of the USDA chemist included analyses of both sugar beets and sorghum. However, it was in the postbellum period, during the career of Harvey W. Wiley, that the USDA attacked the problem of sugar most energetically.[2] Wiley, who oversaw the experiments in Florida, stands out among those who were attempting to expand and modernize the U.S. sugar industry, and, more generally, he played a central role in U.S. food politics at the turn of the twentieth century. In aiding Wiley's research agenda by establishing a station in Florida, Disston had captured the attention of one of the first "bureaucratic entrepreneurs" to emerge in the USDA. President Theodore Roosevelt is reputed to have said that "Dr. Wiley has the grandest political machine in the country" (Coppin 1990, 178).

Wiley, who was chief chemist from 1883 until 1912, is best known for his crusade against food adulteration that led to the Food and Drug Act (1906). In 1912 he joined the staff of *Good Housekeeping* magazine as a contributing editor, with his own Bureau of Foods, Sanitation, and Health in the Good Housekeeping Institute (Mott 1968). Especially during his early career at the USDA, Wiley shared the view that the country should develop a strong domestic sugar industry. To that end, he traveled throughout Europe, studying the technology of beet and cane processing. His commitment to U.S. sugar self-sufficiency is apparent from the variety of solutions he sought, which included experiments with sugarcane, beets, and sorghum, and involved both agricultural as well as industrial innovation. In support of his work, Congress allotted $424,500 to the USDA for sugar experiments during the decade ending in 1890, a decade in which sugar boosterism became infused with the spirit of agrarian populism. Wiley felt that departmental research should identify where best to grow particular crops. By the 1890s, the USDA was concentrating its efforts on sugar beets, publishing a 250-page bulletin entitled *The Sugar-Beet Industry, Culture of the Sugar-Beet and Manufacture of Beet-Sugar.* Readers of this publication were particularly intrigued by a map that Wiley had prepared, showing a belt approximately one hundred miles wide where beets could best be cultivated, "crossing the country in a sinuous manner, depending upon average temperature for the months of June, July and August of seventy degrees Fahrenheit" (Wiley, quoted in Fox 1980, 523). This technique of mapping optimum yet imaginary "belts" was later used as the most graphic weapon in the Florida sugar boosters' arsenal.

Sugar was unusual in the agro-development politics of its day because it was the only staple food for which the United States relied heavily on imports. In addition, it was the only major imported commodity that was also produced domestically, unlike silk, rubber, and coffee (Dalton 1937). Postbellum sugar boosters were thus able to make their arguments for increased sugar production at three levels. At the national level, they argued that the United States should not spend millions to buy a product that could be produced domestically. At the regional level, they argued for the development of an agro-industry that would provide the economic underpinnings for the rapidly expanding farm sector. And at the household level, they deplored the costs to farm families. Advocates for domestic sugar production sought to reposition the United States in the global sugar market.

The example of the European sugar beet industry inspired them. Whereas cane accounted for 95.1 percent of world supply in 1839, with the political support and economic encouragement of European governments, beets supplied more than half the world's sugar by 1883, and nearly two-thirds by 1889 (Deerr 1950; Prinsen Geerligs 1912). Advocates believed that farmers could grow beets throughout much of the United States and thus make the country, which was importing almost 90 percent of its supply, self-sufficient in sugar. While U.S. boosters admired the European beet industry, they decried the bounties that had fostered its development, especially the German system, which provided a strong economic incentive to export rather than to produce for domestic consumption. The nineteenth-century world sugar economy was entirely restructured by the rise of the bounty-fed beet industry, which set the stage both internationally and domestically for the activities of U.S. sugar boosters (Albert and Graves 1984).

With the world awash in sugar, the mid-1880s were a period of crisis for the global industry. Yet overproduction and low prices did little to deter U.S. policymakers, politicians, and farm leaders from promoting domestic production. As early as 1861, some European states began to negotiate international trade agreements on sugar, and in 1863 the first international conference on sugar was held in Paris. However, these were limited measures. "For more than forty years from 1861 to 1903, international sugar diplomacy went from failure to failure" (Chalmin 1984, 14). Finally, the Brussels Convention of 1902, signed by representatives from Germany, Austria Hungary, Belgium, Spain, France, Great Britain, Italy, the Netherlands, Sweden, and Norway, pledged to abolish all direct or indirect bounties on the production or exportation of sugar for five years. Signatories considered the treaty successful, so it was extended for another five years, with Russia also signing (Prinsen Geerligs 1912). Thus, sugar diplomacy did not bear fruit until after the turn of the century, and until that time, U.S. boosters and the U.S. government were "exasperated" by the aggressive export strategies of European beet producers (Chalmin 1984, 16).

Much of the boosters' rhetoric was tinged with the "moral note, which enjoyed great vogue" immediately after the Civil War, when they could contrast U.S. production with the "slave-grown cane sugar" from, for example, Cuba and Brazil (Williams 1984, 388). When slavery was outlawed there as well, boosters adapted their discursive strategies by making more detailed comparisons of labor conditions at home and abroad, or by identifying other social categories in the international division of labor (e.g., "coolie labor"). Boosters elaborated a discourse that made capital investment and government support for U.S. sugar production the morally correct position. In the competition for state support among different sectors of capital and among nationally and regionally based producers, such efforts to define regions as relatively favorable or unfavorable, deserving or undeserving are a familiar practice (Harvey 1996; Paasi 2002). In later decades the competition between Florida's sugarcane region and its rivals, especially Cuba, would be discursively framed as a matter of national security. In the postbellum United States, boosters echoed a discursive strategy used in various other places and times. In France, for example, advocates of the beet-sugar industry in the 1830s invoked a "powerful propagandist appeal against the cane sugar industry on the ground that cane sugar required slave, 'coolie,' contract, or some form of degraded labor, black, brown, or yellow, whereas the beet sugar industry was the product of white and free labour" (Williams 1984, 388).

Herbert Myrick's publications exemplify nineteenth-century agricultural boosterism. Myrick was the editor of the *American Agriculturalist, Orange Judd Farmer, New England Homestead,* and *Farm and Home,* and the treasurer of the American Sugar Growers' Society. Well-known and widely regarded, he was also the director of the Good Housekeeping company from 1900 until 1911, and in that capacity interacted extensively with Wiley, who publicized his crusade for the Food and Drug Act in the magazine. The title of Myrick's 1897 publication, *Sugar: A New and Profitable Industry in the United States for Capital, Agriculture and Labor to Supply the Home Market Yearly with $100,000,000 of Its Product,* presents the gist of his argument concerning that commodity (fig. 2.2). He begins by noting the economic absurdity of U.S. sugar imports: "It required every pound of the wheat and flour exported by the United States during the fiscal year 1896 to pay for the sugar imported. The total value of all live and dressed beef, beef products and lard exported during the past year barely equaled the amount paid for imported sugar." He

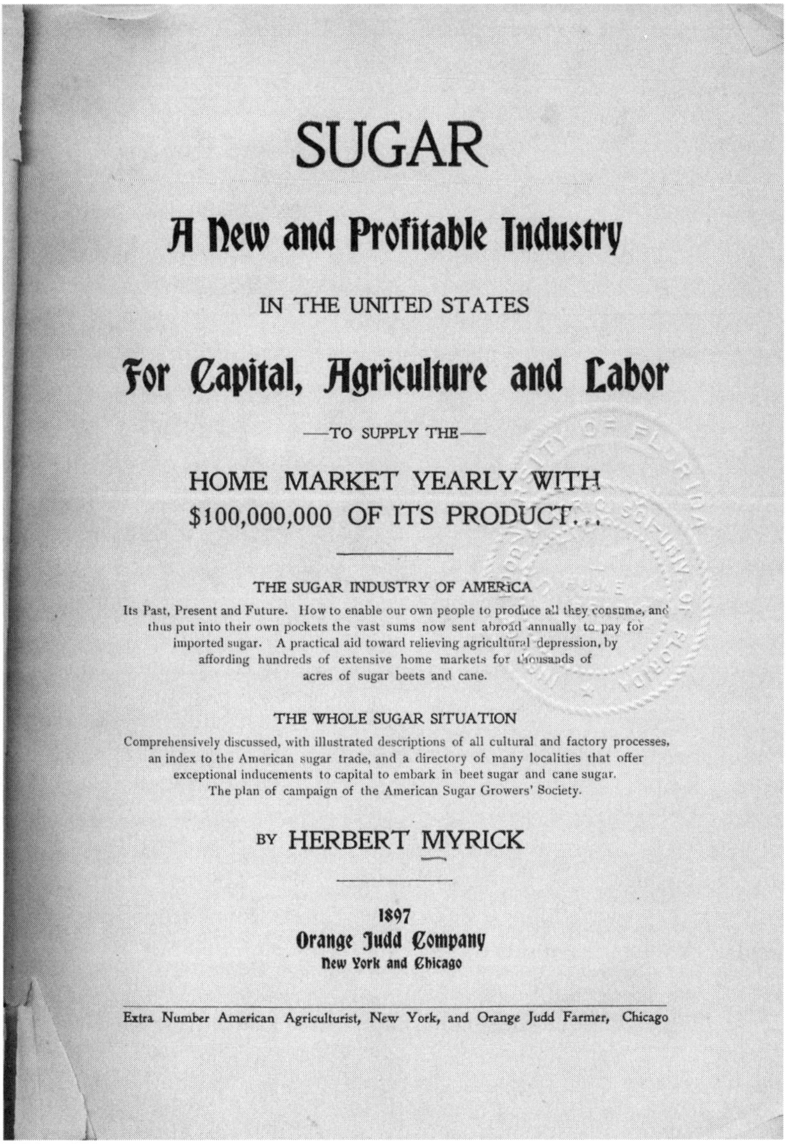

SUGAR

A New and Profitable Industry

IN THE UNITED STATES

For Capital, Agriculture and Labor

—TO SUPPLY THE—

HOME MARKET YEARLY WITH
$100,000,000 OF ITS PRODUCT...

THE SUGAR INDUSTRY OF AMERICA

Its Past, Present and Future. How to enable our own people to produce all they consume, and thus put into their own pockets the vast sums now sent abroad annually to pay for imported sugar. A practical aid toward relieving agricultural depression, by affording hundreds of extensive home markets for thousands of acres of sugar beets and cane.

THE WHOLE SUGAR SITUATION

Comprehensively discussed, with illustrated descriptions of all cultural and factory processes, an index to the American sugar trade, and a directory of many localities that offer exceptional inducements to capital to embark in beet sugar and cane sugar. The plan of campaign of the American Sugar Growers' Society.

BY HERBERT MYRICK

1897
Orange Judd Company
New York and Chicago

Extra Number American Agriculturist, New York, and Orange Judd Farmer, Chicago

Figure 2.2. Promotional publications such as this one by Herbert Myrick on sugar were typical of nineteenth-century agricultural boosterism. Courtesy of the University of Florida Libraries, Special and Area Studies Collection.

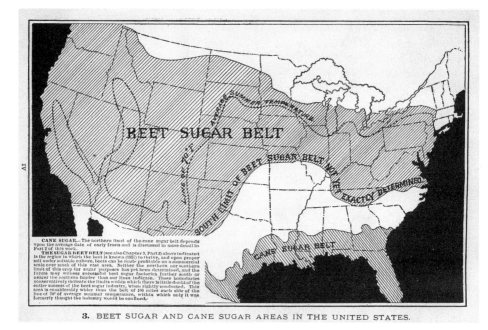

3. BEET SUGAR AND CANE SUGAR AREAS IN THE UNITED STATES.

Figure 2.3. Herbert Myrick's 1897 version of the two sugar belts that boosters claimed would obviate the need for imported sugar in the United States. Courtesy of the University of Florida Libraries, Special and Area Studies Collection.

calls U.S. sugar imports "an economic crime" because they "compel American farmers to raise staples in competition with the cheap-land-and-labor countries, with which to pay for imported sugar" (Myrick 1897, 1). He is particularly vexed by a fivefold increase in sugar imports from Europe and by substantial increases in imports of "the coolie-grown product" of "the Orient." The "worst and most inexcusable phase of the sugar situation is the unjust, unfair, illegal and unbusiness-like competition of sugar from the Hawaiian Islands" where "coolie labor is employed to raise this cane" (3).

Myrick's boosterism included both beet and cane sugar, and was therefore not regionally specific. He used several maps to support his central argument that the United States could produce its own sugar. The first delineated the "probable possible distribution of these commercial crops"—sugar beets and sugarcane—in the United States, and the second showed "the counties that have already started a movement to secure a sugar factory. In some of these counties several towns are aspirants for the factory." A third illustrated an exaggerated "beet sugar belt" and an imagined "cane sugar belt" (fig. 2.3). The primary impediment to the development of the U.S. industry was European beet-sugar imports, which increased a hundredfold in fifteen years: "She has developed her beet-sugar industry by a liberal system of direct subsidies, high protection and export bounties, until the European beet-sugar industry has practically ruined the cane-sugar industry of the tropics and monopolized the sugar market of the world" (Myrick 1897, 10). Therefore, argued Myrick, a duty on imported sugar was necessary so that the domestic industry could supply the U.S. market, the largest in the world.

Though critical of foreign competition, U.S. sugar boosters were informed and invigorated by international developments. The reports of U.S. consuls in Europe on the German, French, Russian, Bohemian, and Austria-Hungarian beet-sugar industries included detailed descriptions of methods of cultivation and processing, as well as illustrations of architectural plans for factories. Another source of technical information was the Cuban industry. In the later nineteenth century, as changes in the sugar market and refining industry drove the price of raw sugar down, Cuba restructured its sugar industry. "Cuba encountered an industrial revolution in the sugar industry in the same period that the abolition of slavery and an oppressive fiscal system caused labor costs and tax burdens to rise" (Jenks 1976, 30). At this time Cuba had gained "preeminence in the world sugar industry largely on the basis of domestic capital and expertise long before North American investors began to cast their eyes in Cuba's direction" (Dye 1998, 3). This restructuring meant the decline of the *ingenios*—the smaller, integrated plantations and mills—and the rise of the central factory system—*centrales*—or as Florida boosters would refer to them, "centrals." With this development came the institution of the *colono* system, "a set of arrangements by which outside growers supplied cane" to the central factories (95). Then, after 1898, an influx of North American capitalists and industrialists contributed to the second industrial revolution in the Cuban industry. Thus, American engineers and machinists were involved with their Cuban counterparts in the ongoing process of modernizing and redesigning mills and transportation links in Cuba. They transmitted their experience in the Cuban industry through trade journals and agricultural publications, which educated southern, and especially Floridian, boosters concerning the efficient organization of the relationship between growers and mills. The idea of the central would figure prominently in the Florida industry's promotional discourse and imagined economic geography.

The industrial revolution in the Cuban industry coincided with that of the Louisiana industry, which, after 1877, began to undergo "a revolution in organization and methods of production" (Sitterson 1953, 252). Changes included improved cultivation and manufacturing techniques, the introduction of the central factory system, and the establishment of the Louisiana Sugar Planters' Association, the Audubon Sugar School, the Louisiana Sugar Exchange, a privately funded research station, and a publication, *Louisiana Planter and Sugar Manufacturer*. These scientific, technical, educational, and business institutions, "at the heart of the late-nineteenth-century modernization of the Louisiana sugar industry . . . were a direct response to the challenges posed by a dynamic international market" (Heitmann 1987, 68).

The bounty provided by the McKinley Tariff of 1890 provided economic stimulus for this established cane industry as well as the nascent beet industry, which, between 1891 and 1894, expanded output from 3,874 to 23,344 tons per year. Between 1861 and 1897, annual per capita sugar consumption rose from 18 to 78 pounds, while between 1865 and 1897, the domestic share of this expanding market rose from 2 to 13 percent (Williams 1984). Thus, during the late nineteenth century, the efforts of Wiley and other domestic sugar promoters were rewarded by an increase in U.S. production and in the domestic market share. We can attribute the expansion of the beet-sugar industry not only to the bounty, but to the efforts of the USDA, which had "preached beet sugar in season and out of season; spread broadcast pamphlets dilating on the advantages of beet growing for the farmer and giving minute directions on cultivation; maintained a special agent who kept in touch with the manufacturers and farmers, and annually reported on the progress of the industry" (Taussig 1931, 80). Looking back after more than a century, we can see that the beet-sugar industry took root at a particularly propitious moment, just as the political geography of the United States was about to be transformed. This timing would have critical implications for sugar in Florida and for the transformation of the Everglades.

The Sugar Question, Circa 1900

The McKinley Tariff of 1890, which had so encouraged Hamilton Disston's doomed efforts to raise sugarcane in the Everglades, marked a distinct juncture in U.S. sugar policy. From 1789 until 1891, when subsidies were first paid, the justification for the tariff on sugar was to raise revenue for the U.S. Treasury. During the nineteenth century, when import duties provided about two-thirds of government receipts, sugar accounted for 20 percent of duties. Tariff structure differentiated between raw sugar and refined sugar. Higher duties on refined sugars protected U.S. sugar refiners. Tariffs on raw sugars protected domestic growers, namely, the Louisiana industry, with varying degrees of effectiveness. The McKinley Tariff was a departure because it placed raw sugar on the "free" list, and then, to make up for the loss of protection, provided a direct subsidy—a two-cent per pound bounty—to domestic sugar producers. This two-cent bounty certainly encouraged the expansion of domestic production. Both the cane and beet industry expanded during this time, with Disston's venture as one example.

Whether that was the sole intention of its supporters is more debatable, for the McKinley Tariff did more than encourage domestic production; it entered into the struggle between East and West Coast refiners over market share. This struggle was largely a consequence of the emergence of the American Sugar Trust in 1887. The trust was initially composed of seventeen East Coast sugar refineries, but soon added the American Sugar Refining Company of California as part of a complex battle over control of the U.S. sugar market between the trust and Claus Spreckels in California (Ayala 1999). Spreckels dominated sugar refining on the West Coast, but refused to join the new consortium. Enactment of the McKinley Tariff coincided with the ensuing price war, and its application produced greatly different effects on the two sides.

By placing raw sugar on the free list, the McKinley Tariff dispensed with the advantage given to West Coast refiners by the special status of Hawaiian sugars, which had been admitted to the United States duty-free since 1876. Now Cuban sugars were no longer subject to duties and therefore entered U.S. markets to supply East Coast refineries (Ayala 1999). Thus the bounty provided support for domestic cane and beet growers and encouraged the Cuban industry to expand. The "loser" was the Hawaiian industry, which included Spreckels. Without tariffs, Hawaii lost its preferential status in the U.S. market vis-à-vis other foreign producers and did not receive the bounty given to domestic producers. The result was general deflation of the island economy and then political discontent, which contributed to the overthrow of the monarchy by American settlers and the establishment of the Republic of Hawaii. With the passage of the McKinley Tariff, the Hawaiian industry "was left exposed to the 'winds of international competition.' For the first time, Congress juggled the interests of the various sugar-producing groups" (Dalton 1937, 24).

The ostensible reason for doing away with the sugar duty was surplus revenue in the treasury. Established Louisiana planters did not want duties to be reduced and were wary of the bounty, "which they feared would not be retained long" (Sitterson 1953, 327). The difference in opinion between Rufus Rose and Hamilton Disston on this score might reflect the fact that Rose, previously from Louisiana, was better in touch with the sugar politics of the time. As it turned out, the bounty was paid for four years' crops, those planted from 1891 until 1894. The *New York Times*, which opposed the bounty, made an example of Disston in an editorial column:

> We have received, in an envelope bearing on the outside the name of Mr. Hamilton Disston of Philadelphia, a copy of an article recently published in Florida relating to the sugar industry. It is urged in this article that the sugar bounty should be 'guaranteed for fifteen years,' or a duty of 2 cents a pound on all raw sugar should be imposed. We find also the remark that Mr. Hamilton Disston has "reclaimed 2,000,000 acres" of Florida land, "of which 500,000 acres are proved to be the richest land for sugar or any other purpose on the face of the earth." We understand, from this, that Mr. Disston is growing sugar, or desiring to do so, and would like to have the wheat growing and all other producers keep on paying the bounty which sugar growers now enjoy. (January 8, 1894, 4)

The 1894 bounty payments were disputed and therefore substantially delayed. Though Congress finally appropriated money for their payment in 1895, the comptroller of the United States refused to make payments until a Supreme Court order of May 1896. Payments were made in the summer of 1896, by which time Disston had already taken his own life.

The 1894 Wilson-Gorman Tariff marked a return to the pre-1890 system, with reciprocity for Hawaiian sugar and duties on all other foreign sugars. As Sitterson notes, despite the efforts of Louisiana planters and allied industries, the tariff, a 40 percent ad valorem duty that kept the half-cent differential on raw and refined sugars, was not a victory for domestic growers. This he attributes in part to the fact that the "sugar interests of Texas and Florida were not strong enough to control their members of Congress, who made no fight for sugar" (Sitterson 1953, 330). The Wilson-Gorman Tariff had especially harmful consequences for Cuba, because sugar was no longer on the "free" list and therefore subject to import duties (fig. 2.4). The result was massive deflation, which added to political unrest against the Spanish colonial regime and made the sugar industry a target of the ensuing revolution (Ayala 1999). Meanwhile, in the presidential campaign of 1896, further protection for U.S. domestic sugar production was a feature of the Republican platform, which stated that the party favored "such protection as will lead to the production on American soil of all of the sugar which the American people use" (quoted in Williams 1984, 388). With the return of Republicans to power, the Dingley Tariff of 1897 strengthened protection of the domestic industry by raising duties. The act was intended to nullify foreign bounties by providing a countervailing duty equivalent to what-

Figure 2.4. Political cartoon, circa 1894, depicting Cuba's entry into the protected U.S. sugar market through the reciprocity agreement. The caption reads, "Cuba—Why not let me in? Porto Rico is inside. American Sugar Grower—She didn't come in this gate. She went in the other one—and I can't control *that!*" Courtesy of the Library of Congress.

ever amount another government's bounty happened to be. The tariff remained in place for the next seventeen years, a period of remarkably stable U.S. sugar prices during which the beet-sugar industry flourished, growing from six factories in 1896 to fifty-one in 1904, while acreage increased from 22,948 to 252,100 (U.S. Senate 1906).

The immediate impact of the Dingley Tariff on Cuba's sugar industry was quite the opposite of its effect on the U.S. industry. At the beginning of the decade, Cuban sugar production had been expanding rapidly. The 1890 McKinley Tariff allowed Cuba's raw sugar free entry into the U.S. market, which translated into an increase in the country's sugar output from 632,000 long tons in 1890 to 1,054,000 in 1894 (McAvoy 2003, 18). In 1895, following the passage of the Wilson-Gorman Tariff, the rapid expansion of sugar production in Cuba ground to a halt. The political-economic repercussions were "a disaster" for Cuba (21). The Spanish government responded against U.S. imports, so living costs soared in Cuba even as producer prices for sugar plummeted. At the end of the 1895 harvest season, dismissed sugarcane workers joined the growing ranks of rebel forces in the revolution for Cuban independence. The War of 1895 was a continuation of the ongoing struggle for independence from Spain and was also—as inspired by the leadership and writings of José Martí—waged to defend against potential annexation by the United States. The war meant "the nearly complete paralysis of the sugar industry" (Zanetti and Garcia 1998, 177) in Cuba. Noting the relationship between the fortunes of Cuba and Florida, the *New York Times* speculated that "the destruction of the Cuban crop this year should greatly stimulate the sugar industry in Florida and other States" (*New York Times* 1896). The 1897 Dingley Tariff further disadvantaged Cuban sugarcane producers by providing incentives for U.S. beet-sugar producers to supply the domestic market. The resulting downward spiral of living conditions added fuel to the revolutionary fires in Cuba, the world's largest sugar exporter. Eighteen months after enacting the Dingley Tariff, the United States established a protectorate over Cuba in the aftermath of the Spanish-American War. From New Year's Day 1899 until May 1902, a U.S. military government administered Cuba (Thomas 1971).

Though complex, contentious, and often opaque, domestic and international sugar politics at the turn of the last century were not arcane. Debate on the sugar question was not relegated to the agricultural press, but raged through the popular and the prestigious publications of the day. An editor of the *Forum* noted, in 1898, "a deepening of the popular interest in political economy and politics" (Mott 1957, 155), and, at that time, "no politico-economic question occupied more space in the magazines and reviews than the tariff in its many phases" (165). In general, during the 1890s "on the subjects of foreign trade, international relations, and political crises abroad, there was a constantly increasing number of articles in the leading magazines" (223). The sugar question brought all of these popular concerns together. In the period between 1897 and 1902, the discourse on the sugar question shifted from considering it as an issue of primarily domestic interest—albeit engaged with the international sugar market—to seeing it as an explicitly international problem concerning the rights of new territories and the responsibility of a colonial power.

In 1897, the *American Monthly Review of Reviews* published "Sugar—The American Question of the Day," by Herbert Myrick. "To sugar or not to sugar," wrote Myrick, "seems to be the present issue in the United States Senate" (673). The Dingley bill under debate, though not as favorable to agriculture as some had hoped, would give the farmers "a reasonable chance." Farmers wanted a tariff rate that would encourage domestic sugar cultivation, he argued. The profits to be made in beets were promising, while the primary expense would be labor, "largely a class that is now unemployed—children, unskilled help, etc." (676). Myrick asserted that a single state such as California or Iowa, if devoted to beet production, could alone provide the entire U.S. sugar supply. "To this end our farmers are justified in asking as much help as Europe has given her beet sugar industry. But authorized as I am to speak for two millions of them, let me say that our farmers ask hardly one-third as much as Europe has done" (677).

Secretary of Agriculture James Wilson took up Myrick's call in the journal *Forum,* in an article entitled "Should the United States Produce Its Sugar?" (Wilson 1898). He described the efforts of the USDA in "experimentation until a belt is established across the continent determining where conditions are most favorable" for sugar beets and sorghum, noting that "while men of forethought and decision are carrying these investigations forward" the question has been raised "whether it is a wise policy for the people of the United States to produce their own sugar" (1). Wilson linked arguments concerning the material advantages of beet production—that it depleted the soil less than did grain cultivation, was useful for rotations, and provided fodder—to issues of trade, arguing that dependence on grain exports to pay for sugar imports led to a net loss of soil fertility that would ultimately reduce the country's productive potential. Also in the *Forum,* USDA Chief Chemist Wiley published "The True Meaning of the New Sugar Tariff." In contrast to political critiques—Democrats said the sugar bill was dictated by the American Sugar Trust, Republicans claimed that it dealt the Trust a staggering blow—Wiley's critique was scientific. He noted that the tariff rate, based on a color standard, was subject to fraud because a nearly pure sugar could be tinted to avoid duties. In addition, the tariff as structured provided incentive to import only lower grade raw sugars, keeping from U.S. consumers the "yellow, crystal cane-sugar much in demand for table use among the richer classes in London. These unrefined sugars, so desirable and palatable, might be brought into more general use, except for the differential duty, which tends to exclude them from our markets" (Wiley 1898, 694).

Around 1900 the sugar question took on a new wrinkle, with numerous reports in magazines such as *Current Literature* and *Scientific American,* as well as a USDA Farmers' Bulletin, which linked the nutritional qualities of sugar to issues of national security. Titled "Sugar as Food," these reports conveyed the findings of experimental work done in Europe to test the effect of sugar in lessening fatigue, first undertaken in Italy in 1893 and then by the Prussian War Office, with results published in 1898. Sugar "has been advanced to a pedestal in the dietary list" since German army authorities "proved conclusively that it is an adjunct to the soldier's diet of almost inestimable value. The British government followed the example set it by the Germans. The result of the experiment has been most satisfactory in the South African campaign. Our army, too, in its revised ration scale, is allowed a generous amount of sweet food" (*Scientific American Supplement* 1901). The German government's motivation for this research was not only to sustain soldiers, but also to spur demand for sugar. Producing one-fourth of the world's sugar, Germany was dependent on foreign demand at a time of increasing protectionism and supply. To maintain the beet-sugar industry, the Germans focused on domestic consumption, seeking to increase "the amount of sugar used by individuals, especially in the army, where increased consumption may be made compulsory" (*Scientific American Supplement* 1900).

More broadly speaking, the sugar question was reconfigured at this time by geopolitical events, with two distinct trajectories. One remained domestic, but concerned the United States as an emerging imperial power. As described by a government economist of the era, Frank R. Rutter, "The problem presented by the sugar tariff—one purely of protection and revenue—has been greatly complicated by the legacy to this country from the Spanish War of new responsibilities over important tropical sugar-producing areas. A colonial policy, as well as a tariff policy, is now involved in the sugar question" (1902, 44). The second was at the global level, where sugar stood as a synecdoche for the whole of international trade relations: "The public prominence that questions relating to sugar now occupy through-out the world is remarkable. It is a most curious and interesting fact that no other food product enters so largely into the domain of state and international politics" (Crampton 1901, 283). Or, more succinctly, "It has been said repeatedly that the sugar question *is* the economic question" (Meyer 1910, i).

In the debate over U.S. "colonial policy," the greatest conflict concerned trade relations with Cuba. Cuba was the world's largest sugar producer, and the relation of the Cuban industry to the United States was of critical concern to the domestic industry. Charles A. Crampton—then chief chemist of the Internal Revenue Bureau and previously assistant chemist to Dr. Wiley at the USDA—argued that the "solution of the problem of successful colonial expansion by the United States will be found in the rehabilitation and development of the tropical sugar cane industry" (Crampton 1899, 276). He advocated protection for colonial sugar. "This must not be objected to on the ground of its being political, and therefore artificial, aid. Fire must be fought with fire, and sugar has been entangled with politics from the time of the first Napoleon down to the present day" (282). He also echoed Wiley's concerns, suggesting that U.S. consumers should use a "high grade raw cane sugar, which is practically identical with the granulated article except for a slight tinge of color, the removal of which furnishes our refineries their immense profits" (281). In a later article, Crampton reviewed the struggle over Cuban reciprocity. He advocated that Hawaii, Puerto Rico, the Philippines, and Cuba should be treated equally, free of tariffs. The opposition of domestic interests to free trade with Puerto Rico, which "surprised the entire country even to the Chief Executive," was really just a skirmish in preparation for the bigger battle over Cuba. Since then, "their political strength has been much augmented by means of the growth of the beet industry and its extension into new States. Thus an alliance of agricultural interests—always powerful with Congress—of widespread geographical extent, and well organized and equipped for such a contest, will oppose during the coming session any reduction of duties upon merchandise imported from Cuba or the Philippines" (Crampton 1901, 285–86). Secretary Wilson's claim that in a short time the domestic beet industry would produce all the sugar consumed by the United States, "rendering us independent of foreign-grown sugar" was, in Crampton's view, a "most remarkable prediction" since USDA statistics showed that consumption far outpaced increases in domestic production. "There is so much said in the daily press about the importance of the beet-sugar industry in this country that the general public has come to regard it as a large and growing institution, but it plays a very insignificant part both as to size and growth" (290).

Opposed to reciprocity was *Gunton's Magazine,* a Republican monthly "subsidized by the Standard Oil Company and frankly the spokesman of trusts, a high protective tariff, and what the Democrats called 'special privilege'" (Mott 1957, 171). The first question it addressed was "Can we raise our own sugar?" Not only could we, argued the author, but we should, because the beet industry "is the finest type of an 'infant industry' developed in this country since the civil war. Shall this 'infant industry' be sacrificed by annexing the island of Cuba or admitting Cuban sugar duty free?" (Crawford 1902, 58). The following month an article entitled "Prospects of Domestic Sugar" claimed that it would be possible to supply half the U.S. demand domestically within a decade. When sugar was restricted to the South it was not adequately protected. "But now that sugar is grown North, East, West and South, there will probably be sufficient political strength to maintain the present generous protection of about 75 per cent" (Wilkinson 1902, 135). That *Gunton's*—the spokesman of trusts—had become so solicitous of beet farmers is best explained by the Sugar Trust's purchase of a large interest in beet factories in 1901 and 1902 (Ayala 1999).

George Gunton, author of *Wealth and Progress: A Critical Examination of the Labor Problem,* weighed in on Cuban reciprocity with an article entitled "Some Free Sugar Fallacies." He noted that the primary beneficiaries of free sugar would be the Cuban planters and American refiners. The Cuban planters, he thought, deserved some measure of support, but U.S. investors in the Cuban industry should not expect the same level of protection afforded domestic producers.

> How charming!—to encourage a state of affairs in which American capitalists, sugar refiners, and what not, could go to Cuba or any other foreign country and use the equivalent of slave labor and be exempt from duty in the United States. When American capital goes to a foreign country in pursuit of cheap labor, it loses all claim to protection or privilege in the American market (Gunton 1902, 144).

George Kennan, who was at that time the Washington correspondent for *Outlook* magazine, reflected on recent Congressional hearings on proposed tariff relief for Cuban sugar. Although people expected that the Cuban sugar manufacturers would make a strong showing and that "their most formidable antagonists"—the beet producers—would find it hard to resist their appeal for tariff relief, this was not the outcome. A key problem, according to Kennan, was that the Cuban representatives were not accustomed to American Congressional methods and had difficulty with English, so that it was not clear that they spoke "not only for the Cuban planters, but for municipalities, boards of trade, bankers, merchants, labor organizations, stevedores, and the people of Cuba generally, even to the firemen of the city of Havana" (Kennan 1902, 367). Even though they presented petitions and appeals from all these groups, some still wondered if any representatives of the Cuban people were present. In contrast, Mr. Henry T. Oxnard, president of the American Beet-Sugar Association, demonstrated the widespread economic impact of the U.S. beet-sugar industry when he stated that there were forty factories in eleven states producing 150,000 tons of sugar and paying seven million dollars annually to farmers. Oxnard claimed that in ten years the beet-sugar industry could furnish all of the 1.5 million tons that the United States imported.

"The question of Cuban reciprocity involve[d] the whole commercial policy of the United States towards its dependencies" (Rutter 1902, 78), and factoring in the interests of consumers strengthened the case for reciprocity. National interest in Cuba was expressed in and promoted by *National Geographic,* which published ten articles on the island between 1898 and 1905, having not previously done so (Schulten 2001). However, even with

U.S. public opinion and the administration, including President Theodore Roosevelt, on the side of reciprocity for Cuba, Congress decided in 1903 to leave 80 percent of the tax in effect. We can interpret this decision as "a historical victory for the beet interests," which now included among its members the American Sugar Trust (Ayala 1999, 61). Nonetheless, because Cuba received a 20 percent tariff preference over other countries and in return gave U.S. shippers significant tariff reductions, the reciprocity treaty—ungenerous to Cuba as it may have been—was seen as "a signal for big further investment" (Thomas 1971, 469). That investment was primarily in sugar. Contrary to the hopes of domestic growers, President Roosevelt wanted to give "Cuban mills a chance to expand production till they supplied all the sugar the U.S. needed" (469).

The outcome of the Spanish-American War reconfigured the political geography of the United States. As a consequence, the political economy of the U.S. sugar supply was transformed by the rise of the "American Sugar Kingdom" (Williams 1984). The American Sugar Kingdom took shape after 1897 with "the enormous concentration of latifundia in the Caribbean under the stimulus of American capital investment" (429). While the geographic sources of sugar remained roughly the same, the boundaries determining which producers were inside U.S. territory and therefore considered "domestic" were redrawn. As evidence of this transformation, consider that full-duty sugar, 53 percent of U.S. supply before 1898, accounted for less than 1 percent by 1913. As John E. Dalton, chief of the Sugar Section of the Agricultural Adjustment Administration in 1934 and 1935, observed, "With the acquisition of new territory following the Spanish-American War, the protective principle was applied over an increasingly larger area and our new sugar policy recast the economic physiognomy of Hawaii, Puerto Rico, Cuba, and the Philippine Islands" (Dalton 1937, 39). Sugars from Hawaii, Puerto Rico, and the Philippine Islands were admitted duty-free. Cuba, outside the free-trade area, received varying levels of trade reciprocity. Regardless of the outcome of tariff battles, the formal political ties established under the Platt Amendment led U.S. capitalist investment in the Cuban sugar industry to increase exponentially (Jenks 1976). Some of this investment went to the sugar region of western Cuba, an area of independent growers (*colonos*) supplying cane to centralized mills (*centrales*). However, the east, characterized by cattle ranges and subsistence farming, was the primary destination for the horizontally consolidated U.S. sugar-refining interests, which now sought vertical integration through development of offshore corporate plantations (Dye 1998). As sugar refiners moved toward vertical integration, they balanced their interests geographically, with investments in the protected domestic beet industry and the efficient Cuban cane industry (Ayala 1999).

Thus, after 1898, Florida's significance as a subtropical frontier was greatly diminished. Agricultural boosters recognized this, and so we see a rivalry between Florida and the new American Sugar Kingdom, discursively constructed as populist protectionism versus capitalist imperialism. Myrick, writing in 1907, argued that free trade "would permanently blight Florida and almost annihilate her agricultural industries, making Florida only a way station to the tropics. After having given freely of our blood to drive out their Castilian oppressors, . . . proprietors in the East and West Indies now seek a yearly bonus of untold millions" (Myrick 1907, 9–11). Against this international backdrop, different development interests in the U.S. South envisaged contrasting roles for sugarcane. They promoted these competing visions at meetings held by the Interstate Sugar Cane Growers Association, a group that, despite the rise of the American Sugar Kingdom, persisted in seeing in the establishment of a southern cane belt hope for the depressed rural economy of the postbellum South.

Imagining a Southern Cane Belt

Though sugarcane cultivation had a long history in the U.S. South, commercial sugar production was still geographically restricted at the turn of the century when southern agriculturalists and political leaders organized to promote the establishment of a regional cane belt. They intended to create something akin to the Midwestern Corn Belt. However, the project itself suggests very different historical and material circumstances than those that attended the emergence of the Corn Belt. Mid-nineteenth-century Midwestern development was at once the story of "new land" being brought into production and of new institutional structures being developed that would serve to mediate the relationship between farmers, agricultural industry, finance capital, and the state. Late-nineteenth-century southern boosterism occurred in the context of regional agricultural decline, and, by that time, encountered an established institutional structure, the USDA.

William Cronon (1991) and John Hudson (1994) each recount the historical process through which the Midwestern agricultural region materialized. Cronon focuses on the formation of urban-rural linkages that transformed the environment, considering how commodity production alters and is altered by technological change, socioeconomic institutions, and natural conditions. He highlights the pivotal role of boosters, who articulated a theory of regional development focused on "the symbiotic relationship between cities and their surrounding countrysides" and epitomized by the growth of Chicago, "Nature's Metropolis" (Cronon 1991, 34). Though some of its methods were similar, cane-belt boosterism differed in its focus on a single commodity from the Midwestern boosterism recounted by Cronon. Sugarcane boosters had a much more modest vision of the future of southern hinterlands that did not necessarily entail intensive urban development. Instead, they emphasized an extensive rural agro-industry centered on a network of regional mills and small towns.

Hudson (1994) is specifically concerned with the question of how an agricultural region emerges. He describes an iterative and interactive process in which a few commodities, corn and livestock, became economically and geographically dominant throughout a self-defined region, the Corn Belt. That is not what southern agriculturalists had in mind, because it would not have worked. Creating an economically competitive cane belt would not have been an iterative process. Instead it would have required substantial capital investment to launch an agro-industrial, regional plantation economy. The material properties of sugarcane—its bulkiness and the fact that it must be milled within twenty-four hours of harvest—meant that commercial sugarcane farming required the establishment of mills close to the cane fields and a destination for the raw sugar once it was milled. This was the problem confronting southern boosters, and, maddening though it was to see the emergence of the western beet-sugar belt under the tutelage of the USDA, the success of the beet-sugar industry also gave them hope. Even more than hope, the tariff that enabled the emergence of the beet-sugar industry "made protection to raw sugar a national instead of a sectional issue; for it gave promise of an ultimate domestic sugar product equal to the entire demand" (Rutter 1902, 47). Secretary Wilson's declaration that the "island possessions" of the United States would raise coffee and not sugar probably fueled the optimism of southern boosters, especially when he announced that he had dispatched "experts of the Agricultural Department" to Puerto Rico, Hawaii, and the Philippines to teach "the natives improved methods in [coffee] cultivation" (*New York Times* 1905, 8).

A notable characteristic of sugarcane boosterism was how closely it could be modeled on the beet-sugar industry's propaganda, for which the USDA provided a veritable blueprint. Sugarcane boosters had the advantage of yearly USDA reports, begun in 1896, that outlined not only the successes of the beet industry, but also "mistakes made, and obstacles met and overcome" (Sayler 1905, 13). One of the chief obstacles was labor supply: the success of the European beet industry was based on an agrarian structure and settlement pattern that was not replicated in the "frontier" regions of the United States where beet cultivation was being promoted. The USDA suggested that city-dwelling "foreigners who had experience in beet growing in their native countries" were a potential source of labor, as were the Salvation Army, reformatory institutions, asylums, and public schools (37). A second obstacle was the difficulty of obtaining seeds, which at that time were imported from Europe, where there was "a sentiment not conducive to [U.S.] interests in securing seed of the best quality" (23). Yet what the reports emphasized most was "the influence the beet sugar industry is exerting for development along all other lines in the States where it has been established" (34).

This latter observation inspired advocates of regional development in the South, but sugarcane farmers faced an even more bedeviling situation than did beet farmers with respect to seed supplies. William Carter Stubbs, director of the Louisiana Sugar Experiment Station from 1885 until 1905, asserted that the South held tremendous potential for sugar because the area capable of growing cane was extensive, including everything below a line drawn west from Savannah, Georgia, through Texas and into Arizona and New Mexico; just within Louisiana alone, he asserted, acreage could be increased tenfold. But, as explained in detail later, cane propagation was inherently difficult. Therefore, he cautioned, regional expansion would necessarily be gradual. "Cane . . . has one peculiar feature, not possessed by hardly any other plant cultivated in the United States. The large amount of cane necessary to plant an acre (from four to six tons) makes it necessary to go slow[ly] in the establishment of a large plantation" (Stubbs 1907, 20). For the first few years, the entire crop must be used for replanting. Consequently, three years are needed to get into production, and so expansion of cane production occurs gradually, unlike that of beets. Stubbs identified the catch-22 of founding the sugarcane industry: "Cane plantations must be established before the factory will be secured, and farmers are slow to establish a crop which requires three years of work and patience, unless they have 'an assurance doubly assured' of a factory" (26).

By the time of his 1901 report on U.S. sugar production to the Commission on Agriculture and Agricultural Labor, Chief Chemist Harvey W. Wiley was primarily interested in beets. He discussed his research in France, which suggested a model to him for the U.S. beet industry (Wiley 1901). He described a large factory being built in Colorado, the location of which was chosen by the USDA on the basis of their data: "That is where the practical part of such investigation comes in—pointing the investor to where the agricultural conditions are favorable and warning him from regions unfavorable" (648). Wiley delineated the proper role for the USDA regarding industry development: "The men who are putting their money in this in-dustry today come to the Department of Agriculture to ask where to locate their factories, because we have collected the data and studied the subject from a purely scientific point of view" (648). His description of the poten-tial for increased sugarcane production shows that his interest had declined both in the crop and in the muck lands of south Florida. Louisiana's output was fluctuating greatly, and he wrote that "it is not probable that Louisi-ana will ever produce any more sugar than to-day" (649). Wiley argued that the future for Louisiana was in rice, and that the future of the U.S. sugar supply was not in domestic cane. "We need not expect any large increase in our cane-sugar production. It is more likely to begin to decrease than to increase. Our production in Texas does not cut any great figure, although Texas has produced considerable cane sugar. Florida is likewise hardly to be considered as a cane-sugar country until at least the swamp lands are freed of water" (649).

As U.S. agricultural policies and promotion became de facto regional development policies, the issue of organizing to develop a southern cane belt took on greater urgency. Taking exception to Wiley's predictions was the Interstate Sugar Cane Growers Association (ISCGA), first convened in 1903 in Macon, Georgia. The meeting was proposed by Capt. D. G. Purse, of Savannah, to bring together political representatives, farmers, business people, and manufacturers of sugar-processing machinery for the purpose of resurrecting and developing the southern sugarcane industry. That is, the sole purpose of the ISCGA was mainland sugarcane boosterism. "After the lapse of near a half a century a revival has taken place in this nearly forgot-ten crop, as a commercial industry, and it is . . . an opportune time for the entire cane belt of the United States, and those interested in it, to assemble and consider plans and methods" (ISCGA 1903, 9).

A sugarcane belt was appealing in light of the economic decline of two other regionally important commodities, cotton and lumber. Cotton was "king" in the antebellum South: it was the most valuable American com-modity sold on world markets (Mann 1990). However, during the Civil War, major international buyers such as the British textile industry began sourcing elsewhere, for example, from India. After the war, domestic U.S. cotton cultivation expanded from its regional concentration in the South-east to include the Southwestern states. Thus, southern cotton cultivation was already struggling with low prices and loss of markets when, in 1890, a thirty-year infestation of boll weevils began, and cotton hit a thirty-year low (Ayers 1992). Timber fared little better. In the early 1880s, northern tim-ber companies had begun buying up southern tracts to replace exhausted northern timber lands, and over the next twenty years they vastly increased the rate and scale of southern lumbering. Many rural areas experienced a cycle of boom and bust. Towns in the timber belt rose and then vanished within a decade, leaving behind a restructured, relocated, underemployed, or unemployed rural proletariat. At the turn of the century, the southern landscape—cultural, economic, and natural—needed a boost. As Capt. Purse explained, it was an especially critical time to reestablish the sugar-cane industry "because of the large areas adapted to the cultivation of sugar cane, more profitably than anything else, as the mill men . . . are clearing the timber from these areas and opening them up for agricultural purposes" (ISCGA 1903, 10).

Addressing the convention was Rufus E. Rose, now Florida's state chemist and previously associated with Disston. Rose began by noting the tremendous and increasing American appetite for sugar, which "is the only agricultural product which the United States imports. Of all other crops we export enormous quantities. Few realize how large a part of our exports is required to pay for the sugar we import" (Rose 1903, 41). Although, for example, wheat exports paid only half the U.S. sugar bill, sugarcane cultivation had not been promoted. "While beet culture has had the intelligent, fostering care of our own and the various European governments, with government bounties, rebates, tariffs and drawbacks, cane sugar has had neither; on the contrary legislation has always been adverse to cane sugar, in our own country and in England, the two largest sugar consumers in the world" (41). Thus, whereas "vast sums" had been spent on the science of beet culture, little effort had been made to improve the quality of tropical cane or the manufacture of cane sugar.

Rose claimed that if a fraction of the sum used to establish beet sugar in America were spent in demonstrating cane production in the South, the United States would be "the principal sugar producer of the world" (Rose 1903, 42). Florida, Rose noted, was especially suited to cane: "Thousands of 'patches' scattered over the State from the extreme Northern Counties to the Keys, attest the fact that sugar cane is easily and successfully cultivated throughout the whole State, and requires but the joint effort of the capitalist and farmer to make it one of the leading if not the foremost industries of the State" (42–43). He suggested that the entire state be devoted to sugarcane, arguing that the "area of lands suitable for cane culture is practically unlimited. There are few townships in the State not capable of furnishing a mill with a capacity of 5,000,000 to 10,000,000 pounds of sugar per season" (48). To achieve this goal, he advocated that "[a]ll low and flat lands must be thoroughly drained. Cane will grow in moist but not in wet lands. Low hammocks, swamps and saw grass marshes, thoroughly drained make the best of cane" (49). Once drainage was accomplished, *centrales* rather than small mills should be set up to achieve an efficient industry.

Attendance, both numerically and in terms of geographic diversity, was much higher at the third annual ISCGA convention, held in Montgomery, Alabama, in 1905. Two factions emerged in the debate concerning how to proceed with industry development: syrup advocates versus sugar supporters. To some extent the division had to do with climate since, in the northern reaches of the envisioned cane belt, sugarcane would not always reach maturity, in which case it would not crystallize into sugar. The other aspect was socioeconomic; syrup-making was a cottage industry that was already widely practiced. Farmers with cane patches could imagine better distribution networks being developed for syrup, whereas establishing a nearby *central* seemed less likely. Within the unified purpose of promoting the cane industry, tensions and confusions were evident. The Hon. John Thomas Porter, speaking on behalf of West Florida, noted that many farmers raised cane in his area and had contemplated buying a smaller, displaced factory system from Louisiana. Their main problem was lack of capital: "We are all too poor in Florida to build a central factory. . . . I think we will solve this question by having central sugar factories, and, if we will use our influence, I think there are capitalists outside of our territory, who can be interested" (Porter 1905, 45–46).

Another Florida representative, Mr. N. H. Fogg, echoed these remarks in his paper "Sugar Cane in Florida." Fogg imagined that refineries "would make Florida a vast field of sugar cane and the largest sugar and syrup producing section in the world." Although the outside capitalists, "the men behind the refineries, would get the lion's share of the profits, the cultivators of the sugar cane will make a very comfortable living and become prosperous and happy" (Fogg 1905, 102). Mr. R. A. Ellis, also from Florida, explained that although sugar was just being recognized as a staple crop, it was widely grown throughout the state for syrup production: "Few of those who raise cane in this way have any accurate knowledge of the amount of land planted in it, or the amount of cane raised per acre, yet farmers will tell you that these little patches of cane pay them better than anything else that they raise" (Ellis 1905, 87). Thus the experience of the South, and specifically Florida, was quite different from that of the Northern regions, where beets had to be introduced and farmers had to be coaxed, cajoled, or at the very least tutored to grow them. The latent potential for the sugarcane industry was apparent to these men. Purse, for instance, argued that a syrup industry based on central mills would provide an alternative industry for the South without having to compete with beets, since beet processing lacks "the intermediate stage of a palatable syrup as in the case of cane" (Purse 1906, 32).

A syrup industry with limited market potential, however, would not save the rural South from economic decline. The problem remained to develop either standardized, marketable syrup or to develop milling capacity to produce raw sugar on a commercial scale. By the time of the fourth convention of the ISCGA, held in Mobile, Alabama, in 1906, the interests of Florida and the rest of the South had clearly diverged. Florida's representatives played a double-edged role, promoting the potential of sugar for the South and simultaneously extolling Florida's superior climate for cane production, suggesting that all domestic shortages could be met by developing a Florida industry.

The first speaker at that year's conference was the Governor of Florida, Napoleon Bonaparte Broward, a fierce advocate of Everglades drainage. Governor Broward's address left no doubt regarding his goal for Florida's ranking in national sugar production and offered a challenge to President Purse's hopes for the South. "I can see all the sugar necessary to make up the deficiency in the production in this country to meet its consumption can be grown in one corner of the Everglades of Florida," he observed, thus implying there was little need for a southern cane belt (Broward 1906, 90). The crop Florida could produce, he continued, is one "that requires $137,000,000 of imports to supply the deficit, before you come in competition with one pound of similar goods raised in your own country" (91). The aggregate value of corn, wheat, flour, beef and naval stores exports, totaling $144 million, was only slightly higher than the cost of sugar imports. "I want to impress every one of you that the future holds in store a great work for those engaged in the sugar industry, as we ask you to turn your eyes to the Everglades of Florida. Three millions of acres of almost level land, the highest point in it being twenty-one feet four inches above sea level and located within twenty-three miles of tide water" (92–93).

After describing preparations undertaken by the state of Florida for Everglades drainage, Broward considered the syrup versus sugar question. If the intention was to farm on a small scale, he concluded, farmers should be content with the manufacture of syrup, but it was clear that his vision was grander.

> Now, if we are going into the money-making business, we must stop grumbling and organize along sane lines. Cultivate the ground, find some method for gathering it up and selling it to central mills and have the cane juice there converted into sugar, if we want to go into business on a big scale. If you are

contented to raise syrup, then get in touch with each other and find some cheap way of transporting it to the common markets of the country. (Broward 1906, 95)

Capt. Rose returned to address the 1906 ISCGA meeting, making points similar to those of Governor Broward. He also addressed the question of syrup versus sugar, maintaining "that until we produce commercially, scientifically, and economically, a pure granulated or crystallized sugar, as do our friends, the beet sugar manufacturers, we will not become a factor in furnishing the sugar of America" (Rose 1906, 129). A carload of syrup could not be sold readily, whereas there would be a market for tons of granulated sugar, and because sellers of raw sugar faced only one buyer, he argued for modern, central factories where growers could sell their cane. "Make none but the pure granulated goods, such as the American public demands, cease the insane effort to produce a cheap raw article, in competition with the raw sugars of Cuba, and other foreign countries. Understand that the tariff is wholly in the interest of the American refiner" (130). Thus, when the South,

> with her superior climate and soil, builds central mills or factories, she can produce sugar at a profit, in spite of free raw sugar from Cuba, as she will have the assistance of the 'sugar trust,' and the beet-sugar grower, in maintaining the price of refined sugar. . . . [W]hen our farmers begin to think, and then combine their practical knowledge, and labor, with capital and skill, now seeking profitable employment, the question of the American supply of sugar will be solved by the cane belt of the United States making the necessary amount to supply the demand. The beet grower will soon discover that he cannot compete with cane, and will naturally gravitate into the cane belt, where his profits will be greater, and his crops more certain. (131)

Rose sketched out the economic geography of this new cane belt: factories costing $75,000 would produce fifty thousand pounds of sugar per day. "There is not a town or village in the 'cane belt,' from Mobile to Jacksonville to Tampa or Miami, that cannot furnish within a short distance twice the required acreage for such a mill. A thousand such mills would be required to produce the 5,000,000,000 pounds imported annually" (Rose 1906, 132). Yet, the state chemist of Florida noted, there were certain drawbacks to extensive sugarcane-growing at the northern reaches of the belt. In the concluding section of his paper, "Florida's Climate Superior for Cane," he explained that it was superior to that of any other state because the rainy season occurs during the growing months, and the dry season during the harvest period. Even in northern Florida, the growing season was thirty days longer than Louisiana's, while in southern Florida it ranged from forty-five to sixty days longer.

From these differences, Rose drew conclusions regarding the feasible scale of the industry. The organization of the mode of production suitable for cultivation was, he argued, in part climatically determined. North of the twenty-seventh parallel, the central factory system—"similar to the beet factory systems of Germany, Austria and the West"—would be appropriate, comprising numerous small fields of ten to forty acres, where each farmer, in case of freezing weather, could windrow his crop to prevent frost damage. Below the twenty-seventh parallel, in Dade and Lee counties, "where vast areas of rich land, in large bodies can be had, the plantation, or 'gang system,' will prove the most satisfactory, where the planter owns the factory, and cultivates the cane also. This system is applicable only where there is no probability of killing frost, where large fields can be safely allowed to stand till wanted by the mill" (Rose 1906, 133).

Two contradictory images emerge from these conferences. One is of sugarcane promoters, regionally unified around the hopeful prospect of agro-industrial development and buoyed by the evident success of the beet-sugar industry. The second, which becomes more striking over time, is of a group that was divided on key points, such as whether the beet industry was so formidable an opponent that a syrup industry would be advisable or so truly uneconomical that it was only a matter of time until the cane belt supplanted it. These divisions were most evident between Floridians and the rest of the South for reasons having to do with both climate and settlement. While other southerners had a populist vision of regional development based on small farmers and dispersed mills, Florida officials were looking to an unsettled frontier with a subtropical climate, which they regarded as a blank slate on which to construct a highly capitalized, centralized, industrial plantation system similar to those recently developed in eastern Cuba. Thus the regional vision of a sugarcane belt gave way to the rhetoric of Florida's "sugar bowl" boosterism. Floridians were confident in their ability to compete with the beet belt. However, at the turn of the century, the USDA had enlarged its area of interest, heading offshore: "No sooner had American troops landed in Puerto Rico and the Philippines during the Spanish-American War than scientists were looking over their shoulders to see what opportunities were available for study. None were more interested than members of the U.S. Department of Agriculture" (Overfield 1990, 31). In 1900 funds were allocated for research stations in several of the newly acquired insular territories, which Secretary Wilson expected would provide the tropical and subtropical products that the United States was not yet producing sufficiently on a commercial basis, such as coffee, tea, and, of course, sugar. As a result, Florida promoters faced new and formidable competition that would call forth new discursive strategies in the rhetoric of sugar-bowl boosterism. They would have to contend with and somehow fit into this "Conquest of the Tropics" that was leading U.S. agronomists, traders, engineers, and investors to ever more southern "frontiers" (Tucker 2000).

CHAPTER THREE

Securing Sugar, Draining the 'Glades

Climate, soil, and square mile after square mile of empty land led Florida's business and political leaders to look toward agro-industry as the engine of regional economic development. The only way for Florida sugar boosters to make their imagined Everglades sugar bowl a concrete reality, however, was through direct federal involvement in removing the two biggest obstacles: water and cheap, foreign sugar. Though it had the requisite climate, the south Florida region faced a fearsome competitor in Cuba and required significant state commitment to drain the Everglades in order to realize the potential productivity of its rich soils. Furthermore, world prices for sugar fluctuated wildly in the early twentieth century, undermining the economic rationale for heavy investment in drainage to make way for agriculture. Given this political-economic context, boosters needed to tie drainage and protective tariffs to issues and interests beyond the quest for profit among Florida's agricultural investors and real estate speculators. They found in national concerns over flood control, navigation, and food self-sufficiency a powerful rationale for promoting a sugarcane industry. Nevertheless, it required events beyond their control, including a world war and two devastating hurricanes, to give adequate weight to their campaign.

At its core, this chapter addresses an apparent economic contradiction: Why were the Everglades transformed into an agro-industrial complex for sugarcane at great environmental and monetary cost during a period of oversupply and depressed prices in the global sugar market? Not only were world prices greatly depressed, but also the largest and most efficient sugarcane producer in the world, Cuba, lay just ninety miles across the Florida Straits and had been a dependable supplier for many years. The answer lies beyond the reach of the market's "invisible hand." World War I had a profound impact on the geography of the global sugar trade, wiping out much of Europe's beet production while creating boom conditions for Cuban cane. Wartime disruptions of the market raised the issue of the vulnerability of the U.S. sugar supply, which had become a national security issue because of its importance in keeping armies on the march. The prospect of war-induced sugar shortages provided boosters with a potent argument in their struggle to bring long-promoted ideas about Florida's potential to fruition. The story that unfolds here brings into focus the relationship between commodity production and place construction and demonstrates how, once established, the "Nation's Sugar Bowl" became central to political struggles over domestic and international sugar policy.

"Drainage Would Cause the State to Prosper"

Governor Broward inherited from his predecessor, Governor William S. Jennings, the issues of Everglades drainage and land "reclamation" and distribution. Jenning's administration (1901–1905) marked a shift in the state's approach to public lands, influenced by rising populist sentiment (Blake 1980).[1] Jennings sought to confirm state ownership of swamplands and to reduce the conflicting claims that continued to plague the IIF. He took special interest in the possibility of draining the Everglades, compiling data on topography, rainfall, the elevation of Lake Okeechobee, and soil fertility (Manuel 1942a). When the legislature's generous land grants to railroads came under increasing criticism during the 1890s, one crucial question was whether the land that had been granted was really "swamp." This in turn raised the issue of the disposal of remaining wetlands held by the federal government, approximately four million acres in the Everglades. Then, as now, the struggle over the swamp was one of definition; only then, as opposed to now, development interests very much wanted the lands deemed "wet and unfit." The outcome of the four million acre question was of interest to several groups; the railroads, which claimed it was theirs; developers, who claimed it on the basis of drainage contracts; various canal companies; and the trustees of the IIF, who maintained that it should be distributed as land for farmers. Whereas it is possible to identify the institutional interests at stake, the interests of individuals became as complicated as their affiliations. For example, in 1898 Rufus E. Rose and James Ingraham, former president of South Florida Railroad, signed a contract with IIF trustees to reclaim 800,000 acres, which they could then purchase at twenty-five cents per acre (Blake 1980). On that basis they organized the Florida East Coast

Figure 3.1. By the turn of the century, land was put up for sale as it was drained. Here a Seminole Indian woman poles past a billboard advertising real estate. Courtesy of Historical Museum of Southern Florida.

Drainage and Sugar Company with the stated intent of making land available to farmers (fig. 3.1).

Rose and Ingraham's experience is illustrative of changing politics. Unable to fulfill the terms of the contract with regard to drainage within the time specified, they sought an extension, which was denied in 1901 after Jennings took office. During Jennings's administration the IIF preferred to make deals with land developers who seemed ready to begin reclamation rather than railroads or speculators. However, according to the confused records of land deals dating to 1879, the legislature had already granted more land than the state owned. When various railroads brought suit against the state for deeds to this land, Jennings took the position that these grants were invalid because they failed to conform to the specifications of the Swamp Act, the sole purpose of which was to reclaim land. Railroad claims remained unsettled when, following severe flooding in 1903, Jennings asked the federal government for one million dollars in aid. Theodore Roosevelt denied his request, instead deeding 2,862,080 acres of Everglades to the state. For this reason, the prominent issue on the eve of the 1904 gubernatorial election was the state's newly acquired land, to which the railroad companies were prepared to lay claim (Blake 1980).

Restricted to a single term by the state constitution, Jennings gave his support to Broward, whose stance on Everglades land and railroads was most like his. Broward, a Democrat, fit the mold of a progressive politician, campaigning for tax reform, state support for education and life insurance, and game conservation laws, and against railroad interests. Opposed by most of the Florida papers, Broward published and distributed an *Autobiographical Sketch,* a "backwoods" narrative that overlooked his planter-origins and stressed the hardships of his youth, when he was orphaned by the Civil War. Two aspects of Broward's campaign highlight unique characteristics of Florida politics. First, an attempt by his opponents to use his filibustering activities during the Spanish-American War against him backfired, winning Broward the support of Cuban nationalists, who outnumbered Spanish loyalists in Florida. Second, the centerpiece of Broward's campaign was Everglades drainage, which he promoted with the aid of what came to be known satirically as "Broward's map" depicting the Everglades region (Proctor 1993). At the time of his campaign, Florida's population was 528,000, making it the smallest of any southern state. It was also one of the poorest, with an estimated annual per capita income of $112 compared to the national average of $202. Broward's focus on land reclamation and his interest in attracting population addressed the critical issue of economic development for a state that at that time "had more acreage and more frontier characteristics than any state east of the Mississippi" (ix).

Broward used the map as well as graphs and charts as he toured the state lecturing on the benefits of drainage: "land salvaged to men who wanted to build homes, plant crops and tap the wealth of the fabulous muck." He argued that Floridians should simply "knock a hole in a wall of coral and let a body of water obey a natural law and seek the level of the sea" (quoted in Proctor 1993, 191). Every voter in the state received a copy of Broward's map, on the back of which was printed an explanation of the issue of Everglades drainage. Bear in mind, however, that progressive was not synonymous with inclusive; Broward was an unreconstructed southern politician, elected in the aftermath of reconstruction when widespread disfranchisement of black voters had been realized through poll taxes, economic coercion, and violence such as kidnappings and whippings (Dovell 1952, vol. 2).[2]

Samuel Lupfer, who, as noted in chapter 2, was Disston's superintendent in earlier drainage endeavors, hailed Broward's election in the pages of *Engineering News* in an article entitled "The Florida Everglades: Their Legal

Status, Their Drainage, Their Future Value" (Lupfer 1905). East of the Mississippi River, noted Lupfer, Florida is the state "whose agricultural possibilities are least developed. But Florida has one possession, the Everglades, of which no other state can boast and which Governor Broward and a number of his friends believe will yet make her one of the richest states in the Union" (Lupfer 1905, 278). Broward's election had removed political obstacles so that reclamation could proceed.

> Every member of the Board of Trustees is with Governor Broward in his fight for these lands, and each one realizes how their drainage would cause the State to prosper, as the Everglades would be cut up with canals . . . bordered with vegetable farms and immense sugar cane plantations; . . . then the whole country would be dotted with immense sugar refineries and . . . these lands would support a happy, contented agricultural population of half a million souls. (280)

Lupfer argued that the task of drainage would be fairly simple. His opinion on the feasibility of draining land for sugar cultivation in Florida was informed by his work on Disston's St. Cloud plantation. Because Disston had been nationally prominent and his suicide was national news, it was essential to the booster effort at this time to attribute the failure of Disston's sugar plantation to factors other than its geographic location in Florida. Lupfer stated that the St. Cloud enterprise had failed because Disston neglected to manage it, "and the sugar plantation was used as a speculative investment to catch 'suckers'" (279). Similarly, Rose prepared a "short history of that enterprise," published in numerous Florida newspapers in 1905 and 1906, in which he attributed the failure of St. Cloud to mismanagement while lauding the ability of Florida's climate and soil to "produce more and better sugar cane than any lands in the United States." In addition, he argued, "they will produce cane equal to Cuba, or Mexico in tonnage of sacharine [sic] quality."[3]

One month after taking office, Governor Broward made his second trip to inspect Lake Okeechobee to evaluate drainage prospects, accompanied by former governor Jennings and State Chemist Rose. They concluded that the lake could be lowered six feet without hampering navigation, to reveal six million acres, "an area capable of producing the entire tonnage of cane sugar used in this country, a crop which alone would be of untold value to the State" (Broward, quoted in Blake 1980, 96). Since the IIF held only half of the six million acres, and railroads and canal companies held the rest, Broward submitted to the legislature a proposal to create a drainage commission with the power to tax. A constitutional amendment subject to ref-

Figure 3.2. A bucket dredge at work in the Everglades (c. 1915). Courtesy of Historical Museum of Southern Florida.

erendum was approved, which set up a board of drainage commissioners composed of the five IIF trustees, with the power to build canals, drains, levees, ditches, and reservoirs, to establish drainage districts, and to levy an annual tax. Using IIF monies on hand, Broward ordered surveys undertaken and two dredges built (fig 3.2). The two dredges—ironically christened *The Everglades* and *The Okeechobee*—began work in summer 1906 and spring 1907, respectively. With funds running low and in need of tax revenue, the IIF designated the Everglades Drainage District, comprising five counties and extending sixty by one hundred fifty miles (Manuel 1942a). Since nonresident owners (that is, the railroads) held by far the largest portion of the land in the district, more than four million acres, drainage taxes were set so that owners would pay higher rates than residents (Blake 1980).

The effort to bring the state into reclamation and to establish collective institutions for drainage was not unique in the nation. The Everglades were exceptional, but the desire to drain them was not. In the late nineteenth century, most states with "overflowed lands" organized drainage districts, and "the peak years of drainage came immediately following the turn-of-the-century" (McCorvie and Lant 1993, 33). Political activities aimed at securing federal support for drainage also flourished during this time. Broward was active in the National Drainage Congress, an organization devoted to exchanging drainage information and promoting federal drainage legislation. Following his presentation on Everglades drainage activities at the 1907 national conference of the Congress, Broward was elected as its president. Earlier that same year, he was invited to accompany President Theodore Roosevelt on a trip down the Mississippi to inspect drainage projects along the way. Broward felt that Roosevelt was sympathetic to his own "progressive" vision of resource development (Proctor 1993).

Closer to home, Broward's drainage campaign had not fared as well. Everglades landowners successfully fought the drainage tax, which the Federal courts declared unconstitutional (Manuel 1942a), and in the fall of 1906, voters rejected the drainage amendment (Blake 1980). What had worked elsewhere was not yet going to succeed in Florida. The collective institutions necessary for drainage had been defeated, and the IIF was faced with a conundrum. The land they held was worth only twenty-five cents per acre wet and up to twenty dollars an acre dry. The money to dredge was running out, but without dredging, they owned nearly worthless swampland. Further, during the ongoing litigation that surrounded Everglades lands, plaintiffs challenged the feasibility of the drainage plans; in response, Governor Broward asked the USDA for engineers to investigate and report on the plans for the purpose of "silencing political opponents who otherwise would have complained that it was 'Broward's engineer and Broward's report'" (Manuel 1942a, 12872).

In an ironic way, the populist-cum-progressive impetus to drain the Everglades ultimately gave rise to the "first Florida land boom," which straddled the administrations of Broward and his successor, Albert W. Gilchrist, and brought land speculators between smallholders and the state. The IIF needed to make land deals to remain solvent, and, in light of the outcome of several suits against the state, had to honor some of the claims of litigants. Broward is credited with making compromises advantageous to the state, which salvaged some two million acres of Everglades land for the IIF (Blake 1980). Drainage was progressing very slowly and in a piecemeal fashion: only four miles of canals had been dug by the end of 1907 and, at the most, twelve thousand acres of land reclaimed. To speed things up, the IIF contracted for two more dredges financed by selling Everglades land, some of it in large parcels. In June 1908, R. P. Davie bought approximately twenty-seven thousand acres for the purpose of general farming, vegetable production, and for the establishment of an experimental cane farm for large-scale sugarcane cultivation as well as sugar mills. Richard J. Bolles of New Mexico purchased five hundred thousand acres in Dade and Lee Counties; his representative for this transaction happened to be the general counsel of the IIF trustees, Jennings (Manuel 1942a).[4]

By 1909 five large land companies, the Florida East Coast Railroad, and Richard Bolles wanted to negotiate with Governor Gilchrist regarding drainage taxes. Granted their request for oversight of the work and the use of private contractors, they agreed to pay drainage taxes for the period 1907–12. For the next two years dredging proceeded apace, financed in part by taxes and payments made by the speculators, who in turn received payments from the approximately twenty thousand people who hoped to become Florida landowners. Bolles's sales methods became the prototype of "land by the bucket" real estate promotion, using gimmicks such as lotteries to lure prospective buyers. Meanwhile, Broward, who had failed in his bid for nomination to run for the U.S. Senate, was engaged in various business schemes, one of which involved "working for the Bolles Everglades Land Company at a reported salary of four hundred dollars a month" (Proctor 1993, 292). The Everglades Land Company pamphlet included an article by the "ex-Governor," in which Broward told prospective buyers that there were "in this submerged Florida, thousands of acres of land suitable for the cultivation of sugar cane," supporting his claim with extracts from Wiley's 1891 report.[5] When, in response to voluminous queries regarding Everglades land, the USDA in 1910 composed a cautionary form letter regarding issues such as soil quality, drainage progress, and immediate agricultural potential, Broward intervened successfully to have the bulletin suppressed (Dovell 1952).

1911 was the critical year for the Everglades drainage project for several reasons (fig. 3.3). First, dredging was now well underway, and it was becoming apparent that the simple methods that had been envisioned simply did not work. In retrospect, an engineering feat on the scale of the Panama Canal would be required to successfully complete the largest reclamation project in the world—without the corresponding economic benefit of international trade and still leaving much of the Everglades inundated. Second, during the winter and spring of 1911, thousands of smallholders who took cheap excursion trains to Florida to see their homesteads were disappointed and outraged by what they saw.. Third, news of the scandal uncovered by congressional investigation of a USDA drainage engineer, James O.

Figure 3.3. A souvenir pamphlet commemorates the opening of a major drainage canal in 1912. Courtesy of Historical Museum of Southern Florida.

Wright, was published by newspapers in forty-three states. When assigned the task of assessing the feasibility and method of Everglades drainage, Wright "acted not as a neutral expert but as a booster for the reclamation program" (Blake 1980, 107). On receiving Wright's 1909 "Report on the Drainage of the Everglades of Florida," USDA Chief Engineer Charles G. Elliott, Wright's supervisor, edited it and "toned down its optimistic findings" (Blake 1980, 115). However, Wright's original, unedited report found its way into the U.S. Senate Document, *Everglades of Florida* (U.S. Senate 1911). The document, a compilation of fifty years of sanguine reports on the drainage and development prospects of the region, of which Wright's report was the most recent and, apparently, authoritative, had been hurried to press following U.S. Senator Duncan Fletcher's resolution calling for its publication. *Everglades of Florida* ignored balanced and realistic assessments of the difficulties of drainage in order to provide boosters and developers with the imprimatur of the U.S. government. Running throughout the document were visions of the extraordinarily fertile and profitable sugar lands soon to be revealed by simple drainage techniques. Wright, who in 1910 became chief engineer of drainage for Florida, resigned in disgrace in 1912 when the committee investigating Everglades land transactions found that his cursory report had been used to great profit by land sales companies (McCally 1999).

In 1912 an independent engineering study commissioned by one of the large land companies found that the state's approach to drainage was actually contributing to flooding through half-finished canals. The report threw cold water on the populist dream of ten-acre farms because it concluded that "private owners would have to build their own levees and ditches to supplement the main canals of the state system. Only large operators could invest the capital required to make Everglades agriculture profitable" (Blake 1980, 121). Thus, Lupfer's prediction in *Engineering News* that the region would support one million souls happily engaged in agriculture was less prescient than Rose's vision of sugarcane cultivation below the twenty-seventh parallel. When it materialized, Rose's vision of gang-system plantations would be based on the labor of poor southern blacks.

With the collapse of land sales, the revenue from installment payments ceased, and money for drainage was running out. In September 1912, Governor Gilchrist went to New York to confer with "gentlemen representing the large landed interests, situated in the Everglades. At this conference, the captains of industry and their representatives were anxious for the great work of the reclamation of the Everglades to be continued" (Trustees IIF Minutes, quoted in Dovell 1952, 2:747). Anxious as they were, the gentlemen were not willing to continue making payments to the IIF unless the state had a clear plan for progress. Gilchrist proposed that the legislature enact a law authorizing the trustees to issue bonds "for a sufficient amount of money to drain the Everglades" (Trustees IIF minutes, quoted in Manuel 1942a, 12875).

Gilchrist's successor, Park Trammell, had been an IIF trustee and came into office ready to build on Gilchrist's suggestions and to rationalize the drainage program. Under his leadership, the State of Florida contracted in the spring of 1913 with the J. G. White Engineering Corporation of New York City, which appointed an engineering commission to conduct a new survey of the Everglades terrain and provide recommendations concerning drainage. The members of the Everglades Engineering Commission, Isham Randolph, Marshall O. Leighton, and Edmund T. Perkins, "represented some of the most prestigious engineers in the United States" (McCally 1999, 111). *Florida Everglades: Report of the Florida Everglades Engineering Commission* (U.S. Senate 1914) is known as the Randolph Report, a testament to the reputation of Randolph, who had supervised the digging of the Chicago drainage canal and was appointed by Theodore Roosevelt to the Panama Canal Commission. The report outlined the problem as follows:

> Lake Okeechobee, the great liquid heart of Florida, which, with the exception of Lake Michigan, is the largest body of fresh water wholly within the United States, lies at the focus of the greatest agricultural drainage problem in our country. This basin receives the floods from a watershed 5,366 square miles in area. When that capacity is exceeded, the excess waters flow southward over an area of 4,000 square miles known as the Florida Everglades. (7)

Having defined the problem in this way, the Randolph commission concluded that "the drainage of the Florida Everglades is entirely practicable and can be accomplished at a cost which the value of the reclaimed land will justify. The solution of the Everglades drainage problem is primarily dependent upon the disposition to be made of the flood waters entering Lake Okeechobee from the north" (5).

The commission disagreed with the conclusion of the privately commissioned engineering study of 1912 "that the whole area must forthwith be covered by a great independent system of canals. We believe this to be an erroneous idea, and that the Everglades can be reclaimed progressively, as is now planned by your board" (U.S. Senate 1914, 8). They proposed that the cost of building the canal system, specifically the Okeechobee-St. Lucie Canal, not be entirely "charged against the draining of the Everglades" because the canal would serve three purposes: first, to control the level of the lake and therefore drainage; second, to provide a navigable canal; and third, to develop hydroelectricity, a "water power of primary capacity of 5,000 horsepower which will return to the district an income that will contribute largely toward the future maintenance of the drainage systems" (5). Thus the Randolph report introduced the idea that the Federal government should cooperate in providing funds for the development of a navigable waterway and proposed another method of financing the project through the sale of hydroelectric power (fig. 3.4).

The report contained candid observations that, considering the prestige and combined expertise of its authors, are all the more striking. One

Figure 3.4. The Florida Everglades Engineering Commission's survey of existing and proposed drainage canals (1913). Source: U.S. Senate 1914.

was their statement that there "is probably no more difficult place on this continent in which to determine run-off than in the Everglades" (U.S. Senate 1914, 42). This was due to a variety of factors including its "inaccessibility, its lack of perceptible grade, its practically unknown geology, the variability of its muck cover," and numerous unknowns with respect to underground water flow. A central and critical issue concerned the relation between runoff and muck soil. In reporting their findings on this matter, they were especially frank: "This commission in its goings about the Everglades has gathered from old residents and from apparently reputable observers and experimenters more contradictory information about muck than the commission's members have confronted about any other subject in all their professional lives. Confusion seems to be unbounded, and about every disputed point there turns a factor affecting run-off" (42).

Despite these uncertainties, the Randolph report provided a generally optimistic blueprint for developing the state's overflowed lands. In June 1913, Governor Trammell signed a bill establishing an Everglades Drainage District, which would levy taxes in accordance with anticipated benefits. What seemed to be a workable relation between the state and large landowners was hammered out during Trammell's administration concerning the economics and engineering of drainage, though drainage progressed slowly from 1914 until 1918. During World War I, "the lake shore gained national recognition for the large food crops produced in that section of the state" (Dovell 1952, 2:749). In 1915 the Southern States Land and Timber Company, organized in 1902 by New York and New Orleans investors, undertook experiments with sugarcane planted in four locations along Lake Okeechobee: Canal Point, Loxahatchee, Indiantown, and along the St. Lucie Canal. Jules M. Burguieres, whose Louisiana experience had been augmented by a year's research in Cuba for the Cuban American Sugar Company, supervised these experiments. However, the "second land boom" would not occur until after the war, driven by interest in the use of Everglades land for sugar production.

Meanwhile, the promotional literature of the nineteenth century gave rise to early twentieth-century pragmatism. In a book titled, *What About Florida?* L. H. Cammack noted that "unfortunately the Everglades got into politics. Campaigns have been lost in a most unscientific way, the fight hinging on the practicability of their drainage" (Cammack 1916, 127). For Cammack, the better question was, who would want to live there? Those who argued that the Everglades could not be drained might be in error, but those "who have maintained that this will make one of the garden spots of the universe, inhabited by a thrifty and satisfied people, are equally in error. The Everglades is no place for the independent man, the man with limited means, the man with a family to raise, with children to school and dependent on his own exertions for a living" (130). Cammack attempted to counter the relentless optimism of Florida boosters, but the questions he raised were ultimately beside the point. Populist rhetoric aside, economic development in the Everglades would not depend on an influx of yeoman farmers, but on the establishment of a large-scale, corporate-owned, agro-industrial complex.

The Sugar Question in National Security: World War I Raises the Stakes

While World War I temporarily slowed the drainage of the Everglades system, in the longer term it gave the sugar industry a powerful discursive tool with which to leverage favorable trade agreements from the federal government. War-induced shortfalls had made it clear "for the first time that America could not engage in an international conflict without finding herself confronted with a sugar shortage" (Dalton 1937, 38). Government planners designated the sugar industry "as essential in war time" and undertook to manage production, transportation, and distribution. Distant offshore sources were problematic because the critical shortage throughout the war was not sugar itself, but shipping tonnage (Bernhardt 1920). Geographic proximity was thus key to securing sugar, which gave Cuba a favorable new status in its quest for access to the U.S. market. World War I thus transformed the geography of the U.S. sugar supply into a question of national food security and gave Florida boosters an opportunity to discursively construct their imagined sugarcane region as a national security issue.

As noted earlier, scientific studies conducted at the turn of the century claimed that soldiers had more endurance when given sugar (*Scientific American Supplement* 1900). Thus by the time of World War I, military planners saw sugar as a strategic commodity. The wartime mobilization of the U.S. sugar industry began in 1917, with the passage of the Food and Fuel Control Act, which gave President Woodrow Wilson the authority to create the U.S. Food Administration, and he appointed Herbert Hoover to the position of U.S. food administrator. To serve as chief of the Sugar Division, Hoover appointed George M. Rolph, general manager of the California and Hawaii Refining Company. During U.S. involvement in the war, discussion of the sugar question was less abstract than it had been: it no longer addressed issues such as "free trade" or colonial responsibility but focused instead on the concrete concerns of proximity of supply, availability of shipping tonnage, price, and distribution. Though less abstract, these problems were also com-

Table 3.1. Production of raw sugar in selected regions, 1913–20 (1,000 tons)

Year	Europe	Cuba	Java	United States	Philippines	Other Countries
1913–14	9,043	2,909	1,549	2,009	408	5,236
1914–15	7,598	2,922	1,454	1,966	421	6,514
1915–16	5,434	3,398	1,797	2,106	412	5,738
1916–17	5,194	3,422	2,009	2,279	425	5,263
1917–18	4,594	3,890	1,960	2,042	475	7,330
1918–19	3,611	4,491	1,473	2,062	453	6,514
1919–20	3,278	4,184	1,681	1,905	467	6,474

Source: Ballinger 1975.

plex and political-economic in nature. William Clinton Mullendore, who served on the staff of the Food Administration during the war and afterward became its official historian, noted that "the problem of sugar control was probably more complicated than that of any other commodity because of the widely different character of the many sources of supply and the varying seasons in which they reached the market" (Mullendore 1941, 170). When the war began, Cuba was providing approximately half of the U.S. sugar supply. As the war destroyed the European beet-sugar industry, Cuba also began to supply the allies, especially the United Kingdom, the only close rival to the United States in terms of national sugar consumption.

The Sugar Division was created to deal with the combined problems of shortages and rising prices for U.S. consumers, who were now competing with the Allied countries, primarily the United Kingdom, for supplies. The United States and the United Kingdom, with annual consumption levels of 83 and 91 pounds per capita, respectively, were by far the largest users of sugar. Before World War I, more than 70 percent of the sugar consumed in the United Kingdom came from European beets, which were unavailable or destroyed during the war, forcing the United Kingdom to look to tropical suppliers, namely Java and Cuba (table 3.1). The Cuban industry, which benefited from extensive U.S. capital investment, had by then become the largest and lowest-cost producer in the world (Dye 1998). Cuba was the geographically logical choice since, technically, the problem of sugar during the war was primarily a shortage of shipping tonnage (Bernhardt 1920). Thus, U.S. administrators felt it necessary to intervene in the international market to secure access to foreign supplies, as well as to coordinate supplies from the domestic industry.

In 1917, Hoover advised the senate that sugar problems were confounded by the fact that

[at] the present moment our sugar refiners are competing with the allied sugar commission for the purchase of Cuban sugar. It must be patent that if we create a sugar commission and if that sugar commission cooperates with the Allies and the Cuban producers to take over the Cuban crop at the fixed price, that we can effect a considerable saving on the present inflated price of raw sugar, and we can stabilize the price of sugar throughout the whole of next year. (quoted in Bernhardt 1920, vii)

However, the Sugar Division did not have the authority in 1917 to fix prices or purchase raw sugar abroad; instead, it had to rely on devising strategies to "mobilize" the domestic industry for the crop year 1917–18. Encouraging increased sugar production through price mechanisms was tricky. For example, a price that would stimulate domestic production was well above the costs of production for Cuba. As Joshua Bernhardt—chief of the Statistical Department of the United States Sugar Equalization Board (USSEB) and sugar statistician for the U.S. Food Administration—explained,

The problem would not have assumed such complexity if the entire sugar supply of the United States had originated in one or two localities. But of the total annual sugar consumption of the United States in prewar years, 49.8 per cent was shipped from Cuba, 13.66 per cent from Hawaii, 8.01 per cent from Porto [*sic*] Rico, 6.27 per cent came from Louisiana, 15.97 per cent from the domestic beet crop, and the balance, about 7 per cent, from miscellaneous foreign sources and the Philippines. Each of these regions differed in the average costs of production, and reliable statistics of costs in most of these regions were meager and unsatisfactory. (22)

Given the complexity of the problem faced by the Sugar Division and its limited powers, it is no wonder that it proved inadequate to the task. The Sugar Division had to rely on voluntary cooperation to curtail domestic sugar consumption and to achieve increases in production while maintaining stable prices. Both missions were difficult. First, to curtail domestic consumption required reversing a century-long trend, during which U.S. annual sugar consumption per capita rose from about 10 pounds in 1810 to 75.4 pounds in 1910 (Woloson 2002; Ayala 1999). Annual consumption per capita was 83 pounds just before World War I, and the prosperity of wartime tended to increase the consumption of sugar (Mullendore 1941). Second, it would have been difficult enough to stimulate domestic production due to wartime shortages of fertilizers and labor, but the concomitant mandate to keep prices from rising made it especially so. In an effort to control prices, the division suspended futures trading in sugar, expanded export controls, and "propagandized" about fair sugar prices. As it was, the average price paid for a ton of beets rose from six to ten dollars from 1916 to 1918. However, in an August 1917 memo, Hoover blamed the increasing price of sugar not on strong consumer demand nor on domestic producers, but on the fact that "certain Cuban sugar producers (who are out of our reach), have combined to force up the price of the remaining 1917 Cuban crop" (quoted in Bernhardt 1920, 11). In August 1917, representatives of the beet industry met with the Food Administration and agreed to a fixed price; later that month an embargo was placed on exports from the United States, and in October the Cuban government placed an embargo on all sugar exports except to the United States and Allies. Cane refiners agreed to purchase raw cane sugar only through the agency of the Food Administration, which formed the International Sugar Committee, composed of two British representatives and three U.S. representatives, the latter all from the refining industry. The committee negotiated with representatives of the Cuban industry for a set price on the entire 1917–18 Cuban crop, which was to be divided among the Allies.

Sugar shortages were anticipated for 1918 for several reasons: stocks were depleted, the Allies were increasingly dependent on production from the Western Hemisphere, and, despite appeals to reduce it, sugar consumption was increasing. In January 1918, manufacturers of foods deemed "nonessentials," such as candy and gum, were restricted to 80 percent of their previous year's sugar use. In a memo to President Woodrow Wilson dated June 17, Hoover outlined the problem of shortages, noting that "[s]ugar is one commodity that voluntary conservation does not sufficiently reach. I suppose the great sugar eaters are those of least moral resistance in the community" (quoted in Bernhardt 1920, 44). He suggested that the president authorize a sugar corporation, to be modeled after the Grain Corporation of the Food Administration. Wilson, convinced of impending shortages by Hoover's missives, in July 1918 authorized the formation of the United States Sugar Equalization Board, Inc., which he provided with five million dollars in emergency funding. Essentially, its mandate was to remove geography from production, as directed by the Food Administration: "The purposes of the Board are to equalize the cost of various sugars and secure better distribution" (quoted in Bernhardt 1920, 45), by adjusting for shipping and production costs, and by subsidizing costlier (domestic) sugars with cheaper (foreign) sugars. The USSEB included Hoover as chair and Rolph as president, as well as five directors, among them Frank W. Taussig, a Harvard economist who served on the United States Tariff Commission and who would in the following year play a critical role in the events leading to volatility in the postwar sugar market. In September 1918, the USSEB negotiated to buy the entire 1918–19 Cuban crop.

Also in July 1918, sugar consumption was further restricted for numerous reasons: the previous Cuban crop was smaller than anticipated, French factories had been destroyed, large quantities of sugar were lost in sinking ships, and the canning season had begun. A complex program of rationing certificates was devised. Nonessential food manufacturers were reduced to 50 percent of the previous season's supply, canned goods manufacturers to 75 percent, restaurants and bakers were rationed, and retailers were not permitted to sell more than two pounds of sugar at a time to urban residents or five pounds at a time to rural residents, with the exception of "housewives" with "a home canning certificate which entitled them to buy sugar in twenty-five pound lots" (Mullendore 1941, 112). As a result of these measures, U.S. sugar consumption for 1918 was 230,000 tons less than the average of the preceding four years (Mullendore 1941). In an address to state food administrators in November, Hoover noted the difficulty of forecasting the sugar situation, especially due to the uncertainty regarding Europe's consumption levels. If Europe remained "on present rations," then world supplies were adequate, but if "Europe raises its ration very considerably," Hoover predicted there would be shortages of sugar (185).

Homegrown Sugar: A Nationalistic, Economic, and Geographic Imperative

World War I sugar policies and shortages stimulated an intense period of Florida sugar bowl boosterism, which culminated in several attempts at commercial-scale production. Florida sugar bowl boosters saw their opportunity in the wartime destruction of the European beet-sugar industry and Cuba's new role in supplying the Allies. As a result of wartime shortages and the threat of insecure supplies, they were able to graft onto the nationalistic rhetoric of previous sugar boosterism the strategic significance of U.S. grown sugar. Because in wartime the sugar question became explicitly geographic, Florida boosters made their case using cartographic representations to prove beyond doubt the strategic need to develop an Everglades sugar industry. Although much of their argument was recycled from other places and earlier times, what was especially inventive were the moral embellishments and the meticulous mappings at the national and the local scale, which together provided a thoroughly detailed though imagined economic geography of a future Florida sugar bowl.

The Cuban industry played a double-edged role for Florida sugar boosters. On the one hand, the proximity of the world's largest, most efficient sugar producer seemed to obviate the need to develop Florida's sugar bowl. The economic efficiency of the Cuban industry was well known to sugar enthusiasts. In 1917 the U.S. Department of Commerce published a comprehensive, comparative study of the industries in Hawaii, Louisiana, Puerto Rico, and Cuba. Entitled *The Cane Sugar Industry* (U.S. Department of Commerce 1917), this bulky document gave detailed accounts of organization and expenditures in each location, and, in the case of Cuba, included comparisons between the eastern and western provinces. Though Hawaiian producers excelled in certain aspects of production, the Cuban industry was by far the most efficient, largest, and lowest-cost producer in the world.

On the other hand, Cuba's proximity suggested that Florida shared the geographic characteristics suitable for sugarcane cultivation. By World War I, the period known as the second great sugar expansion in Cuba, which began after the Cuban War of Independence, was well underway. An especially significant feature of this expansion was the contrast between eastern and western Cuba, which illustrated different historical layers expressing uneven processes of regional development. Whereas the plantation system was well developed in the western provinces by the mid-nineteenth century, the sugar industry in the eastern provinces was still small and unimportant at the end of the century. The east, characterized by open cattle ranges and subsistence farming, provided a frontier where, after the turn of the century, "the largest mills on the island were erected on new sites yet unknown to cane cultivation. Land was quickly absorbed into cane. Private company railroad systems were extended throughout the surrounding cane zones for hauling cane" (Dye 1998, 15). The east became the destination for sugar entrepreneurs, backed by North American capital and, with this investment, also became the center of technological dynamism in the industry. Whereas western cane growers were often from the planter class and often owned the land they farmed, "[e]astern *centrales* established themselves on lands previously unoccupied by the planter class by purchasing vast tracts of land in relatively remote areas of the country" (17). This example, which demonstrated the opportunities afforded to the sugar-capitalist on the tropical frontier, was not lost on Florida boosters, who saw under the sawgrass an equally "virgin" terrain. The challenge was to draw out from the Cuban model the lessons of industrial organization and still demonstrate the urgent need to make a Floridian sugar bowl.

Central to the discursive strategies of Florida sugar boosters were comparisons with other regions that produced for the U.S. market. The most prolific booster was C. Lyman Spencer, a "scientist and successful realtor" who was considered the "man of the hour in the sugar industry" (Emerson 1919). Spencer's booklet, *The Sugar Situation*, provided the economic, social, political, environmental, and, above all, geographic, arguments for developing south Florida as "America's Sugar Bowl" (Spencer 1918), making extensive and detailed reference to statistics compiled in *The Cane Sugar Industry* (U.S. Department of Commerce 1917).[6] On the front cover was a map of the southeastern United States and Caribbean region, delineating places suited to sugarcane according to climate and distinguishing between the syrup and granulated sugar regions (fig. 3.5). Above the map, Spencer posed the following question: "The Sugar Situation: About one-half the world's surplus sugar has been cut off. If sugar consumption increases proportionately for the next ten years, where will you get your sugar in 1923?" He answered it cartographically: "The sugar must be produced in the section of the world shown below. Beneath, it read, "The world's largest, most accessible, and cheapest 'sugar-bowl' is shown on the above map." The accompanying map

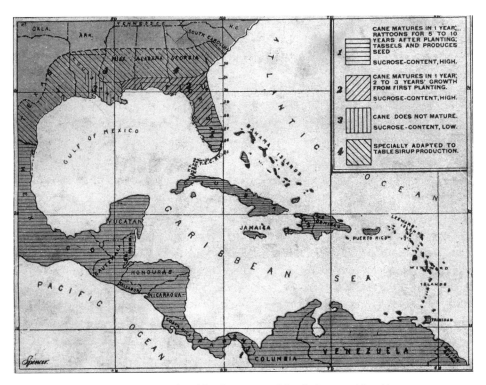

Figure 3.5. C. Lyman Spencer's publication presented detailed geographic evidence on the climatic limits of sugarcane production for granulated sugar. Courtesy of University of Florida Libraries, Special and Area Studies Collection.

graphically demonstrated that south Florida was the only area within the Caribbean basin that could produce domestic sugar for U.S. markets without naval shipping. Echoing Herbert Hoover's memo, Spencer emphasized that "in Cuba, beyond our governmental control, 50 Cuban sugar factories, prior to the European War, produced about one-half of the Cuban sugar crop at an average cost of $1.45 per 100 pounds, f.o.b. factory. It was good food, 96% pure. How much did you pay?" And anticipating Hoover's concern for the postwar sugar supply, he asked, "When peace is declared and all Europe is dipping into this 'sugar-bowl,' what are your prospects for low-priced sugar?" (Spencer 1918, front cover)

Spencer dismissed Florida's two primary competitors for the U.S. sugar market, the domestic beet industry and Cuba. According to him, the future of the U.S. beet-sugar industry was questionable. It had emerged as "a *scientific method of marketing* the agricultural product of a section located a long distance from market" and had served, "when labor, land and fuel were cheap" to transform the products of the Western irrigation farmer "into a non-perishable, easily marketed crop—beet sugar." However, the beet-sugar situation was changing because western lands were "needed for the production of large area machine-crops like wheat and corn" (Spencer 1918, 8; emphasis in original). Conceding that Cuba was "the world's richest sugar bowl," Spencer predicted that after the war, when

> the restraining hand of the International Sugar Committee no longer exists, it is more than probable that all of Europe will annually purchase large quantities of Cuban sugar; and thus *our sugar supply will be diminished* . . . No safeguards having been thrown around the future sugar supply of the United States, we have reason to view the future American sugar situation with anxiety and alarm. (6; emphasis in original)

Spencer's arguments regarding Cuba reflect the post–World War I shift in booster discourse that amplified the nationalistic ideas of self-sufficiency with the issue of food security to lobby for the construction of the Florida industry on a "war sugar basis" (44).

Reliance on Cuban sugar, according to Spencer, also brought the danger of vertical integration of an oligopolistic industry. "American capital is developing a great sugar industry in Cuba, *beyond our governmental control,* notwithstanding the fact that *in the same section of the world,* but within our own borders, sugar may be produced cheaper than elsewhere on the Globe" (Spencer 1918, 40; emphasis in original). Resurrecting populist alarm concerning the Sugar Trust, Spencer argued that recent restructuring of the Cuban industry put "large business units," which are "in control of the larger part of our sugar supply, as well as its *distribution"* in "direct or indirect control of the *production* as well" (40; emphasis in original). Centralized sugar distribution would open the door to trade control, and American sugar farmers might therefore "find themselves in the same predicament as the tropical sugar farmers, with but a single American buyer for their output— the American sugar refiner" (41). Although the largest refiners took credit for the low price of sugar in the United States to justify centralization, one "ought to take into consideration the fact that America is *located* in that part of the world where climate and soil make *possible* the production of sugar at lower costs than elsewhere on this Globe" (41–42; emphasis in original).

Beet sugar posed a different problem. "The modern sugar cane crop may be described as a *machine made* crop adapted to American conditions; while the sugar beet may be said to be a *hand labor* crop, for which purpose a very large army of men and women must be assembled each year" (Spencer 1918, 18; emphasis in original). Accompanying photos vividly illustrated the arduous labor associated with beet cultivation: "Polish Women Thinning Beets," showed twenty or so babushka-clad workers kneeling under a foreman's gaze, with the caption, "It would seem that the *labor problem of the American beet sugar industry is incapable of solution.*" In arguing against beet sugar on the basis of labor conditions and labor intensity, Spencer was reworking and reversing arguments advanced by the beet-sugar producers, beginning in the 1830s in France, who had promoted their crop in contrast to sugarcane produced by slaves. Now, ironically, cane sugar took the high moral ground with regard to labor exploitation.

In both tables and text, Spencer compared the labor and capital efficiency of the Western beet fields to the Southeastern sugar belt, finding the former to be one-sixteenth as efficient as the latter. He delineated the Florida "sugar bowl" with an elaborate and specific, though imagined, geography of production.

> Each side of the million acre cane field will measure 45.52 miles. We can add 25% to the area of that cane field, to provide for rotation crops, roads, ditches, property lines, etc., and distribute the 70 sugar factories regularly throughout the enlarged cane field, and the longest haul of raw material (cane) to any one of the 70 sugar factories will be only 3.30 miles, on a direct line. This concentration of manufacturing capacity and raw material for sugar is impossible in the beet fields, or, for that matter, in any other cane sugar section on the Globe. (Spencer 1918, 14)

Although Spencer characterized sugarcane as a "machine made crop," a supply of hand labor would apparently still be necessary. He thought prisoners well suited to the work of establishing "an American cane sugar industry" and suggested using prisoners of war from Europe or federal and state convicts. Farmers would be contracted to plant cane: "There will be no difficulty in making such contracts with reliable men. Large owners of such sugar lands stand ready and willing to cooperate in every way." Once enough seed cane had been planted, *centrales* could be provided. "All of the essential elements on which to base a profitable sugar industry in Florida are known to the last detail" (44).

Spencer was president of the All-American Sugar League, "devoted to the promotion of modern cane sugar and cane syrup industries in Florida," which published several issues in 1919 of *The Florida Planter.* The cover of the second issue proclaimed, "Florida Must Become a Sugar Producing State Because of the World's Sugar Shortage—High Prices Coming." The title page cried, "America First! Put the American Sugar Peninsula to Work! Convert Southern Sunshine into Foodstuffs for American and European Tables!!!" In the lead article, Spencer offered an alternative explanation to those who blamed trusts or profiteering for the high costs of foodstuffs. A "bird's eye view of the United States" showed that the problem was geographic (Spencer 1919, 1). Using numerous maps illustrating U.S. population distribution, climate belts, rainfall, growing seasons, and crop ranges, he argued that the entire U.S. agricultural system was not optimally located. Because of proximity to population and climate factors, the Southeast was underutilized, especially with regard to sugar production. F. E. Byrant, who would be one of the founding partners of the Florida Sugar and Food Products Company, in his congratulatory letter to the new publication observed,

> Apparently capital is going to be very timid at the commencement, being afraid of the great bugaboo—The American refiner. Once, however, the American public make up their minds that the sugar consumed in the United States should be, as stated by Mr. Spencer, produced in the United States by American labor and American capital, the American sugar refiner or so-called "trust," will have to play fair with the American sugar producer. (Bryant 1919, 5)

In the aftermath of World War I, Florida boosters were thus able to discursively construct an Everglades sugar bowl using a heady mix of late nineteenth-century populism and postwar nationalism (fig. 3.6).

During this period, numerous articles on sugar appeared with increasing frequency in the *Florida Grower,* a much more long-lived journal. An article entitled "Sugar Centrals and Plantations for South Florida" compared the extant and future sugar bowl. The commentary noted similarities of cli-

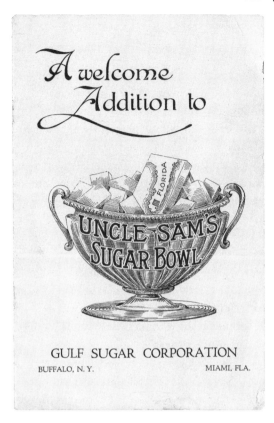

Figure 3.6. A commemorative souvenir from an Everglades sugar producer (1921), stressing the national importance of the region. Courtesy of Historical Museum of Southern Florida.

mate, rainfall, and soils between Florida and the lowest-cost exporter, Cuba. Thus, in "considering Florida as a sugar producing country, it is well to compare the conditions now existing there, with similar conditions in Cuba" (Chance 1919, 6). Labor costs were comparable, if not higher in Cuba, and construction materials were also higher there. Florida, the author claimed, would have several advantages over Cuba, including lower costs of mills, newer, more modern plants, more efficient labor, and the one-cent U.S. duty. With these advantages, Florida could produce sugar more profitably than Cuba if "[l]arge centrals and plantations" were built. In addition, Florida would not use the Cuban *colonos* system "of independent farmers selling their own product to the central under contract," but instead would establish a sugarcane-production region "where the central owns, as it should, the bulk of the cane land" (6). Florida sugar boosters thus took a populist stance toward the "trust," but their brand of populism and vision for a sugar bowl in the Everglades did not include the yeoman farmer.

Managing Deflation and Depression in the Global Sugar Market

Florida's potential was brought to the attention of Manuel Rionda, a preeminent Cuban sugar producer and president of the major U.S. sugar brokerage firm of Czarnikow-Rionda (McAvoy 2003). Among his numerous achievements in the sugar industry, Rionda established the Cuba Cane Sugar Corporation in 1915, and by 1918 it was the largest sugar enterprise in the world (Ayala 1999). Rionda was closely involved with the geopolitics of sugar all his life, and today his grandnephews are among the top producers in Florida. In January 1922 Rionda received a letter from a nephew in the family sugar business, suggesting reconnaissance of "the planting of sugar cane in the swamp-lands that have been reclaimed from the Everglades." Based on reports of the area, he predicted, "successful sugar mills will be erected before long, specially now that they will have the extra protection of the Fordney Tariff." However, a business associate who had actually visited the cane fields of Florida had already written to Rionda about problems encountered by growers on reclaimed Everglades land—such as low yields and pests—concluding "it certainly looks as though Cuba has nothing to fear from Florida competition in growing sugar."[7]

Indeed, against the wishes of Florida sugar boosters and despite the pleas of small Florida farmers, most U.S. capitalists interested in sugarcane during the early twentieth century were investing in Cuba. From 1903 until 1929, Cuba was the major supplier of sugar to the U.S. market, supplying 65 percent at its peak in 1922 (Dye 1998). In general, U.S. investments in Cuba began to increase after the Spanish-American War, accelerating greatly during and immediately after World War I. Specifically, investment in sugar increased more than sixfold from 1911 until 1927. As early as 1913, 39 of 172 working mills were classified as of U.S. ownership, although "nominal ownership meant little; many mills had partially Cuban, partially U.S. capital; many of the so-called Cubans were often more North American than Cuban, and sometimes *vice versa*" (Thomas 1971, 541). Adding to the confusion was the difficulty of disentangling Spanish and Cuban ownership or, moreover, the owners' identity: "A 'Spaniard' such as Manuel Rionda broke all generalizations" (541; see also Dye 1998; McAvoy 2003).[8] Among U.S. interests invested in Cuban sugar were some of the major shareholders of the American Sugar Refining Company.[9] Refiners were already heavily invested in domestic sugar beets; Cuban sugar plantations offered them another opportunity for vertical integration. Beginning in 1915, rising sugar prices increased the pace of foreign and domestic investment in the sugar industry, as evidenced by the fact that 39 new mills were built between 1916 and 1918.[10]

International price controls were in place for the 1918 and 1919 crop years, but at levels sufficient to maintain a prosperous sugar industry. Only months before the end of the war, the USSEB had contracted to buy Cuban sugar and to provide central management of the distribution of domestic sugars, so that 1919 began with contracts in place, prices stabilized, and no restriction on consumption. Thus, in the immediate postwar period, both the U.S. and Cuban sugar systems seemed stable in relation to a profoundly affected world sugar system. Prewar, world sugar sources were approximately half cane and half beet, which had alternating seasons. "Nature had thus established an equilibrium between production and consumption" (Bernhardt 1920, 88). The war cut beet production in half, while cane production increased somewhat, so that seasonal deficits were now more likely. With price controls in place but no demand controls, the result was a "'run' on the nation's sugar bank" (Mullendore 1941, 191). At that time, Cuban representatives, including Manuel Rionda, approached the USSEB offering the American government the entire 1920 crop, which the USSEB advised President Wilson to accept, with the key exception of Frank W. Taussig, who sent a separate letter of dissent: "The regulation of the price of sugar cannot in my judgment stand alone. The whole relation of government to industry in time of peace is involved. If the price of sugar is to be specifically controlled, so should that of bread, of meat, of clothing" (Taussig, quoted in Bernhardt 1920, 115). President Wilson's decision was made by default; when almost two months had passed with no reply from him, the Cuban representatives withdrew their offer, and the USSEB advised American refiners to purchase raw sugars in accordance with prewar conditions.[11] Thus, with Wilson's acquiescence, the "free" market in sugar returned abruptly, and with it came what Havana journalists termed the "Dance of the Millions."

The Dance of the Millions followed the release of price controls. From March through June, prices soared, and wild speculation characterized the sugar economy. Then, from a high of 20 cents per pound, prices collapsed to the prewar level of 3.75 cents, and by the spring of 1921 eleven domestic Cuban banks, including all of the largest, had closed. The Dance of the Millions underscored the significance and centrality of sugar production to the Cuban economy, which was fundamentally altered in its aftermath. According to Dye, "Foreign banks went from having 20 percent of the deposits and making 30 percent of the loans in 1920 to having 79 percent of the deposits and making 84 percent of the loans in 1922" (1998, 63). The low price of world sugar, which bottomed at about 2 cents per pound in 1922, led U.S. beet producers to lobby for tariffs, which were increased in 1921 and 1922. Even so, Cuba was able to produce economically as world sugar prices began to rebound in 1922, to the extent that the 1925 crop was considerably expanded from the previous three years, albeit the product of a restructured industry. As Dalton explained,

> This increase did not come from the locally owned plantations but largely from the properties which had been "taken over" by the large banking interests in New York City after the debacle of 1920. Via Wall Street, a large part of the Cuban sugar industry was sold to "investors" in the United States and sugar production passed largely from the old sugar firms and the Cuban *hacendados* into the hands of absentee bankers and lawyers. (1937, 249)

Cuba was not the only place where sugar production was increasing. Two factors were leading toward a worldwide glut. One was the "wave of economic nationalism" (Dalton 1937, 45) following World War I. Domestic sugar was among the industries that governments sought most to protect. European beet, Australian cane, U.S. beet and cane, and the nascent British beet industries all enjoyed policies that maintained prices above those of the world-market for domestically produced sugar. World production was also increasing as a result of the development and diffusion of technological changes in breeding, cultivation, and processing that led to increases in yields, sugar content, and sucrose extraction. Intensification of production in Java, Hawaii, Puerto Rico, Cuba, and the Philippines brought substantial increases in the sugar produced in these areas during the 1920s.

The postwar boom in sugar prices added an economic logic to populist arguments for increasing domestic sugar production. The boom was brief, however; the price of sugar in the United States jumped from seven to twenty cents per pound in 1920 but had fallen to three cents by January 1922. Cheap sugar was deemed so detrimental to peacetime economic stability that the Cuban industry was "rewarded" for its wartime cooperation by the imposition of tariffs on Cuban sugar. For the U.S. Food Administration, Cuba had been the "swing" producer of an essential wartime commodity. While Louisiana and domestic beet producers had failed to increase production during the war, despite the urgings of the Food Administration, producers in Cuba, who were receiving much lower prices than domestic producers, increased production from 2,922,000 tons in 1914–15 to 4,491,000 tons in 1918–19 to meet the needs of the Allies (see table 3.1). While even higher prices to domestic producers might have stimulated production, "the wiser policy was adopted of assuming a price level which would encourage production in the only source of supply where large increases could be immediately expected in response to relatively small price increases, that is, in Cuba" (Bernhardt 1920, 128). These large increases in

Cuban production were accompanied by social restructuring as more land was brought into production, and more small farmers and workers earned their livelihoods from sugarcane.

As the price of sugar began to climb in 1923, the Tariff Commission commenced a two-and-a-half-year study to determine whether sugar tariffs should be reduced in order to reflect differences in production costs, as intended by the Fordney-McCumber Tariff Act of 1922. In a special session in 1921, Congress passed the Emergency Tariff Act, raising the duty on Cuban sugar from 1.0 to 1.6 cents per pound. In 1922 it was raised to 1.76 cents. In 1925 the commission recommended that tariffs be reduced to 1.23 cents; in explaining his refusal to act on this recommendation, President Coolidge stated that the nation needed to "be independent as far as we may of overseas imports of food" and that the only way to assure a supply of sugar would be to maintain the beet-sugar industry. Coolidge's decision thwarted the purpose of flexible tariff provisions in that "the use of the difference-in-the-cost-of-production theory was doomed so far as sugar was concerned" (Dalton 1937, 62).

Cuban leaders could foresee but not forestall the world sugar depression. In 1925 the Cuban government responded to the risk of worldwide overproduction with a unilateral policy of output restriction that was renewed annually through the 1928 crop year. In addition, Cuba, as the largest exporter to the world market, negotiated with Germany, Czechoslovakia, and Poland to restrict their output in 1928 as well. Meanwhile, Commissioner of Agriculture Nathan Mayo used Cuba's attempted economic diplomacy to boost Florida in a state promotional pamphlet entitled, *Florida, An Advancing State: 1907—1917—1927:*

> Florida can safely consider entering the field of sugar production in a large way. . . . It will be remembered that during the year 1927 an announcement was made affecting the entire outlook of the world in sugar, by reason of an agreement between Cuba, Germany, Poland and Czecho-Slovakia to limit the output of sugar in their territories by government control. The effect of such action would be keenly felt by us unless the United States itself can go into larger production. . . . [T]he simplest answer to the problem lies in Louisiana and Florida. (Mayo 1928, 195)

As it happened, sugar prices were hit hard by the decline in consumption accompanying the general economic depression of 1929. In the context of world depression, Cuba negotiated the First International Sugar Agreement, otherwise known as the Chadbourne Agreement, which was signed in Brussels in 1931. The agreement limited exports from nine signatories—

the major beet- and cane-sugar exporters—for a period of five years.[12] However, "surplus sugar in warehouses from Havana to Batavia remained," and the agreement was "between producers, not governments, and therefore was not binding" (Thomas 1971, 562). Moreover, this effort to reduce world supply was undone by areas outside the agreement, primarily the British Empire and the United States and its insular territories, which over the same period increased production by nearly 20 percent. Between 1930 and 1933, the Cuban share of the U.S. market dropped from 49.4 percent to 25.3 percent, while the domestic share rose from 31.8 percent to 47.9 percent (Thomas 1971).

From the point of view of Cuban producers, injury was added to insult with the Smoot-Hawley Act of 1930, which raised the duty on Cuban sugar to two cents per pound in response to general deflation in sugar after 1929. The act "had a devastating effect on Cuban sugar production. In two years, production fell to one-third of capacity" (Dye 1998, 28; and see table 3.2). The tariff offered only a modicum of protection to U.S. producers; by 1932 the pre-duty price of sugar was the lowest in history, the duty was the highest since 1890, the duty-paid price was the lowest on record, and consumption had declined. Yet, for a variety of reasons, from 1929 until 1933, domestic, including insular, sugar production expanded, with a 41 percent increase in sales. Production in Puerto Rico and the Philippines increased as a result of technical improvements in sugarcane cultivation and processing. Western beet production expanded because, low as sugar prices were, beets remained relatively profitable compared to other crops. As Dalton noted not long thereafter, in the "year 1933, conservatively characterized as the most chaotic in American agricultural history . . . in comparison with the prices of other agricultural commodities, which had dropped in the same period by nearly 60 percent, the decline in the price of beets (about 30 percent) left beet producers in a relatively favorable position" (Dalton 1937, 65). The Smoot-Hawley tariff was implicated in this general agricultural decline, because trading partners retaliated with import duties, creating "huge surpluses of cotton, wheat, lard, and tobacco" (Carlebach and Provenzo 1993, 1).

Dye contrasts the 1921 and 1922 sugar tariffs to that of the Smoot-Hawley legislation. The former might be considered an adjustment to Cuba's wartime increase in milling capacity, whereas the latter "was an outcome of political bargaining alone, backed by widespread populist sentiment in the United States that something had to be done for American beet farmers" (Dye 1998, 66). Part of beet's clout might be explained by vertical integration between refiners and processors, and part by the fact that sugar was the

Table 3.2. Contributions to United States total sugar consumption from all areas, 1925–33 and 1934 quotas (short tons, raw value)

	Contributions							
	Continental United States				*Hawaii*		*Puerto Rico*	
	Beet		*Cane*					
Year	*Tons*	*Percent*	*Tons*	*Percent*	*Tons*	*Percent*	*Tons*	*Percent*
1925	1,063,500	16.11	149,500	2.26	763,000	11.56	603,500	9.14
1926	1,046,000	15.39	84,000	1.24	740,500	10.90	551,000	8.11
1927	935,000	14.73	46,500	.73	762,000	12.00	578,000	9.11
1928	1,243,000	18.71	138,500	2.08	819,000	12.33	698,500	10.51
1929	1,026,500	14.74	189,000	2.71	928,500	13.33	460,000	6.61
1930	1,140,500	17.00	197,500	2.94	808,000	12.01	780,000	11.62
1931	1,343,000	20.47	206,000	3.14	967,000	14.74	748,500	11.41
1932	1,318,500	21.10	160,000	2.56	1,024,000	16.39	910,500	14.57
1933	1,366,000	21.63	315,000	4.99	989,500	15.67	791,000	12.52
Quota 1934	1,556,166	24.03	261,034	4.03	916,550	14.15	802,842	12.40

| | *Contributions* | | | | | | | |
| | *Philippines* | | *Virgin Islands* | | *Cuba* | | *Other Countries* | |
Year	*Tons*	*Percent*	*Tons*	*Percent*	*Tons*	*Percent*	*Tons*	*Percent*
1925	485,000	7.35	10,000	.15	3,486,000	52.79	40,500	.61
1926	375,000	5.52	6,000	.09	3,944,500	58.04	47,500	.70
1927	521,000	8.21	6,500	.10	3,491,000	54.99	6,500	.10
1928	570,500	8.59	11,000	.17	3,125,000	47.05	35,000	.53
1929	724,500	10.40	4,000	.06	3,613,000	51.88	17,500	.25
1930	804,500	11.99	6,000	.09	2,945,500	43.89	30,500	.45
1931	815,000	12.42	2,000	.03	2,448,000	37.19	40,000	.61
1932	1,042,000	16.68	4,500	.07	1,762,500	28.21	26,500	.42
1933	1,241,000	19.65	4,500	.07	1,601,000	25.35	8,000	.13
Quota 1934	1,015,186	15.68	5,470	.08	1,901,752	29.37	17,000	.26

Source: Ballinger 1975

primary food staple that the United States both imported and produced domestically. Leading farm organizations, such as the American Farm Bureau and the National Grange, rallied to the beet cause with the slogan "The American Market for the American Farmer" (Dalton 1937, 150). The beet cause served Florida sugar boosters as well, providing a timely window of economic opportunity in which the industry became established. Thus, it was in the context of global overproduction and falling sugar prices that the Florida sugar industry emerged. This "window" would soon narrow to reflect the political-economic ties of the United States to the Cuban industry. With that adjustment, which came in the form of the Jones-Costigan Act (also known as the Sugar Act of 1934), the regional dynamic of the Florida industry became tied to the diverse geographic locales from which the U.S. sugar supply was sourced.

Establishing the "Nation's Sugar Bowl" in Florida

Although Cuba attracted the bulk of investment in the sugar industry, high sugar prices in 1918 and 1919 sparked a flurry of interest and speculation in south Florida. As an example of a small-scale venture, Judge John C. Gamling of Miami managed to plant two hundred acres of cane near Moore Haven, Florida, before his operation succumbed to inadequate drainage and improper equipment (Sitterson 1953). A large-scale attempt was made by the Pennsylvania Sugar Company (Pennsuco), a Philadelphia-based refiner that bought seventy-five thousand acres about twenty miles northwest of Miami and moved a mill from Texas. With twenty tractors and mechanized cultivators, the company was advanced for its time in the southern United States. Under the management of Ernest Graham, who moved his family from Michigan to the Everglades in 1921 to supervise it, the operation was enlarged in 1922 and again in 1923, and the mill began grinding in 1924 (Graham 1998).

We can follow the story of Pennsuco's decline step by step in the telegrams sent by Graham to F. C. Elliott, chief drainage engineer of Florida, such as one of January 16, 1922: "The water has risen ten inches. No secondary dam was put in not withstanding our appeals for protection. Many of our fields are now under water. We were ready to start planting today. There is no chance now of planting before January thirty."[13] Although well-funded, Pennsuco failed because of inadequate drainage. The company president, George Earle, Sr., felt that drainage had been guaranteed by the State of Florida. Earle's scathing, nine-page letter addressed to Governor John Martin, the trustees of the IIF, and the Board of Commissioners of the Everglades Drainage District blamed his company's failure on their ineptitude and deliberately misleading promises: "When we were told that you not only could but were promised that you would give us a sufficient depth of drainage for agriculture, and upon this promise were induced to abandon our plans of leaving the Everglades and spent enormous sums of money, we were told what, if I understand the matter, was an impossibility." Especially frustrating was that, with drainage, Pennsuco "got a tonnage of cane in excess of the average of Cuba at that time" and moreover, "the only way we can know that you even receive our letters is by sending them registered."[14] In 1926 the company gave up on sugar and shifted to large-scale truck farming;

several years later Earle got out of the business entirely and left Graham—who later served in the Florida senate—with a portion of the acreage "as a sort of severance" (Graham 1998, 114).

Another ambitious undertaking was that of the Florida Sugar and Food Products Company, incorporated in Massachusetts to produce sugar and syrup in south Florida. The company bought four thousand acres of land near Canal Point, Florida, and hired Anthony R. McLane, "a sugar land expert and engineer with twenty-four years' experience in Cuba, Puerto Rico, and Hawaii, to make a survey of its holdings and adjacent lands in the Everglades region" (Sitterson 1953, 366). McLane's report could hardly have been more optimistic: he wrote that there were fields producing sugar in 1921 that had last been planted in 1913; that killing frosts had not occurred in the Everglades; that no fertilizer was required, and that the level, treeless land was ideal for mechanized cultivation.

This report together with several other factors—the tariffs of 1921 and 1922 and the establishment of the Canal Point Breeding Station by the USDA in 1921—served to encourage the company's backers. Among these was F. E. Bryant, who had emigrated from England in 1894, originally settling in New Mexico to manage a sheep ranch and to learn about irrigation. He then moved to Florida, "convinced that if scientific procedures could be established, the Everglades were destined to become the source of great food supplies for the nation." With some associates, Bryant—whose letter in the *Florida Planter* regarding the need to secure "sufficient capital" for a modern mill was mentioned earlier—organized the Florida Sugar and Food Products Company in an effort to do just that. In 1923 the company arranged to transport a second-hand mill up the West Palm Beach Canal to Canal Point by barge, which was the first commercial sugar house in the Everglades. Bryant, "invariably wearing a tropical helmet and long boots, . . . became a symbol of the industry" (Hanna and Hanna 1948, 305).

One of the early findings of the Canal Point Breeding Station, of great significance to the locality, was that this vicinity was one of but a few subtropical regions worldwide and the only region in the continental United States where sugarcane would consistently flower and produce seed. The initial purpose of the station was to search for sugarcane genetic material that was resistant to mosaic, a disease introduced to the U.S. South from Asia around 1914, perhaps as early as 1912. The intended beneficiary was the Louisiana industry, which was nearly destroyed by mosaic (Wade 1995). Mosaic was believed to have spread from Louisiana into Florida, where there were widespread reports of infection by 1919.[15] In 1927, Bryant wrote to E. W. Brandes, senior pathologist in charge of sugar plants at the USDA, concerning the presence of mosaic in his fields. Brandes replied that the information was "disconcerting" and that the "situation in the Everglades, insofar as commercial sugar production is concerned, is more critical than it has ever been before." He praised Bryant as "one of the few" who understood the gravity of the situation and was "willing to use heroic measures." He went on to note that commercial cane planters and the state "could well afford to spend a large sum of money in effecting another clean up within the area that is so well adapted to the commercial production of sugar cane."[16] Within ten years of its establishment, the Canal Point Station had developed three disease-resistant canes that both reduced losses to disease and improved yields. In 1921 the state of Florida established an agricultural experiment station at Belle Glade, about ten miles from Canal Point. Though not specifically focused on sugarcane, its contributions to muck soil science would prove invaluable to the industry (Dovell 1947a). Thus in the early 1920s, the region attracted capital investment and gained industrial expertise and scientific knowledge. By June 1925 *Popular Mechanics Magazine* was reporting Everglades sugarcane yields of sixty tons per acre, four times the Louisiana average.[17]

However, high yields were not enough to overcome the problem of insufficient funds.[18] In an effort to secure investment, Florida Sugar and Food Products was reorganized as the Florida Sugar Company, and in 1925 it was incorporated as the Southern Sugar Company when Bror G. Dahlberg, who had emigrated from Sweden as a child, bought control (Hanna and Hanna 1948). The Southern Sugar Company was reputed to have a unique angle on the sugarcane industry. Table sugar was ostensibly a by-product of the company's primary product, "Celotex," a building material manufactured out of bagasse, which are the stalks from which cane juice has been extracted (Johnston 1928). Dahlberg, the new company's president, promoted Celotex as a sustainable alternative to wood. Company publications billed Celotex as "an insulating lumber" and described it as "a material that would supply insulation with structural strength" (Southern Sugar Company 1928, 7).

Dahlberg was head of the Celotex Company of Chicago, which owned several sugar plantations in Louisiana. "When disease hit the Louisiana cane, diminishing the supply of fibre available, it became imperative for the Celotex Company to develop new sources of raw material" (Johnston 1928, 42). In an effort to secure larger quantities of bagasse, after "an exhaustive survey of the sugar cane industry in the United States, and a most thorough study of the soil and growing conditions in the northern fringe of the Everglades, The Southern Sugar Company was organized by northern and western capitalists under the laws of Florida" (Southern Sugar Company 1928, 3). Listed among the directorate were not only "locals" such as F. E. Bryant, formerly of Florida Sugar and Food Products Company, and Jules M. Burguieres, supervisor of the 1915 sugarcane experiments by the Southern States Land and Timber Company, but also George L. Eastman, president of the Los Angeles Chamber of Commerce, W. J. Tully of New York, director of Corning Glass Works, along with a dozen other similarly positioned men.

The directors of the Celotex Company must have been encouraged by the engineering report entitled "Lands of the Sugarland Development Company," which argued that Southern Sugar would become "the greatest and most profitable Cane Sugar producer in the world." Key to predictions of success was the location of their 15,000 acres, just southwest of Lake Okeechobee, where the soil type known as "custard apple muck" was found. Predicting average annual yields per acre that were double those of Cuba or Puerto Rico, the report concluded that

> even without the present Tariff Protection of 1.76 cents per pound for Cuban Sugars and 2.2 cents for all others, Southern Florida will prove itself as much superior to Cuba as a Sugar Cane district as Cuba has shown itself superior to the other West Indies Islands. With the present saving in cost for freights and duties, the business, once properly established, should prove a "bonanza." (Smith and Ames 1925, 24)

By 1929, the Southern Sugar Company controlled approximately 130,000 acres "of the choicest muck lands around the southern shores of Lake Okeechobee or adjacent thereto." Of these, approximately 43,000 acres were "under water control" (Dahlberg Corporation of America 1929, 13), largely because of the eighty-five miles of canals, hundreds of miles of ditches, and extensive levees built by the company. The two large mills in the region—the Canal Point mill of Florida Sugar and Food Products and the Hialeah mill of the Pennsylvania Sugar Company—were moved to Clewiston and combined to make a mill for the Southern Sugar Company with a fifteen-hundred-ton capacity (Sitterson 1953). The Southern Sugar Company was now part of the Dahlberg Corporation of America, which Dahlberg organized to develop and expand the Dahlberg Sugar Cane Industries and "related projects"—including the Celotex Company, the Cypremort Company, and the Clewiston Company, Inc. The last represented the corporation's takeover of a progressive-era town, Clewiston, sited in 1920 and planned by John Nolen, "an international giant in the field of modern city and regional planning" (Low 1998, 312).[19] During the 1920s, Nolen designed model towns throughout Florida. Clewiston, one of his first in Florida, was archetypical of Chautauqua-movement design, which "aimed to enhance and carefully integrate environmental, social, economic, and political conditions and needs to serve all segments of society fairly" (317) (fig. 3.7).[20] By 1929 this progressive town had become a company town, with the Dahlberg Corporation owning in fee the "3,500 acres of land covered by the townsite and numerous buildings" as well as "the public utilities, including telephone, electric light and power, and water systems" (Dahlberg Corporation of America 1929, 16). Boosters saw "a Chicago in the Everglades" arising on the southern shore of Lake Okeechobee, envisioning Clewiston as "the metropolis [at] the center of a great and growing countryside" (Reese 1929, n.p.). Not only did it have an "ideal location" with respect to water transport, but it was also "accessible by two great railway systems" one of which, "the Atlantic Coast line[, had] its terminus at Clewiston" (Southern Sugar Company 1928, 18; see fig. 3.7).

Dahlberg was able to recruit significant agricultural and engineering expertise to his Clewiston enterprise. The professional biographies of the founding members show that the company had the benefit of geographically diverse experience and demonstrate that the region was becoming part of the circuit of this internationalized industry. For example, P. G.

Figure 3.7. The Atlantic Coast Line to Clewiston nearing completion (1921). Courtesy of Historical Museum of Southern Florida.

Bishop, operating vice president, began his career in the sugar business in Puerto Rico and had most recently been vice president of the Cuba Cane Sugar Corporation, which Manuel Rionda had established. The executive vice president, chief engineer, and manager were all veterans of the Louisiana industry. B.A. Bourne, originally from Barbados, had been assistant director of agriculture there and subsequently was head of the Plant Pathology Department at the University of Puerto Rico. Bourne arrived in Florida as assistant pathologist and cane breeder at the USDA experimental station in Canal Point, and Dahlberg hired him from there in 1929 to head the research department of the Southern Sugar Company. He was joined by R. V. Allison, who had also worked at the Canal Point station and would later head the University of Florida's Everglades Agricultural Experiment Station. Soon thereafter, Harry Vaughn, a sugar chemist from Louisiana with further experience in Cuba and Haiti, was hired. A final example is that of the mechanical engineer, N. C. Storey, who had earlier been in charge of mechanical maintenance for the Panama Canal (Reese 1929; *Sugar Journal* 1961; Heitmann 1998).

The Dahlberg Corporation's approach to developing a sugarcane industry in Florida was thus part of a geographically diverse, vertically integrated plan for interregional agro-industrial development, and it was recognized as such at that time. The opening of the Clewiston mill on January 14, 1929, marked the "beginning of a new industrial era" with headlines such as, "Early faith that Florida potentially is sugar producing region realized after many years" (Reese 1929). In an astute discursive move, the company marked the event as the opening of the "Nation's Sugar Bowl," thereby linking its profits to the national interest (fig. 3.8). Telegrams from Florida's U.S. congressional representatives assured the company's owners that the "entire Florida delegation are working in and out of season to secure the passage of a bill" that would provide federal support for drainage and reclamation of more wetlands for sugar (Sears 1928, n.p.), while the new Florida Governor, Doyle E. Carlton, proclaimed that "the term 'Sugar Bowl of the Nation' . . . is no misnomer" (Carlton 1928, n.p.). Several thousand people—including John Hays Hammond, former minister to Great Britain, Thomas Meighan, a film star of the day, and Glen Curtis, aviation pioneer—attended the dedication ceremony, at which Governor Carlton threw the switch that started the mill machinery (Sitterson 1953). "Florida Will Fill America's Future Sugar Bowl: Organized Industry Taps Pristine Resources of Long Neglected Empire" headlined the February issue of the *Florida Grower*. The rhetoric of the ensuing article was inflated even by booster standards: "Florida . . . is now incubating a 'billionaire baby' which

Figure 3.8. Southern Sugar Company appeals to nationalist sentiments in opening its new mill (1929). Courtesy of United States Sugar Corporation.

even in adolescence eclipses its world-wide rivals. Introducing the epochal Everglades in a new and different role—future sugar bowl of these United States. Not a new idea, but an idea which has been handicapped in hatching by that formidable menace which for centuries has held the 'Glades in bondage—too much water" (Dacy 1929). The roles would soon begin to reverse, however, as more and more of Everglades water was subjected to the bondage of canals and check dams in order to free the 'Glades fertile soils for agro-industrial development.

It remained for the Southern Sugar Company to "inaugurate bonanza sugar growing in the 'Glades on a scientific and practical scale with the avowed purpose of exploiting that spacious territory acclaimed by the USDA as 'the best resourced locality on the earth's crust for sugar production'" (Dacy 1929, 1). Southern Sugar also had the advantage of entering production after the Everglades region had eradicated mosaic virus. One consequence was that the expansion of cane plantations had to proceed geometrically, since, to prevent recurrence, the company was restricted to using only its own seed cane grown from newly developed resistant canes, the POJ (Proefstation

Oost Java) varieties. By 1929, eleven thousand acres had been planted to cane, with the goal of reaching twenty-five thousand acres as rapidly as possible. Thus, nature provided at least two formidable barriers to the sugar industry. One was inherent in the crop itself: expansion depended on the propagation of seed cane, which meant "making haste slowly" (Dacy 1929, 8). The second was geographically specific: the Everglades drainage district was "the largest drainage enterprise in the world" and an unwieldy and misunderstood one at that. By 1930, Southern Sugar had reached its goal of twenty-five thousand acres and had increased its factory capacity to four thousand tons per day. An equipment inventory listed 144 tractors, 408 cane wagons, 13 cultivators, 235 cane cars, and 4 locomotives (Sitterson 1953). However, though Southern Sugar had from the outset approached the drainage question with engineering expertise and capital investment, the company still found the task of flood control daunting. "Not only was the company liable for taxes on its land, but it had to drain whole areas for the sake of small arable sections contained within them" (Manuel 1942b, 12959).

"A Modern Giant Factory": Reclaiming the Everglades for Agro-industry

In November 1923, the *Ft. Pierce News-Tribune* reported that the Fort Pierce Chamber of Commerce had invited Dahlberg to inspect the local port. The fact was newsworthy then and noteworthy now because it reflected the political maneuverings of local boosters whose aim was to link drainage and transport issues once again, as they had been in the pre-railroad days and later in the Randolph report. The chamber also "endorsed the work of the Florida Flood Control association and recommended that the county commissioners contribute to its support." The purpose of the Florida Flood Control Association (FFCA) was "to furnish data and information in support of federal aid for the Everglades." Though flood control was its primary goal, "it had been necessary to tie navigation development into the project" because federal appropriations would be based on navigation, and thus, "having one of the ports in nearest proximity to the developed Everglades region, Fort Pierce is of necessity concerned in any proposal" (*Ft. Pierce News-Tribune* 1923).

A meeting concerning a proposed "cross-state waterways project . . . [drew] prominent representatives from all of south Florida" (*Ft. Myers News Press* 1926, n.p.). The most prominent was Fred Williamson, general manager of the Clewiston Development Company, whose views were quoted exclusively in the local press. Still working for the Dahlberg Corporation, by 1929 Williamson was both vice president of the Southern Sugar Company and president of the FFCA. In this dual capacity, Williamson served as a key spokesperson in the effort to garner federal support for Everglades reclamation in the guise of navigation. Making the request for government aid to canalize the Everglades, Williamson judiciously linked flood control and transport in articulating his machine-age vision of the future of the vast wetland.

> The present status of Everglades reclamation and development might well be compared to a modern giant factory building, with all needed raw materials available and with machinery to manufacture that raw material, but standing on a temporary foundation, and reachable only by foot paths. Permanent strengthening of the foundation must come from carefully prepared plans for adequate control of excess waters. Adequate transportation will only come from complete utilization of available water routes. When these two necessary objectives have been attained, all South Florida can develop, and the Everglades will truly be like a great factory creating new wealth for Florida, out of Florida raw materials.[21]

Ultimately, the FFCA and those who shared its sentiments succeeded in garnering federal aid after overcoming two types of political obstacles. One concerned the role of the federal government in providing "local" infrastructure; the second was the question of how to divide costs between the state and federal government. Overcoming the first difficulty required redefining the federal government's role in local flood control. The conclusion of the chief of the Army Corps of Engineers in the aftermath of the 1926 hurricane was that diking Lake Okeechobee and draining the surrounding landscape were projects that should be undertaken using local and state resources. The hurricane of September 1928 again demonstrated the vulnerability of the lakeside population, taking more than two thousand lives in Belle Glade and its environs. In his January 1929 report, the chief of the Corps presented a $10.7 million dike and drainage project with the recommendation that the federal government contribute $4 million to its completion. This was insufficient federal involvement from the point of view of Florida officials.

When President-elect Herbert Hoover visited Florida, one of only four southern states that had supported him in the election, Governor Carlton led him on a tour of the south shore of the lake. Hoover's trip to the devastated region—and the two hurricanes that prompted it—is considered a turning point in federal involvement there (Blake 1980). Shortly thereafter, Carlton asked the state legislature to create a state agency that would be empowered to cooperate with the federal government in Everglades reclamation. The result was the Okeechobee Flood Control District, which included all of south Florida from just north of the lake, excluding the Keys. The board, headquartered in West Palm Beach, was empowered to build flood-control works, enter into agreements with the federal government, issue bonds, and impose acreage taxes. Thus, it was more extensive, geographically and politically, than the Everglades Drainage District. In February 1930 the newly created district sent to the Corps an engineering report that outlined a $29 million program of levees and drainage throughout south Florida. In March the Corps recommended an almost $10 million project, nearly two-thirds of which would have been federally funded (Blake 1980).

At the time, the sugar region of south Florida was already viewed as an industrialized landscape and production system. Among the reports submitted to Corps Chief Major General Lytle Brown was E. R. Lloyd's, "On the Possibilities of Agricultural Development in the Everglades District of Florida," which had been transmitted to the division engineer in August 1929 (U.S. Senate 1930). An introductory note from Secretary of War Patrick J. Hurley explains that the "report was made . . . in connection with proposed river and harbor and flood control improvements in the State of Florida." Lloyd, an agronomist and director of the Mississippi Agricultural Experiment Station, spent three weeks observing crops and agricultural practices, and interviewing workers at the station, farmers, and "those in charge of corporations operating large holdings devoted to special crops" (4). In discussing the economic merits of the "Florida saw-grass Everglades," Lloyd emphasized sunk costs as evidence of an emerging agro-industrial region. The focus of the report is "farming on a large scale, which," Lloyd explained "is more economical than small-scale farming. The large companies have well-trained men, the latest and best farm machinery, and they make a careful study of the best methods of treating and conserving the soil" (14).

In 1930 the federal share of the cost of infrastructure was significantly increased when the Okeechobee project bill was moved from the Committee on Flood Control to the Committee on Rivers and Harbors to be considered as a bill to aid navigation. In July 1930 it was signed into law, having passed through Congress as part of the general Rivers and Harbors Act (Blake 1980). The FFCA had succeeded in its objective: the federal government would construct and maintain the Hoover Dike, using land contributed by the state of Florida in addition to a cash contribution from the state of a mere five hundred thousand dollars, roughly 2.5 percent of initial costs. When completed, the Hoover Dike would be a phenomenal earthwork, 85 miles long, 125–150 feet at its base, rising 34 to 38 feet above sea level to girdle and obscure the lake. It represented a significant alteration in both the political-economic and ecological dynamics of the region. It was the beginning of tremendous federal investment to restructure the landscape of south Florida, of which the completed Hoover Dike project, representing an expenditure of $19,146,000, was just the start. It also was fundamental to the transformation of what is now termed the K-O-E ecosystem, by making immutable the previously protean lake. Encased in rock and concrete, Lake Okeechobee could no longer reform itself seasonally by transgressing its littoral boundaries. The political-economic processes of the latter half of the decade, which had been influenced by cataclysmic natural events such as storm surges, thus, in turn, fundamentally altered the physical geography of south Florida, creating a "second nature" (Smith 1990; Cronon 1991). The development of a "modern giant factory building" for the production of tropical commodities required not only the management of seasonal precipitation but also the reduction of the threat posed by an extensive inland body of water in the hurricane-prone subtropics.

Though Lloyd had highlighted the Southern Sugar Company in his report extolling the virtues of large-scale Everglades farming, the company was to be short-lived. In the fall of 1930, as construction of the Hoover Dike began, Southern Sugar was forced into receivership. Among stockholders filing suit against the Celotex Company was William L. McFetridge, who alleged that Dahlberg and associates had received $10 million in "secret profits" and had "unloaded" sugar mills and plantations in Puerto Rico, Cuba, and Louisiana onto the corporation "in the face of a 'ruinous sugar' market" (*Minneapolis Journal* 1930, n.p.). Whether or not Dahlberg was guilty as alleged, it was true that Celotex faced a "ruinous" sugar market and was overcapitalized. For the 1930–31 season, the company was operated by court-appointed receivers as ownership was transferred to its creditors, notably Charles Stewart Mott, vice president of General Motors Corporation. Mott is credited as the founder of the United States Sugar Corporation (USSC), for decades the largest and today the second largest of the Florida sugar companies. The company's naming—combined with the previous designation of the region as the "Nation's Sugar Bowl"—was part of an effort to discursively construct Florida's cane plantations as vital to national interests (Hollander 2005). Indeed, in the case of the strategically named USSC, the distinction between national government initiatives and private enterprise would become blurred in the general public's mind in the New Deal era (Goluboff 1999).

In April 1931 the courts approved the Southern Sugar Company's reorganization as USSC. Reorganization was formalized in December when "Florida's largest corporate farming enterprise—the Southern Sugar Company with Everglades cane fields and grinding mills valued at $15,000,000—went on the legal auction block. . . . It was sold at a bid of $900,000 to the United States Sugar Corporation, recently formed by Bitting, Inc., of New York, in behalf of certain groups of creditors and stockholders in the old company" (*Florida Grower* 1932). The seemingly low price reflected the fact that USSC was assuming the debts of its successor. Also in 1931 there were reports of a second sugar company being organized in Fellsmere, Florida, under the direction of Frank Heiser. Heiser and other farmers in the area had been experimenting with sugarcane since 1918, when they had formed an informal sugar-cane producers association. By 1931, Heiser's Fellsmere Company had sufficient seed cane to plant 1,000 acres and enough capital backing to begin assembling a mill. In 1933 the mill began grinding, and Fellsmere became a depression-era boomtown, outside of which were signs "warning that the mill was not hiring outsiders" (Patterson 1997, 419).

By the early 1930s these reorganized companies—USSC and Fellsmere Sugar—had the benefit of the hard-won experience of previous years. The idiosyncrasies of the physical landscape—of the muck soils, drainage, and microclimatic variations—were increasingly understood and manageable. The less tractable problems were political-economic. The immediate problem was the low price of sugar in the world market, and, more important, in the United States. As noted, USSC executives quickly set themselves to work lobbying Congress for protection from "foreign" competitors through higher tariffs. However, by June 1932 "the price of sugar before duty was the lowest in history, the duty was the highest since 1890, and the duty-paid price was under 3 cents, the lowest on record" (Dalton 1937, 63). In the longer term, the solution to the price problem posed the greater difficulty to the nascent Florida industry. Florida sugar interests and boosters would be exceedingly frustrated by the Jones-Costigan Act of 1934 and its subsequent revisions, which marked the point at which the de facto regional development policy implicit in U.S. sugar policies became internationalized (see table 3.2). Protection from competition was a double-edged sword when it came in the guise of maintaining the status quo, which in 1934 included a small Florida industry that had attempted to expand as rapidly as possible within the agro-ecological constraints of seed-cane propagation and under the hydrological challenge of drainage.

CHAPTER FOUR

Wish Fulfillment for Florida Growers: Managed Market, Disciplined Labor, Engineered Landscape

The decade leading up to U.S. involvement in World War II was a critical period in the establishment of the sugar agro-industry in south Florida. The federal government became increasingly involved with—indeed complicit in—sugar investors' imagined economic geography. This was the New Deal era, when federal programs proliferated and brought government regulation into all aspects of sugarcane production in Florida, including labor, markets and trade, and environmental management. By this time it was clear that the U.S. sugar industry would receive some kind of protection from foreign producers, but the form and extent of that protection was still a point of political struggle. There was also widespread concern over conditions for migratory agricultural labor, including the migratory labor stream flowing through Florida's sugar plantations, though again the form and extent of federal involvement were subjects of intense debate. This was also the heyday of large-scale, federally directed environmental engineering projects. These were the salad days of federal agencies such as the Bureau of Reclamation and the U.S. Army Corps of Engineers, whose projects transformed arid lands and wetlands alike into phenomenally productive agricultural regions, albeit at phenomenally high environmental costs. Finally, it was a transformative period for the U.S. South, including Florida, as northern industrial capital flowed into the region seeking cheap, principally black agricultural labor, which in turn was increasingly flowing northward in search of better employment opportunities as the Depression wore on.

These political and social trends were part of the sugar question in the 1930s and early 1940s when the federal government established labor, market, and water control structures that were fundamental to accumulation and expansion in the industry for the next half-century. This chapter explores the machinations of sugar lobbyists, the social welfare visions of America's capitalist class, and the dreams and desires of Florida boosters, as well as the roles that each played in the formulation of the federal laws and policies that established these structures. The chapter's narrative thread is woven through the lives of key individuals, beginning with Bror Dahlberg, whose behind-the-scenes efforts to shape U.S. policy on Cuban sugar have been largely unrecognized in the historical literature. Though Dahlberg's business enterprise in the Everglades failed, his political efforts bore fruit for his successors, the shareholders of USSC. The company, especially its vice president, Clarence Bitting, took the initiative in defining the sugar question for the American public and its elected officials. In this new corporate discourse, Cuba, as it had since the Everglades sugar region was first imagined, played the foil to Florida's nascent agro-industry. To understand why Bitting thought such a corporate propaganda campaign was both necessary and effective, we need to return to the debates surrounding the Smoot-Hawley tariff and the national political context in which they took place.

"Little Cuba Tried to Do Her Share"

If New Deal sugar policies seemed to mark the denouement in the ongoing drama of depression-era sugar politics, then the earlier debate over the Smoot-Hawley tariff was the critical first act. The House and Senate hearings preceding the passage of the 1930 Smoot-Hawley Act revealed how different competing interests were formulating the sugar question in 1929. Historians commonly attribute the success of U.S. sugar interests in securing this protective tariff, which punished Cuban producers, to the populist appeal, nationalist rhetoric, and political power of the domestic beet industry (see Thomas 1971; Benjamin 1990; Dye 1998; Ayala 1999; McAvoy 2003). None have noted any role for Florida interests in the political lobbying surrounding the debates on tariffs. However, the nascent Florida sugarcane industry was much more politically proactive than historians have recognized. Already by 1929 Florida sugarcane interests were becoming a formidable force with a particular penchant for obstructing the access of Cuban sugar to U.S. markets. In order to understand the emergence of south Florida in the 1930s as a region of agricultural production, we first need to consider the broader outline of the sugar question that emerged from the 1929 Senate hearings and then to look specifically at the political maneuverings of the Florida industry at that time.

Senator Reed Smoot of Utah, a "fervent protectionist determined to attain for the United States a 'high degree of self-sufficiency'" (Schlesinger 1957, 164) presided over the Senate hearings. Smoot, who chaired both the Committee on Finance and the subcommittee that held hearings on "Sugar, Molasses, and Manufactures of," would later serve on the Executive Committee and as director of the Utah-Idaho Sugar Company from 1935 until 1941. Among the four other members of the subcommittee was James E. Watson of Indiana, Republican leader of the Senate. Dozens of witnesses appeared before the subcommittee, among them representatives of national cane or beet associations, local cane or beet organizations, domestic growers, Americans invested in the Cuban industry, Puerto Rican producers, industrial users of sugar, sugar refiners, and trade associations. By far the largest number came from the first three groups. Bror Dahlberg—who was also working effectively behind the scenes to garner national political support for his Florida plans—testified as a representative of his Louisiana and Florida sugar companies.

On balance, domestic sugar growers faced in the Senate a sympathetic audience. Like other agricultural producers, farmers from sugar-producing states counted on their representatives in Congress to defend their economic interests. Beet farmers' political influence was significant—in some years as many as twenty-two states produced beets for sugar. Domestic sugar producers felt that they were integral to the U.S. economy, helping to stabilize farm incomes, and therefore deserved protection. Tariff proponents also argued that sugar was a strategic good, deserving of protection for reasons of national defense. They stressed the importance of U.S. independence from foreign supplies and that "Army and Navy authorities in the United States give sugar a place in the forefront of military rations" (U.S. Senate 1929, 50).

Counterpoints of disagreement and difference arose regarding the possible undesirable impact of tariffs and the character of the U.S. industry. One objection to tariffs was that consumers would be forced to pay artificially higher prices for sugar. Also, the national security argument could cut two ways, because existing tariffs had already begun to have the effect of displacing Cuban sugar from the U.S. market in favor of more distant but "free" sugar producers such as the Philippines and Hawaii. Tariff opponents portrayed these distant supplies as vulnerable in time of war. As Rudolph Spreckels, representing Spreckels Sugar noted, "Contemplate what would happen to us if we were dependent upon the Philippine sugar in the event of war. We could not get a pound of it. It would be a distinct menace to our nation" (U.S. Senate 1929, 177). The same logic held in arguing for policies that would contribute to the stability of the Cuban industry (i.e., no or low tariffs), which was a proven wartime supply ninety miles from the continental United States. (Hollander 2005).

Beyond the arguments over protectionism versus free trade typical of any tariff debate, the Smoot-Hawley hearings dwelt on two issues that were or would become emblematic of Florida sugar politics. The first revolved around questions of labor in the U.S. sugar industry, including racial and national origins and the conditions for domestic workers relative to those in other producing countries. At the time of the hearings, labor had yet to become an issue for Florida, but the testimonies foreshadowed a political and economic predicament that would dog Florida sugar growers for decades. The issue of labor figured significantly in the debate on sugar tariffs, manifest in several ways. One was the question of whether tariffs could be justified to protect "American" sugar if production relied on "foreign" workers. A second issue was the condition of workers' lives in the United States, whether they were earning enough to live decently and whether, if workers were attracted by seasonal sugar work, they would remain and take jobs from local workers. All of these were domestic concerns, but when the committee heard from representatives of Cuban producers, the line of questioning shifted to comparisons between U.S. and "foreign" standards of living.

For example, the committee, especially Smoot and Watson, questioned Edwin P. Shattuck extensively on the issue of comparative labor conditions. Shattuck—who had been one of the directors of the USSEB—was retained as a lawyer to represent, as he explained, "the United States Sugar Association and its membership, who are Cuban producers of raw sugar." In his opening remarks, he prefaced the remainder of his testimony with two assumptions. First, that American investment in the Cuban sugar industry was twice that in the protected industry—which included producers in the U.S. mainland, Hawaii, Puerto Rico, and the Philippines—and second, "the fact, which I think these hearings have brought out, that there are no real differences, no real inequalities in the type of labor employed in the different agricultural regions of sugar production, cane or beet" (U.S. Senate 1929, 191). Shattuck was aggressively questioned on two claims he made with regard to labor: that Cuban workers were paid as much as workers in the domestic industry, and that "the labor in the agricultural fields in the United States is largely foreign" (197). The hearings also probed the racial identity and national origin of labor in U.S. domestic sugar production. For example, Senator Harrison, a Democrat from Mississippi, questioned a beet-grower representative by quoting the statement of another beet spokesman "that there is not a white man of any intelligence in our country that will work an acre of beets" (43) and a Department of Labor report that "75 to 90 per cent of the workers in the beet fields of Ohio, Michigan, Minnesota and North Dakota were Mexicans" (45). As the committee wrestled with the question of whether sugar produced in the United States using foreign labor was "American," Montana sugar beet farmer O'Donnell noted that "a great many of these Mexicans are native born" and suggested another potential and indubitably American labor source: "Montana has lots of Indians. We are trying them out in the beet fields and they are making good" (127).

The nascent Florida industry made its presence known at the hearings primarily through its relationship to Cuba's sugar growers. This was the second salient issue in the testimonies: what did the United States "owe" Cuba, and how would different tariff structures hurt or help Cuba. Shattuck, the Cuban industry's able advocate, attempted to give the Senators a history lesson on the relation between the United States and Cuba, punctuated with quotations from Presidents McKinley and Roosevelt. He appealed to the idea of moral obligation and the reciprocal relationship between the two countries, testifying that the tariff they were contemplating would "be her destruction." Smoot replied, "It will not be her destruction. And I know this, Mr. Shattuck, that Cuba brought it upon herself by the amount of sugar she produced." Shattuck responded that Cuba had cut production by two million tons in three years, while other regions had increased output: "Little Cuba tried to do her share in bettering the situation, but nobody else in all this big world did a thing to help her. And you accuse Cuba of being the cause of your distress" (U.S. Senate 1929, 217).

It was Bror Dahlberg, heavily invested in the Southern Sugar Company of Clewiston, Florida, whose testimony brought the hearings' attention to the intertwined destinies of the Florida and Cuban industries. Dahlberg portrayed the southern U.S. sugarcane industry as resurgent, fully capable of expanding production to meet the nation's sugar needs. At the same time, however, he spoke of its vulnerability to Cuban competition. Speaking of Louisiana and Florida, he argued for higher tariffs "so that . . . we can have adequate protection against the competitor who is now dumping sugar on us at ruinous prices. That is Cuba." Cuba, he claimed, "enjoys no advantages whatsoever, as far as sugar making is concerned, above Florida" (U.S. Senate 1929, 133). Furthermore, and in contrast to the domestic beet industry, "no foreign labor is employed in the cane fields or the sugar mills at any season of the year" (14). Like many U.S. producers of agricultural commodities who came before and after him, Dahlberg portrayed his business operations as robust while asserting that without an increased tariff they would fail: "We

are operating with as improved methods, with as competent an organization, with as up-to-date machinery as any place in the world. On the present
sugar situation it is impossible to support ourselves" (136).

In the end it was Dahlberg, not Shattuck's clients, who won this early argument between the rival industries of Florida and Cuba. Without doubt
Dahlberg and his fellow investors rode the coattails of the beet-growers'
lobby, so capably shepherded by Senator Smoot. The testimony transcripts
show that Smoot, representing one the largest beet-producing states in the
West, wielded his power as chairman of the hearings to assure a tariff favorable to U.S.-based sugar growers. His strategy seemed to be one of badger-
ing into submission any witness who testified against the tariff. Toward the
end of his testimony, Shattuck, apparently worn down by the effort, declared, "I am trying to represent in my feeble way a very large investment in
Cuba" (U.S. Senate 1929, 214). The fix for U.S. sugar growers, one might conclude from the hearings' testimony, was in. As it happened, however, back-
stage of the public theater of the tariff hearings, Dahlberg was orchestrat-
ing a much more complex and comprehensive political deal for his Florida
sugar enterprise.

"Everything That We Are Asking For": Sugar Investors' Political Payback

Dahlberg's testimony was only the tip of the iceberg with respect to his political activities on behalf of promoting a higher sugar tariff. The extent of
Dahlberg's political machinations was not clear until after the company had
gone into receivership, when, in response to a newspaper article alleging improper lobbying activities, a subcommittee of the Senate Committee on the
Judiciary initiated an investigation (U.S. Senate 1931). What came to light
as a result of this investigation contradicts contemporary scholars, who as-
sign credit—or blame—primarily to U.S. beet interests in the struggle over
sugar tariffs in the late 1920s. Indeed, the Florida industry is rarely, if ever,
mentioned in the scholarship concerned with the relation of the Cuban industry to the U.S. sugar system at that time (e.g., Thomas 1971; Benjamin
1990; Dye 1998; Ayala 1999; McAvoy 2003). Yet, as Dahlberg developed his
Florida plantations, he was actively lobbying for tariffs higher than any advocated by domestic beet-sugar producers. Dahlberg's vertically integrated
agro-industrial empire in Florida was his launching pad into national politics, and Florida development was his raison d'etre. Whereas he ostensibly
moved to Florida to gain a source of raw material—bagasse from sugarcane
for his Celotex factories—the Senate investigation into Dahlberg's political activities reveals a more ambitious agenda, including repositioning Flor-

ida in national politics with the purpose of breaking the Democratic block on the South. While that would have been difficult to do in Louisiana in the 1920s, Florida's relative "emptiness" provided Dahlberg with a political frontier. We can see that he was intent on restructuring not only the physical landscape of the Everglades but also the political-economic landscape of Florida, and more generally, of the nation. While he had broad national political objectives, it was his ambition specifically with respect to Everglades development that provided the impetus for his lobbying efforts and behind-the-scenes campaign activities at the national scale.

During the time period under the subcommittee's investigation, in addition to Celotex, Inc., Dahlberg was president of Dahlberg & Company, a holding company, and of four operating companies, including two for the production of sugar—Southern Sugar Company in Florida and South Coast Company in Louisiana—as well as Cypremort Corporation, which was organized to take over several sugar plantations in Florida, and his real estate company, Clewiston Company, Inc. In 1929 he organized a holding company to control all of these, Dahlberg Corporation of America, with Dahlberg as president. He was also president of Almar & Company, a private, personal corporation of which he owned all the stock, which was succeeded by a similarly structured corporation, Bromar & Company. John G. Holland, counsel for the subcommittee, acted as the investigator for the Judiciary Committee's lobby investigation. Holland discovered that both Senator Watson, who, as previously noted, was one of the five members of the subcommittee overseeing the hearings on sugar tariffs, and Senator James J. Davis had received stock in Southern Sugar Company in exchange for promissory notes that were then cancelled and returned to the senators, along with "profits" from the sale of the stock and shares in the Dahlberg Corporation of America. That is, the records showed that neither of them had paid for stock from which they received income, and furthermore the books of Almar & Company showed matching expenses labeled "B. G. Dahlberg special account" (U.S. Senate 1931, 5044).

Holland found that Dahlberg had been active with regard to the sugar tariff "long prior to the hearings on the bill before the Ways and Means Committee," at least since 1927. In a confidential memorandum dated June 25, 1927, to the head of his public relations department, Dahlberg wrote, "For some time I have been wondering if it would not be a good thing for us to begin to stir up the question of the import duty on sugar. Have some facts looked up on this tariff subject. Let us start something to get started." He added, "If we do anything on this subject it should be handled strictly on its own merits without any reference to the Celotex Co.

or to myself. It should be handled as an academic situation and as a natural issue, which undoubtedly can be fathered by the American Sugar Cane League and by the farmers associations in the different sections affected" (U.S. Senate 1931, 5032).

Dahlberg fashioned a network of alliances and lobbyists to further his goals, and assigned political activities to his employees. His executive vice president in charge of the Florida operations was Jules Burguieres, whose brother, Ernest Burguieres, was president of the Domestic Sugar Producers Association of Louisiana. Jules was to represent Dahlberg's companies with that organization, as well as the American Sugar Cane League and the Southern Tariff Association. He designated Fred Williamson (president of Clewiston [Florida] Building Materials and vice president of Southern Sugar Company) to act as his representative in relations with Glenn B. Skipper, Republican national committeeman of Florida. He sent D. C. Goodwin, director of public relations for Celotex, to Washington D.C., where he remained for months in order to serve as Dahlberg's personal representative on sugar tariff matters. Dahlberg owned several airplanes, which he used to travel between Chicago, New York, Louisiana, and Florida, with frequent stops in Washington, D.C. in connection with the sugar tariff. Correspondence indicated that Dahlberg, his brother C. F. Dahlberg, and Goodwin kept in close touch with Senator Smoot through frequent telegrams and personal visits. As Holland explained, "He was really interested in a higher tariff than most of the other American producers" (U.S. Senate 1931, 5033).

In September 1928 Dahlberg sent a personal check for five thousand dollars to the Republican National Committee. In examining the books of Celotex, Southern Sugar Co., and South Coast Sugar, Holland found that Dahlberg had reimbursed himself for this contribution with checks totaling five thousand dollars combined from the three companies, which amounted to a violation of laws regarding corporate campaign contributions at the time. Meanwhile, Dahlberg was in touch with the Republican state committee of California, through an intermediary, Kernan Robson, a stockbroker in San Francisco who was handling the sale of securities in Dahlberg's corporations there. In August 1928 Robson wrote to Dahlberg, informing him that there was "a considerable revival of speculation interests here in California" and enclosing a letter from Mark Requa, executive campaign director of the Republican state committee. Requa acknowledged that he was writing, "as requested," to outline a campaign strategy: "The bringing in of outsiders in the Southern States is not desirable. A strong local man experienced in politics is infinitely to be preferred. There are, in the South, numerous centers of anti-Smith propaganda among the Democrats and some of these groups might well be contacted with and their experience used in work in Florida" (U.S. Senate 1931, 5061).

Dahlberg replied to Robson that Requa's letter was "interesting. I have had our people go over the subject with the local Republican organization in Florida and they are working along the lines suggested by Mr. Requa" (U.S. Senate 1931, 5061). In his response, Robson noted that

> Mr. Hoover's closest attachments are here in California, and the men who will have the most influence in his activity are here. I want to keep your name and the names of your company in the minds of these men. The feeling seems to be growing in San Francisco that the solid South is going to tumble. I am very anxious to have the electoral votes of Florida recognized as coming through your efforts to the Hoover column. (5061)

In reply, Dahlberg sent Robson copies of a new advertising circular for Southern Sugar Company and assured him that with respect to Hoover's campaign, "our boys are putting their shoulders to the wheel in good shape in Florida" (5062). For example, Jules Burguieres wrote to the editor of the *Clewiston News,* which relied on Dahlberg's advertising for revenue, "I had a further talk with Mr. Dahlberg concerning the policy of the paper and he has again suggested that in matters of national politics we do nothing that would hurt the Republican nominee. In fact, we should show him up in a favorable light. In Mr. Dahlberg's own words, it is all 'right for us to be Democratic locally but we must be Republican nationally'" (5063).

Dahlberg used the relationship that he had established between his employee, Fred Williamson, and Glenn Skipper, Republican national committeeman of Florida, to funnel money from his companies to the Republican party and on behalf of lobbying activities for sugar tariff and flood-control legislation. In a manner similar to Dahlberg's contribution to the Republican national committee, Williamson would write personal checks that were then reimbursed, "charged to expenses on the books of Southern Sugar Co." (U.S. Senate 1931, 5062). Through Williamson, Dahlberg gained access and influence at the highest level of government. After Hoover's election, Skipper was headquartered in Washington, D.C., and "correspondence passed frequently between Mr. Skipper and Mr. Williamson, at Clewiston, Fla., at such times when Mr. Williamson, himself, was not up in the Capitol in connection with flood-control legislation" (5064). Their efforts appeared to pay off when Hoover visited Clewiston, where he toured Southern Sugar Co. More important, as Skipper wrote in January 1930 to Williamson, "I have succeeded in getting the administration forces lined up behind the Okeechobee project. It will go through, and in my opinion, we will get everything that we are asking for, and if the engineer's report comes through in the right shape there is quite a probability that we will put it through without having hearings either here or in Florida" (5076).

This was part of the patronage that Dahlberg expected, which he had made perfectly clear in a letter written in late November 1928 to Williamson while he was staying at the same hotel as Skipper in Washington:

> The Republican administration should certainly take their coats off and do everything reasonable that they can for Florida going Republican, not only as a reward for the Hoover vote, but in order to indicate to the other so-called solid South States, that it is best to get on the Republican band wagon. There are three ways in which the Republican administration can help in Florida:
> 1. By assisting in the control of Lake Okeechobee.
> 2. By assisting in the development of waterways.
> 3. By putting on adequate import duties on sugar so that the American sugar business may be developed and the hundreds of millions of dollars now annually being shipped over to Cuba for sugar, diverted to our own citizens. A tariff of 4 cents should be established. (U.S. Senate 1931, 5077)

Ultimately Dahlberg got his three wishes, though the sugar tariff was not as high as he wanted. Dahlberg's geographical strategy was to take the tariff fight from Louisiana to Florida and aid in delivering a southern state to Republicans in the national elections while establishing in the Everglades an agro-industrial complex that would attract investments from capitalists as distant as San Francisco. He structured his corporations so that he could offer lucrative stock options to men of power and influence, such as Senators Watson and Davis. Dahlberg met with Senator Smoot "every time he came to Washington" (U.S. Senate 1931, 5039) and kept close track of Cuban representatives, such as Shattuck, whom he assumed were working against his interests "or they would not be here [Washington]" (5023). Thus, it is fair to conclude, Senator Smoot had anticipated and was well-prepared to attack Shattuck's testimony at the tariff hearings.

Likewise, Dahlberg had identified Senator Harrison, a Mississippi Democrat, as one of the "free-trading Democrats of the South who are the greatest enemies of sugar" (U.S. Senate 1931, 5036). Therefore, to gain further protection for sugar, Dahlberg and his cronies sought to break the hold of the Democratic Party on the South. Florida was key to this project. The critical point then, is not that the beet cause was unimportant in the politics of the sugar tariff, but that Florida was much more important than has been previously recognized. Florida's invisibility in the historical record is to Dahlberg's credit; he was a strategist who saw that in Washington it had to be a "farmer's fight," which gave beet farmers a particular cachet in fighting for the cause of tariffs (5023). Although Dahlberg succeeded in altering the physical and political landscape of Florida, his enterprise there was soon in receivership.

Cuba and Florida Compete for Sugar Quotas

Dahlberg illustrates and exemplifies the complexity of political and economic interests that President Franklin D. Roosevelt's first administration faced as it sought to revisit the sugar question. The fact that beet-sugar interests were geographically diverse and had a populist political profile had served the Florida sugar industry well when the issue was tariffs to protect the domestic industry from "foreign" sugar. However, by 1933, when it was apparent that tariffs were failing to avert a "national sugar depression," the proposed remedies pitted various producing regions against one another (Dalton 1937, 57). The political clout of the beet interests no longer seemed to be an asset to cane producers. Initially, beet-sugar interests requested that sugar be declared a "basic commodity" so that it would be included in the Agricultural Adjustment Act of 1933 (AAA). Introduced by Senator Edward L. Costigan of Colorado, the amendment that would have achieved this passed the Senate but not the House. Instead, the AAA permitted sugar producers to develop a voluntary marketing agreement, which they undertook to do. Under the aegis of the newly established Agricultural Adjustment Administration, industry representatives formed a subcommittee to draft a Sugar Stabilization Agreement. Five months of negotiation resulted in an industry plan that "pleased no one" because in "the badly fragmented industry, every division believed the other segments were out to destroy it" (Heston 1975, 100). In October 1933 the secretary of agriculture rejected the industry's Sugar Stabilization Agreement, giving as his reasons "uncertainties in Cuban production," "difficulties of operation" of the proposed plan, and that it would have "tended to increase rather than remove present disparity in agriculture's purchasing power" (quoted in Dalton 1937, 84). The last point referred to the fact that the $10 million given to 42,000 sugar producers would cost six million other farmers more than $14 million in increased sugar prices.

John E. Dalton, identified previously as chief of the Sugar Section of the Agricultural Adjustment Administration, and who afterward served as executive secretary of the United States Cane Sugar Refiners Association, lists several reasons that the plan failed. Foremost was that "there was no organized 'sugar industry'" (Dalton 1937, 85). The Agreement "attempted to weld the interests and compromise the demands of the widest-spread agricultural industry in America" (86). If widely spread, it was actually well organized by sector, including two beet farmers' associations, a beet factory association, the American Sugar Cane League, representing Louisiana planters and processors, the Hawaiian Sugar Planters' Association, the Philippine Sugar Association, the Cuban Sugar Stabilization Institute and, for the nine seaboard refining states, the U.S. Cane Sugar Refiners' Association and the Sugar Workers' Alliance (Dalton 1938). Florida was represented by USSC. In order to reconcile competing interests, regional quotas had been increased to the extent that the purpose of the agreement — to restrict production in order to increase price — was nullified. Second, the agreement did not provide sufficient economic assistance to Cuba. Cuba, the only affected area that lacked a vote on the plan, was given an allotment based on its 1932 (post-Smoot-Hawley tariff) crop year, which was less than half its record-high. That Cuba was slighted was unsurprising, given that the plan's signers held "to the fundamental principle of preserving the domestic market for the products of domestic agriculture or industry" and intended that "no benefits shall accrue under said Sugar Marketing Agreement to the Republic of Cuba . . . except that the President of the United States shall from time to time determine" (quoted in Dalton 1937, 88).

Indeed, President Roosevelt's views regarding Cuba differed sharply from those of the industry's subcommittee. In August 1933 Cuban students and workers had revolted against the regime of President Machado. The U.S. government actively opposed his successor, Ramón Grau San Martín, and was instrumental in replacing him with Fulgencio Batista, who supported U.S. business interests. Because the "special relationship" between the two countries depended on Cuba's access to the U.S. sugar market, another way to regulate the U.S. sugar supply had to be found (Markel 1975). The large size of the 1933–34 domestic U.S. crop and low sugar prices demonstrated the inadequacy of the tariff as a means to both manage the market and achieve foreign policy goals. In the face of worldwide sugar depression, U.S. insular and continental production was stimulated by the tariff, resulting in overproduction and the displacement of Cuban sugar from the U.S. market. In January 1934 Costigan again introduced a bill to declare sugar a basic commodity. In response, on February 8, 1934, Roosevelt presented to Congress his plan, a quota system that he hoped would serve to maintain the domestic industry but also restrict its expansion; would prevent inordinately high sugar prices; and would aid the economies of Cuba, the Philippines, Hawaii, Puerto Rico, and the Virgin Islands (Gerber 1976). Roosevelt's comments reflected his concern that the domestic industry should not be stimulated by protectionism and that the Cuban industry should not be further harmed by it. He noted that Cuban purchases of U.S. goods had dwindled steadily as sugar exports to the U.S. declined.[1] The rate of Cuba's consumption of U.S. goods would become a hotly contested issue in the debates over sugar policy in the succeeding decades.

If beet interests thought they detected an unsympathetic aspect to the Democratic president's proposal, they were right. Roosevelt's private correspondence recorded his desire to place sugar on the "free list" as well as the fact that he had "discussed the possibility of wiping out the beet sugar industry over a series of twenty years" (Harold L. Ickes, quoted in Heston 1975, 102). Indeed, his cabinet members said as much publicly. During the Jones-Costigan hearings, the prominent New Dealers Henry A. Wallace and Rexford Tugwell, then secretary and assistant secretary of agriculture, respectively, proclaimed sugar to be "a parasite industry" (Ickes 1954, 269). Wallace, whose father was secretary of agriculture under President Warren Harding, had abandoned the Republican Party in 1928 to support Democratic presidential candidate Al Smith. Wallace left his family's party because he strongly opposed Hoover's agricultural policies and viewed Hoover as his father's main antagonist in the Harding Cabinet (Schlesinger 1959). In his "secret diary" Secretary of the Interior Harold L. Ickes expressed a general agreement with Wallace and Tugwell's publicly expressed sentiments, but feared the ramifications for the 1940 presidential election. "God knows this is true, but it will be used against Wallace now. I do not believe that Louisiana will go Republican next Election Day, but it will be a tough fight in the beet sugar states of the West" (Ickes 1954, 269).[2] The sugar question thus persisted as an important consideration in national political campaigns throughout the Great Depression.

One of the principal aims of Wallace's New Deal agricultural policies was to limit output and reduce surpluses. In exchange, farmers would receive price subsidies. The devil was in the details, however, as producer interests, predictably, pushed for the fewest limits and highest prices. Thus the marketing quotas for sugar as recommended by Roosevelt to Congress were significantly different from those proposed in the industry plan. As originally drafted by Representative Marvin Jones of Texas and Senator Costigan, the House and Senate bills attempted to balance regional interests by adding sugarcane and beets to the basic commodities of the AAA, with production quotas to be set by the secretary of agriculture, based on regional averages of any three years from 1925 until 1933 (see table 3.2). The strongest opposition to the bills came from beet interests, who wanted no restrictions on production and protection against imported sugar (Heston 1975). The result was that the Jones-Costigan Act of 1934 contained critical revisions: minimum continental quotas were set, with cane producers receiving Roosevelt's recommendation of a 250,000-ton quota whereas beet producers' quota was 100,000 tons higher than the president wished. Insular area quotas, both domestic and "foreign," were to be determined by the secretary of agriculture in accordance with the original wording of the act. Continental producers were also guaranteed 30 percent of total consumption in excess of the estimated 6,452,000 tons. A last-minute addition to the bill allowed the secretary of agriculture to raise the quota of mainland producers of less than 250,000 long tons (i.e., Louisiana and Florida). Thus the Jones-Costigan Act, which appeared to curtail hopes of expansion, actually allowed room to grow. For example, because beet producers fell short of their quota from 1934 through 1936, domestic cane growers' quota was increased to 392,000 tons in 1936 (Heston 1975). In a sense, the act substituted political risk for economic risk, guaranteeing producers a price and quota but making levels of production a political question to be answered by USDA bureaucrats.

In January 1936 the Supreme Court declared the processing tax and payment provisions of the AAA to be unconstitutional, leading to "the renewal of the sugar fight in Congress" (Dalton 1937). The Jones-Costigan Act, rewritten as the Sugar Act of 1937, assigned quotas on the basis of percentage shares for producing regions. It also included a new category, "direct consumption sugars, defined as all sugars, whether refined, semi-refined, or raw, which entered directly into consumption," that sugar refiners fought successfully to insert into the act (Baldwin 1941, 105). Offshore producers, both domestic and foreign, thus received quotas that restricted "D.C." sugars to a small percentage of their total. Representatives of the Roosevelt administration disputed any inclusion from the refiners—as sugar *manufacturers*—in an *agricultural* bill and argued that the designation of direct consumption sugars in the bill "perpetuates a new geography, creating a continental and an offshore America, where we only know one kind of America" (106). However, Roosevelt signed it, stating at the time that he was "approving the Bill with what amounts to a gentleman's agreement that the unholy alliance between the cane and beet growers on the one hand and the seaboard refining monopoly on the other, has been terminated by the growers" (Roosevelt, 1937, quoted in Baldwin 1941, 107).

The Sugar Act guaranteed mainland cane growers 420,000 tons, a substantial increase over their 1936 quota. Louisianans, who wanted the mainland cane quota to be split on the basis of past performance, angered representatives of the Florida industry. Writing at the time, Dalton saw a stark contrast between the economically frail Louisiana industry and Florida's: "Young, flourishing, and profitable, and with abundant lands in the Everglades, there was, prior to the passage of the Sugar Act, no reason to expect any diminution in its development" (Dalton 1937, 183). He argued that the issue raised by the Louisiana industry was whether to provide federal support to stimulate the growth of an otherwise uneconomic industry, whereas the question posed by the Florida industry was whether to subsidize an industry capable of expansion without benefit payments. The latter question had taken on some urgency, as the public realized that payments to the Florida industry were essentially divided between two companies, USSC and the Fellsmere Sugar Corporation, the former receiving the lion's share. Under the Jones-Costigan Program, USSC had received a total of $1,067,665 by April 1936, by far the largest amount paid to a single producer (Heston 1975).

In the fall of 1937 Manuel Rionda requested and received an annotated summary of the annual report of USSC. From this he learned that USSC sold its entire production to the Savannah Sugar Refining Corporation. Sales figures were presented in comparison to the Rionda family's large plantation in southeastern Cuba, Francisco, along with the comment that USSC "[n]aturally, . . . gets the advantage of duty on Cubas [sic] on all its sugars." USSC's lower costs for cane—a difference of ten cents per twenty-five pounds—were attributed to the fact that "practically 100% of the U.S. Sugar cane supply is administration owned."[3] The comparatively low sucrose content of USSC-grown cane was highlighted, with the comment, "That shows the climate and lands could never compete with Cuba on even bases [sic]." Annual profits for 1936 and 1937 for USSC were reported to be $804,910 and $871,082, respectively, the latter representing the largest crop produced. The final comment was, "The Company wishes to have no restrictions in production in the United States. Of course not!"[4]

The Sugar Act was revised in 1948 and renewed, with modifications, until 1974 for a total of "40 years of a thoroughly managed market for sugar" (Mahler 1986, 167. At the outset creating this managed market was a tremendously complex undertaking. It required that the secretary of agriculture estimate total U.S. consumption, apportion this estimated requirement among producing areas, apportion each area's quota among cane or beet processors, and—as originally written—make benefit payments to producers. This task was only possible because there were fewer than 250 processors and refiners to be regulated, and because "[c]old economic adversity had given the sugar industry in appearance, at least, some unity and cohesion" (Dalton 1937, 114). As of 1935, a hundred thousand cane and beet producers were under federal contract for crop adjustment. The close relationship between the numerous growers and the small number of processors enabled administration of the program through mills or refiners. Even with this structural advantage in administrative efficiencies, the bureaucratic challenges were enormous. For example, the guidelines established by Congress for setting non-mainland quotas resulted in eighty-four possible combinations of three-year periods for each area. To determine these quotas, the State Department, War Department, and Department of the Interior were consulted with respect to Cuba, the Philippines, Hawaii, and Puerto Rico, respectively.

We can trace some of the program's complexity to the Supreme Court decision of 1936, which forbade production control with benefit payments but permitted the limitation of marketing through quotas. The decision in part reflected the political-economic tensions of the Great Depression between, on the one hand, a shift to the centralized direction of national economies and, on the other, a more laissez-faire, market-oriented approach. Sugar was at the center of this debate worldwide. In an attempt to stabilize the world market, representatives of twenty-one countries, including the United States, signed the International Sugar Agreement of 1937, and established the International Sugar Council in London to administer the agreement. The outbreak of hostilities in Europe in 1939 rendered the agreement inoperative, though the formal structure of the International Sugar Council remained.

Sugar policy captured the attention of Oswin W. Willcox, a "onetime professor of soils at Iowa State College," who authored numerous books and was involved in public debates of the time on farming. Willcox advocated "a superintensive, highly technological agriculture," envisioning an "agrobiological utopia" that "resembled Italian fascism" (Stoll 1998, 172). Willcox saw in sugar the ideal commodity to demonstrate the potential of his agro-industrial political vision. In *Can Industry Govern Itself?*—a comparative study of ten national sugar policies—he claimed that sugar had "been brought more completely under production and price control by more variously situated bodies politic than any other major industry" (Willcox 1936, x). Virtually every sugar-producing region in the world employed "proration," the economic principle applied under conditions of overproduction, which created a system of allotment to "prorate" production and distribution within an industry. Proration allowed centralized political control over the allocation of production and distribution without threatening the institution of private property (viii). According to Willcox, the U.S. quota system was unique compared to the sugar programs of other governments, which relied on direct agricultural bounties. Because the U.S. sugar program was designed to maintain existing producing regions, the Supreme Court decision that forbade direct bounties meant that the price of sugar had to be maintained at a level that would protect the highest-cost areas. The cost to consumers of protecting the domestic industry through quotas was therefore higher than if direct bounty payments had been made.

Given its diverse sourcing, proration of the U.S. sugar industry required the most inventive program among the world's nationally directed sugar industries. It was the most geographically complex of sugar programs worldwide and of U.S. industrial programs nationally. The Florida sugar agro-industry, as Dalton observed, was exemplary of the principle of proration during the Great Depression: "There is no better example in the contemporary economic and political scene of the close relationship between government and business than that found in the case of the Florida raw cane-sugar industry. Since its inception in 1929 it has spent three years under a tariff system, two years under a quota system with benefit payments, and one year under a quota system with Soil Conservation payments" (Dalton 1937, 187).

Since the agricultural program assigned quotas regionally—divided among both domestic and foreign producers—Florida investors directed their political and public relations energies toward increasing Florida's quota at the expense of other production regions. Their promotional writings emphasized geographic difference, particularly with regard to labor and housing conditions, quality and scale of infrastructure, and agro-economic efficiency. The Sugar Act thus created the context for a company-generated literature of place-based competition, replete with claims concerning the types of rural society engendered by regional manifestations of the industry around the world. This literature took the form of a distinctive series of booklets written and published by USSC. Its publications praised the advantages of large-scale production units for promoting social welfare. That USSC, whose plantations accounted for 96 percent of 1930s Florida production, favored large-scale production units is no surprise. The company's touting of its benefits to social welfare, however, reflected not only Florida investors' need to favorably differentiate the Everglades from competing regions, but also the general preoccupation with labor union activism among industrialists at the time. Corporate paternalism was their answer to unionism, and there were no better spokespersons for this ideology than the investors who took over Florida's sugar agro-industry from Bror Dahlberg.

"That [Florida's] Life's Blood Might Water the Cane Fields of Foreign Nations"

During the debate over the Sugar Act of 1937, Senator Claude Pepper of Florida named General Motors Corporation as the outstanding stockholder in USSC. More accurately, several directors of USSC were at that time stockholders and officers of General Motors. Modernization in the New South hinged largely on capital investment from the north, and Florida's sugar agro-industry was a typical beneficiary.[5] Agrarian capitalists of the New Deal era in Florida came from the northern auto industry and Wall Street (Heitmann 1998). The professional biographies and philosophies concerning employee relations of these investors are important for understanding the centrality of representations of plantation labor to the development of the south Florida sugar industry. Among the most notable was Vice President Charles Stewart Mott of General Motors (GM). Mott and his family owned 68 percent of USSC's common stock and held several directorships on the board. Mott began his career in industry in a family enterprise, the Weston-Mott Company, which manufactured hubs, wheels, and axles. In 1900, when he became superintendent, the company was shifting from the production of bicycle parts to automotive parts. In 1906, as president, Mott moved the company to Flint, Michigan. In 1908 Flint-based GM acquired 49 percent of the stock of the Weston-Mott Company, and in 1913 Mott accepted GM stock for the remaining 51 percent. He was then made a director of GM, was appointed chief of advisory staff in 1921, and was considered second in command at the corporation in the 1920s (Young and Quinn 1963).

Mott's early career in industry coincided with the period during which "many of the major innovations in American welfare practice originated in the private sector" (Katz 1996, 185, 192). Because Mott was a key figure in developing the largest U.S. corporation (GM) and because of his particularistic approach to "welfare," his professional life illustrates major themes in U.S. social history. Mott held to an ideology of corporate paternalism as an alternative to labor unionism. He strongly opposed unionization and once suggested that the 1930s sit-down strikers in GM's Flint, Michigan, plants should have been ordered to move on by the governor, "and if they didn't they should have been shot" (Mott 1970, 135). His general anti-unionism led him into a brief political career after Flint's "business and industrial community" was "jolted" by the election of a Socialist mayor in 1911 (Young and Quinn 1963, 45). Republicans and Democrats joined cause to nominate Mott as the Independent party candidate, and in 1912 he was elected mayor of Flint.

It was Mott's philanthropic work, however, that served as the most important vehicle for widely disseminating his ideas about workers' welfare. He founded one of the country's largest philanthropic organizations in 1926, the Charles Stewart Mott Foundation, in order to spread his ideas of social engineering. In a comparative study of foundations, Nielsen characterized the "Mott method" as "one of aggressive evangelism and rigidly organized civic uplift, arbitrarily imposed from the top" (Nielsen 1972, 204). In contrast to the informal culture of other foundations, "Mott has applied the techniques of modern industrial management . . . to its philanthropies" (203). Thus, when the Mott family gained majority ownership of USSC, corporate paternalism and industrial managerialism were melded with an agro-industrial enterprise based on plantation production embedded in the racism of the Jim Crow South.[6]

The other key figures in the development of Florida's sugar agro-industrial complex were the Bittings, Clarence and William, who together owned 10 percent of USSC's stock. Clarence was president of Bitting, Inc., a New York–based management company (Manuel 1942b), and chairman of its executive committee from the time of its reorganization until 1946. Early in his career he brought his expertise in financing and managing industrial properties to the New South, managing a large Mississippi cotton plantation prior to investing in USSC (Hanna and Hanna 1948). During his fifteen years as president, Clarence Bitting was the company's key spokesperson, articulating, in congressional testimony, press releases, and company publications an agro-industrial-corporatist ideology to support USSC's expansionist goals. In 1936 Bitting chartered a special train to bring nearly one hundred members of Congress and "other influential persons" to Florida to see USSC's operations first-hand in an effort to win congressional support. Although these visitors "were impressed by the sugar enterprise and surprised at the airport, golf course, and charming inn" there was "no relaxation of the quota forthcoming" (Sitterson 1953, 377).

One of the most powerful components of Bitting's corporate propaganda was the paternalistic characterization of USSC's treatment of plantation labor as more socially progressive than the treatment given in Caribbean plantations, which he portrayed as oppressive. One of his periodically published booklets took the reader on a mock tour of the company's plantations, pointing out the "neat, orderly and well-maintained cottages of the happy, contented plantation workers" (Bitting 1937a, 8). Later in the same publication, he suggested that the company's paternalism was smart business management, explaining, "A well-paid, contented working force makes for efficient operation" (29). In a subsequent publication he argued, "It is poor economy to use shacks for housing employees; the field worker and his family who reside in a good house are healthy and happy" (Bitting 1940, 11). In these and other public writings, Bitting stressed the importance of bringing "to agriculture the viewpoint and technique of the American industrialist" (Bitting 1936, 2). Large-scale, industrialized agriculture was portrayed as morally superior to the system of small farmers because of its socially progressive treatment of labor.

Bitting's presidency of USSC coincided with a crucial decade for the South, 1935–45, when the region was transformed by the Great Depression, the New Deal, and World War II, and his writings are best understood in that context. In some cases, New Deal agricultural reforms made conditions more difficult for tenants and small farmers. Large landowners benefited from subsidies and mechanization, with the result that during the 1930s more than one million people left the South. Second, the South's distinctive identity was under scrutiny as "North-South differences were given more sensational exposure in the popular press" (Grantham 1995, 134). This "othering" of the South by northern writers had been an established practice since Reconstruction, but it took on a particular meaning in the context of the Great Depression and the New Deal. When U.S. Secretary of Labor Frances Perkins in 1933 disparaged the South as an "untapped market for shoes," he represented an administration that had targeted the region as "the Nation's No. 1 economic problem" (165–66). Third, during this time social scientists such as Howard Odum and Rupert Vance developed a scholarly approach to understanding southern regionalism, and various academic disciplines began to study the South, especially race relations.

But ideas of the South were contradictory. On the one hand social reportage in the form of photographic essays and documentaries focused on the South to reveal "not a pretty picture" (Grantham 1995, 134). On the other hand, the most popular southern novel of the decade, *Gone With the Wind* (1936) and the film version (1939) along with films such as *Jezebel* (1938), served to romanticize the Old South for a national audience and contributed to the development of neo-plantationism. That is, the South's history as a slave-based plantation economy was idealized, romanticized, and used as instruction on social and racial relations in the present. In post-Reconstruction Mississippi, performances organized by the descendants of the slave-owning class romanticized life in planters' mansions and idealized the lives of slaves, emphasizing the paternalism of the Old South when "a planter looked after the welfare of his slaves" (1941 pamphlet, quoted in Hoelscher 2003, 659). The ideology of paternalism projected through such performances and historical reconstructions was central to the justification of Jim Crow and the maintenance of racial subordination in the South.

This backward-looking neo-plantationism intersected with the ideology of modernization taking root in the New South—evidenced in various ways under President Roosevelt's New Deal—in the racialization of the southern labor market. One manifestation of the New South's embrace of modernization was the extraordinary popular response to the Tennessee Valley Authority, which "gave rise to a new vision of progress" in the region (Grantham 1995, 156). Another was the widespread enthusiasm for economic development in the South during the 1930s, when, "[i]n their quest for industrial plants and new capital" southerners looked northward for investors (165). This vision of modernization included racial discrimination, which the FDR administration accepted in the operation of its own programs and promulgated through labor legislation. During the New Deal, collaboration between northern and southern Democrats protected the region's particular form of economic and political subjugation of blacks (Linder 1987; 1992). Planters and industrialists "depended on the expansion, consolidation, and enforcement of Jim Crow rule to keep labor cheap and disciplined" (Gilmore 2002, 18). For example, in the South, the 1933 Agricultural Adjustment Act benefited mostly white planters with large land holdings, while tenant farmers (almost all black) were adversely affected (Browne 2003). In many cases in the New South, New Deal federal agencies in charge of regulating employment "continued to segregate jobs according to the employer preference and local custom" (Goluboff 1999). An immediate effect of federal involvement in labor relations was to widen the wage gap between blacks and whites in the South (Wright 1987). In sum, Jim Crow, Old South romanticism, and New South modernization intersected at the historical moment and site of USSC's effort to establish south Florida's sugar plantation complex.

Bitting, as USSC spokesperson and shareholder, encouraged and benefited from both the prevailing (at least among southern politicians and business leaders) sentiment of neo-plantationism and the ideologies of modernization and corporate-driven social welfare. His contribution to the debate regarding the 1937 Hours-Wages bills is telling. Bitting opposed the southern position that small employers should be exempted from its provisions. In a letter to Senator Pepper, which Pepper forwarded to President

Roosevelt, Bitting made the plausible argument that this exemption would lead to the reinstitution of "sweat-shop" conditions by employers less scrupulous than USSC, who would "find it advantageous to break their operations into small units to be operated under some form of contract or agreement with an individual employing one less than the minimum number of employees to which the proposed legislation would be applicable. I can visualize what such methods would mean if carried out in the Everglades. They would mean the utter breakdown of the high standards now established."[7]

Bitting's writings were often published as part of The Little Green Library, a series of green-covered pamphlets used to promote and publicize company interests (USSC 1944). Booklets of the late 1930s for the most part dealt with the issue that concerned the corporation most: the struggle to increase Florida's quota, primarily at the expense of Cuba's. In these writings, Bitting relied on the now familiar discursive tropes of the U.S. sugar producers' lobby: sugar supply as a national security issue, economic nationalism (or, as the booklets repeatedly exclaimed, "American markets for American producers!"), agro-industry as modernization, and, increasingly, a comparative regional discourse focused on labor conditions and social welfare. The pamphlets, distributed to libraries nationwide, were widely available to the various print media and to the public. The green booklets, then, were a key part of USSC's discursive strategy to spread the Florida sugar industry's perspective to a wide audience in order to gain national support for its political agenda.

One of Bitting's booklets, *Some Notes on Offshore Conditions*, comprised of extracts from testimony at public hearings in November 1937 on fair and reasonable wage rates in sugarcane cultivation and harvesting, listed various reasons the United States should not depend for its sugar supply on Cuba, the Philippines, Puerto Rico, and Hawaii. Criticism of the first three was centered on low wages or "deplorable living conditions" (Bitting 1938a, 2). More generally, Bitting argued that offshore production was "uneconomic and could not exist except by reason of the tariff and the exploitation of labor" (22) and that food security—especially in the case of war—was best served by domestic mainland production. *Some Notes on Cuba*, published after the Sugar Act of 1937 gave 29 percent of the U.S. sugar quota to the island, summarized the history of the industry and of U.S. involvement in Cuban politics and economics. Admitting that the climate and soil of parts of Cuba were ideal for sugarcane, Bitting noted "that during the 'Dance of the Millions' there was wanton destruction of irreplaceable natural resources of that nation for the sole purpose of placing many thousands of acres in sugar production that simply contributed to world surplus" (Bitting 1937b, 5). Cuba's production of foodstuffs, "formerly imported from the United States, [which] are now being produced locally" was cited as progress toward a balanced economy there and as reason to develop self-sufficiency in the United States (Bitting 1937b).

The booklet *Florida Sugar*, under the heading "Our Men Wear Shoes," compared the standard of living of Florida cane workers to that of offshore workers (Bitting 1936). Florida cane workers owned autos and appliances; cane field workers had shoes and hats, and "their wardrobe consist[ed] of a great deal more than a second-hand or third-hand pair of overalls" (Bitting 1936). Subsequent booklets develop this theme further, with charts comparing ownership of key consumer items in the United States and offshore. Numerous photographs reinforced the overarching theme of the booklets: that Florida's agro-industry was the most progressive among U.S. sugar sourcing regions (figs. 4.1, 4.2). Facing pages entitled "Living in Cuba" and "Living on an Everglades Sugar Plantation" depict a thatched roof hut and a small frame cottage, respectively (Bitting 1937b, 19, 20). Illustrations of pre-industrial sugar-making in India and Indochina are contrasted with photographs of USSC's mills and trains to emphasize the industrial efficiency of the Florida plantations. The booklets are remarkable for their thorough research, numerous references, and clearly articulated vision: appropriate lobbying tools to evoke the sense of an emerging agro-industrial region. They are also remarkably arrogant; although the Florida industry had benefited greatly from technologies developed elsewhere, Bitting would not praise the attributes of any other producing areas. *Controlled Output*, listing among its sources Willcox's (1936) comparative study (not surprisingly, given Willcox's views), concludes, "Diligent search has failed to disclose a single instance, other than the United States, where the home producer of a necessity of life is discouraged and restricted for the benefit of alien peoples" (Bitting 1938b, 30).

The booklets show clearly that USSC aimed to position south Florida at the top of a moral hierarchy of sugar-producing regions, primarily by emphasizing differences in working conditions and the benefits of American, corporate-driven social welfare. Regional newspapers, whose publishers generally supported economic growth and expansion, contributed to this regional competition. Indeed, journalists and editors were better positioned to carry the nationalistic arguments and moralizing to greater emotional heights. Commenting on U.S. sugar policy in the *Miami Herald's* All-Florida section, the paper's editors wrote:

Figure 4.1. Bitting's green booklets presented the practice of using women's labor to harvest cane as evidence of the backwardness and degeneracy of offshore plantations. Courtesy of United States Sugar Corporation.

Figure 4.2. Bitting's booklet depicted United States Sugar Corporation's practice of using men only for cane harvesting to demonstrate that Florida's plantations were more advanced than those located offshore. Courtesy of United States Sugar Corporation.

For lo these many weary months, ALL FLORIDA has stood by the suffering victim, Florida, and waited and watched hopefully, for some indication of returning sanity to the sadly ailing doctor, the U.S.D.A. We had faint reason to believe
that all of the intelligence of the main stooges, who reflect in their administration the fantastim [sic] hooey of the 100 per cent failure, reciprocal trade treaties, had not entirely waned and an Associated Press report from Washington
now seems to justify the faith and hope to which we have clung since the dark
days of the enactment of the Jones-Costigan monstrosity that preceded the
iniquitous 1937 Sugar Act. Bound to a butcher's block without the courtesy of
an anesthetic, Florida has been bled almost white by these insane wielders of
scalpel-sharp knives, in order that her life's blood might water the cane fields
of foreign nations. (*Miami Herald* 1939)

In one sense, such hyperbolic prose was an expression of a long and gener-
alized history of regional boosterism. In another sense, it was a more recent
and contingent expression of the regional competition for quotas that the
Sugar Act generated and would continue to generate for four decades. In
the spring of 1940, USSC directors attempted to influence public opinion in
favor of a larger quota for Florida producers by supplying political cartoons
to regional newspapers. Personnel Director M. S. Von Mach wrote editors
to ask if they were "interested in the development of the vast Everglades"
and told them that "if you are interested in both local and statewide benefits to be derived from the expansion of the sugar industry in Florida, it will
be necessary for your readers to become conscious of the vast opportunities
that will accrue to them through the lifting of Federal restrictions on the
sugar quota." The corporation would "gladly furnish" a cartoon each week
for publication, samples of which were enclosed (figs. 4.3 and 4.4).[8]

In a last-minute amendment to the extension of the 1937 Sugar Act, signed
into law in October 1940, mainland cane growers gained an unlimited allotment for 1940 (Heston 1975). By this time, USSC's costs of producing raw
sugar had declined from 2.8 cents per pound in 1932 to 2.09 cents per pound
(Sitterson 1953). For the 1940–41 season, Florida's output reached 98,000
tons, surpassing its all-time high of two seasons earlier. At that time, about
90 percent of the cane processed by USSC was "administration" cane — that
is, cane grown by the company on its plantations. The remainder came from
twenty-eight independent growers whose contracts included a "cooperative participation supplement" that entitled them to a share in the profits of
the sugar house. In 1939, Bitting had taken the lead in organizing the Florida Cooperative Sugar Association (Heitmann 1998). Thus in the early 1940s
a relationship between "big sugar" (large-scale corporate operations) and
"little sugar" (small-scale family operations) developed that persists into the

Figure 4.3. Cartoon produced and distributed by United States Sugar Corporation alluding to the importance of Florida's sugar quota for the 1940 presidential election. Courtesy of United States Sugar Corporation.

Figure 4.4. Cartoon produced and distributed by United States Sugar Corporation claiming Congressional favoritism of Cuba over Florida in the allocation of quotas (c. 1940). Courtesy of United States Sugar Corporation.

twenty-first century and provides USSC with an important source of company propaganda and political leverage. As Manuel observed at the time,

> The Sugar Corporation has pointed out that independent growers, without having to invest large amounts of capital, enjoy the benefits of the company's experience with improved methods of cultivation and varieties of cane. The independent growers have supported the Sugar Corporation on local issues and national proposals. In their capacity as producers of vegetables or stock feed, the growers have testified at several fair wage hearings . . . that higher wage rates would cause them to lose their workers to the company or compel them to pay wages so high that they could no longer farm profitably. In behalf of the independent growers and other truck farmers, Mr. Bitting has repeatedly urged Congress to lift the quota on mainland cane sugar, thereby permitting the farmers to diversify their crops and grow cane which they could sell to the corporation. (1942b, 12963)

The emergence in Florida of this new structural arrangement between big and little sugar allowed USSC to enhance its use of Dahlberg's earlier strategy of making the campaign for a favorable sugar trade policy in Washington a "farmer's fight." While the directors and shareholders of USSC and their vision for a sugar-producing region in Florida were anything but populist, they could, and did, affirm that the success of family-owned farms in the region was contingent on a "fair" U.S. quota system. The company's lobbying and its little green booklets provoked a direct response from Cuba. Published in Havana, the book *Azúcar en la Florida* described the history and geography of the Florida industry in general and specifically the *ingenio* at Clewiston, concluding with a chapter entitled "*La campaña de Florida contra Cuba*" (The campaign of Florida against Cuba).

> The Cuban sugar industry, in this year of 1941, wrestles against strong and dangerous rivals who are skillful, dangerous, and determined, and are doing everything possible to manage their respective industries on the highest scientific basis. Some of them, logically, seek a larger share of the United States market, and have decided to obtain this at Cuba's expense, employing all the methods of propaganda against the interests of Cuba. We refer to certain sugar producers of Florida, our newest and most vociferous rivals in the sugar market. (Tejada y Sainz 1941, 8)

The Limits of Neo-Plantationism and Corporate Paternalism

USSC, which relied on a seasonal migratory labor force to do most of the harvesting and planting, encountered increasing challenges to the discourse of neo-plantationism. As already noted, there was during Bitting's presidency a growing national concern over poverty in the South. This intersected with an increasing political awareness at the national level of the social and economic problems of migrant laborers, exemplified by the U.S. House of Representatives' Special Committee Investigating the Interstate Migration of Destitute Citizens.[9] Bitting testified before the committee, presenting the familiar paternalistic arguments about the unique ability of large, corporate agro-industry to meet all workers' social welfare requirements. To redress the plight of migratory agricultural labor, Bitting proposed steps to rationalize agricultural production and labor in the United States. He suggested that the government "First—encourage larger operating units; . . . Second—These large operating units could cooperate with similar units in other parts of the country whose peak labor demands do not coincide; . . ." and "Third—These larger units . . . could afford to undertake private research looking toward lengthening seasonal peak labor requirements; development of subsidiary crops to provide additional employment in slack season; . . . finding and encouraging rural location of small industrial plants which can absorb some labor during slack agricultural seasons" (Bitting 1940, 25).

According to Bitting, the scale of USSC's operations allowed the company to provide for worker welfare in a way that small operations could not. For management purposes, the company was subdivided into a dozen plantations, with eleven plantation villages "strategically located" to "keep the employees close to the center of their activities" (Bitting 1940, 10; see fig. 4.5). The villages included cottages for families, housing for single workers, "office, store, shops and equipment sheds, as well as schools, churches, recreational and first aid facilities" (11). He argued that good housing facilities were good economy: "The field worker and his family who reside in a good house are healthy and happy; illness is common-place in shacky construction; thus two shacks are necessary to house two ill, miserable and unhappy families, as against one well-built cottage to house one healthy, happy family" (12). Plantation life included "movies and home-talent entertainment; . . . plantation boxing and inter-plantation bouts," while "inter-plantation baseball and football leagues make Sunday afternoons a joyous occasion throughout the property" (15).

Some independent observations suggested that Bitting's testimony was not purely corporate hype. Carey McWilliams, who documented the history of the brutal treatment of agricultural labor by California's agribusiness, was wary of the trend toward "factories in the fields" and the appearance, in the eastern United States, of a pattern of migratory labor similar to

Figure 4.5. United States Sugar Corporation plantation village for southern black workers (1939). Courtesy of Library of Congress.

that of the Pacific Coast. Yet he was favorably enough impressed by USSC plantations to write,

> In Florida, the United States Sugar Corporation . . . employs 2,500 workers the year round and about 5,000 during the peak period. The company has a system of retail stores throughout the plantation which do a business with its employees of $750,000 a year. The company furnishes good housing for its employees; maintains a health service; refuses to employ child labour; and has established a system of schools throughout the area. It indulges in paternalism on a large scale. (McWilliams 1945, 173)

However, there was evidence that many local workers actually preferred not to live on the plantations, no matter how neat and orderly the housing. Moreover, USSC's migrant labor force was composed of solitary men whose families resided elsewhere in rural poverty (Manuel 1942b).

When investors planned Florida's sugar-plantation system early in the twentieth century, they envisaged tapping into the South's, low-cost and highly racialized labor market. Before World War II, the South functioned as a regional labor market, separated from national and international labor markets (Burrows and Shlomowitz 1992; Wright 1987). Labor circulated within the region in an east-west direction rather than beyond the region in a north-south direction, a pattern that was reinforced by the existence of a shared regional culture, familial and friendship bonds stretching across the South, and familiar and relatively consistent agro-environmental conditions (Wright 1987). By the 1920s racial wage differentials appeared in the

South among manual laborers, and increasing racial differences in work experience and education meant that blacks were in effect restricted to agriculture. Agricultural wages in the southern states were half, or less than half,
of those in other regions of the country, undoubtedly an attractive situation
for northern investors.[10]

This separate, southern regional labor market was the source of field
workers for USSC's sugar plantations. The U.S. Employment Service
(USES), which directed and oversaw the interstate movement of labor in
the 1930s and 1940s, was central to USSC's labor recruitment. The USES
Farm Labor Report forms from the 1940s listed six southern states — Alabama, Florida, Georgia, Mississippi, South Carolina, and Tennessee — as the
sources of "colored" labor for sugarcane cultivation and harvesting. USES
offices in these states were in constant communication to coordinate the
movement of workers in the region. These black workers were mostly farmers, typically cotton sharecroppers or itinerant farm laborers. Black sharecroppers from the northern parts of the South were recruited during the
slack periods of their agricultural cycle to work on USSC plantations during the peak of cane cutting. The Great Depression initially deepened this
pool of labor, as increasing numbers of sharecroppers were driven out of
tenancy and the proportion of farm laborers in the black rural population
rose from 4.9 percent in 1910 to 25.0 percent in 1940 (Kyriakoudes 2003). In
April 1935, to address the problems of the rural poor, President Roosevelt,
as part of the "second" New Deal, created the Resettlement Administration,
which was restructured in 1937 as the Farm Security Administration (FSA)
and became part of the USDA. At this time, a Migratory Farm Labor Division was created, which was responsible for administering a labor camp program. The first FSA camps on the east coast were built in Belle Glade, Florida, not far from USSC's plantations: the Osceola Camp for white migrants,
the Okeechobee Camp for blacks (Hahamovitch 1997).

Though south Florida was not part of the Old South, it was a Jim Crow
state, and the northern industrialists who invested in sugarcane production
did not question this system. Indeed, they embraced it and took full advantage. There was never any question that the manual labor force would be
exclusively black or that there would be separate "quarters" for blacks on
the company's plantations. USSC recruited seasonal field labor only from
black communities whose members were subject to the strict controls of
Jim Crow and associated vagrancy laws. Company officials emphasized in
their testimonies and publications the modern, paternalistic, and socially
progressive character of USSC's sugarcane production, though the whole
system depended on a racialized labor force with roots in the slave planta-

Figure 4.6. The Clewiston headquarters of United States Sugar Corporation. This building faces the Clewiston Inn, both designed in the "southern plantation architecture" style.
Photograph by the author.

tions of the past. Indeed, USSC tapped the Old South romanticism, building the Clewiston Inn for visiting company executives and dignitaries in the
style of "southern plantation architecture" (Bitting 1937a, 12; see fig. 4.6).
That USSC's corporate propaganda "made sense" was partly due to prevailing white nostalgia for the Old South's plantations and the idealization of
historic race relations between planters and slaves. Mott's beliefs about and
company policies for the social betterment of labor backed the company's
propaganda, yet the recruitment of seasonal workers in a racialized regional
labor market belied the idea of progressive labor relations. The contradictions between USSC's dependence on cheap labor disciplined by Jim Crow
violence and its corporate paternalism were never reconciled and ultimately
would prove untenable.

Indeed, federal investigations contradicted Bitting's rosy portrayal of
USSC's happy plantation workers, finding that "housing on company plan-
tations, relatively good though it is, . . .[is] still below an acceptable standard.
At the 1938 fair wage hearing, Mr. Bitting testified that the average corpo-

ration dwelling was over three rooms. According to subsequent testimony of Sugar Corporation employees, families of six and eight live in two-room apartments" (Manuel 1942b, 12972). Evidence that the company's brand of paternalism was not universally popular came forth in hearings when a cane cutter testified "that many of the sugar company's employees lived in privately owned quarters in Belle Glade, South Bay, and Miami Locks, paying $1 or $2 for a room. He said that the company had houses available but that workers preferred to live off the plantation" (Manuel 1942b, 12972). That workers would opt for housing off the plantation is understandable given the findings of a separate Federal Bureau of Investigation (FBI) investigation that uncovered USSC's abusive labor recruitment strategies.

"Our First Control of Farm Labor"

During the Great Depression, USSC's plantation managers had grown accustomed to recruiting an oversupply of seasonal labor, a strategy that kept wages depressed and also afforded a greater level of labor control. Their seasonal labor force for the tasks of planting, cutting, and loading cane was exclusively black, drawn from the South's ranks of poor sharecroppers and farm laborers and housed in one of eleven plantation villages dispersed among the sugarcane fields. Housing field workers at the site of production gave USSC greater control than would be possible otherwise, with less worry about having its labor practices observed or challenged. In the years leading up to World War II, federal agencies multiplied, and the regulation of USSC's labor supply grew diffuse and complex. The FSA, the USES, the Sugar Section of the USDA, state land-use planning committees, the Extension Service, and the Bureau of Agricultural Economics were some of the agencies and organizations responsible for various aspects of regulating the supply or mitigating the conditions of agricultural labor.

The agencies involved were frequently reformed or renamed, and individuals moved from one to another, creating a complex set of relationships shaped by overlapping jurisdictions, personal ties, and oversight responsibilities in labor market restructuring. Employees of one agency often had to file reports on other agencies. Their reports provide valuable insights into labor conditions in the sugar plantations, the importance of racialization in structuring the labor market, and the close associations between labor agencies and USSC. The case of Allison French is illustrative. According to a 1942 Bureau of Employment Security (BES) report, French developed an intimate working relationship with USSC in the 1930s, first as manager of the Florida State Employment Service (FSES) and later in a position with

USES.[11] Because the USES was the federal agency primarily involved in recruiting agricultural workers for the sugar plantations, French's position
there made him a key player in USSC labor strategy. As an earlier interagency report noted, over the years "a system of cooperation" had been

> worked out between Mr. Allison T. French . . . and Mr. Von Mach of the Sugar
> Company, that seems ideal. When in need of additional workers, the Sugar
> Company advises the Employment Service the locality where they might find
> the required number. . . . If the local supply is exhausted, which is usually the
> case, the Sugar Company advises the Employment Service the locality where
> they might find the required number of experienced sugar cane cutters and a
> clearance order is issued.[12]

The "ideal" cooperation between Misters French and Von Mach reflected
both the fact that USSC made the most frequent and largest requests for
labor and the hand-in-glove relationship between business and the state in
regulating labor markets.

As public interest in the welfare of migrant workers and federal involvement grew in the 1930s, labor conditions on USSC plantations came under
increasing scrutiny, culminating in an FBI investigation. The complaints received at FBI field offices suggested that USSC's strategies for recruiting and
controlling black harvest labor, often aided by local law enforcement, included debt peonage, forced labor, and even killings (Shofner 1981; Jones
1992). Word circulated among black communities in the South that USSC
was running a "slave camp" at its sugar plantations (Goluboff 1999). In letters
to federal officials, concerned family members in Tennessee and Alabama reported that field hands were guarded all night by "armed guards and not
allowed to write home" and that foremen carried "Black Jacks and Pistols
that men have been Killed because the [sic] insisted for the wages" (quoted
in Goluboff 1999, 786, 788). The FBI conducted dozens of interviews with
black agricultural migrant workers across the South in 1942 and 1943. Investigations found that black workers attempting to leave plantations were
"shot at" and "returned to the plantations and forced to work" (FBI report,
quoted in Jones 1992, 181). Interview subjects reported that USSC plantation supervisors wore guns and carried blackjacks, workers were threatened
with death if they tried to leave, beatings were common, and conditions
of debt peonage prevailed (Shofner 1981). In November 1942 a two-count
indictment alleging violations of workers' Thirteenth Amendment rights
was brought against USSC, M. E. Von Mach (personnel director), and three
other employees in federal district court in Tampa. The case was dismissed

in the spring of 1943 when the presiding judge ruled the grand jury had been improperly impaneled.

Acute agricultural labor shortages became the norm for USSC. In 1941, the corporation had twenty-two thousand acres in sugarcane in three counties bounding Lake Okeechobee: Hendry, Palm Beach, and Glades. Sugar growers needed about 4,500 field hands during peak periods, half to cut, trim, and pile the cane, and half to load it into the wagons.[13] The local USES office reported that the labor shortage for sugarcane work had grown from 1,000 workers in May 1942 to 1,418 in April 1943.[14] With labor in short supply, USSC was "making a desperate effort to secure labor to cut the cane which must be cut within thirty days if it is to be saved."[15] "Labor shortages are threatening the Everglades sugar crop, which was expected to produce a record 100,000 tons this year," noted the *New York Times* (1943, 27). As the crisis in Florida's agricultural labor supply unfolded, owners intensified their use of traditional methods to force local black workers to cut cane. Black workers in the local labor market had grown wary of USSC, as the 1942 BES report described:

> When asked if local sources of labor supply had been tapped to meet the company's needs, he [French] explained that no attempt had been made to recruit Florida labor for cane cutting, first, because no definite orders were on hand, and second because *Negro labor in Florida will not work for the Sugar Corporation.* Mr. French could not explain this situation except that certain "rumors" about poor treatment at the hands of Sugar Corporation foremen had always circulated among the Negro population. These "rumors" he explained were unfounded, although it was true that Negroes were occasionally beaten for attempting to leave the job when they owed debts at the company's commissary, and others were sometimes required to work as many as eighteen hours a day at cane cutting.[16]

That a federal labor official could admit that workers were beaten and forced to work eighteen-hour days while denying rumors of mistreatment speaks volumes about the standard of working conditions for blacks on the plantations. Combined with the tight labor market of the war years, such "rumors" probably influenced migrant agricultural workers' decisions to refuse work on the plantations, as the USES reported it was "unable to get any of them to go to the Glades."[17]

Though local black workers apparently avoided working for USSC, vagrancy legislation allowed local law enforcement officials to arrest blacks without restraint (Shofner 1981). French exploited this power in an

attempt to meet the growing agricultural labor gap: "Belle Glade and Pa-
hokee officials have notified a joint meeting of negro ministers and juke
joint operators that beginning Monday a vigorous enforcement of vagrancy
laws will be put into effect that will operate on the slogan 'Work for the
Farmer or Work the Streets.'"[18] Vagrancy laws were one part of an array of
Jim Crow mechanisms that facilitated the control of black agricultural labor in the sugarcane region. Spatial segregation of blacks on the local scale
allowed closer surveillance and monitoring of the pool of agricultural labor.
Located near the shores of Lake Okeechobee, South Bay and Belle Glade,
where a city ordinance required that "all Negroes . . . be off the streets by
10:30 p.m.," were designated as "black" towns (Federal Writers Project 1984,
474). Adjacent to Clewiston, which was owned and controlled by USSC,
black residents resided in a district known as Harlem. Thus a standard tactic
for dealing with agricultural labor shortages was to send local law enforcement officers into the black communities surrounding the plantations and
round up violators of vagrancy and curfew laws. In a February 1943 summary, French was able to report, "Idle labor has been considerably reduced
by the enforcement of vagrancy laws."[19]

World War II began to draw rural southern labor into military service
and industrial production, driving agricultural wages up and transforming
labor surpluses into shortages. By 1942 USSC executives were complain-
ing of "a disappearance of a labor surplus in the Everglades" due to military enlistment and "the lure of the big city and war-industry plant on all
country-bred people" (quoted in Manuel 1942a, 12965). One way to reduce
the labor shortage would have been to mechanize harvesting. Contrary to
the claims of early Florida sugar industry boosters, however, sugarcane was
not, even at the beginning of World War II, "a *machine made* crop" (Spencer
1918, 18) but required a substantial field labor force for planting and especially for cutting. The agro-industrial character of the sugarcane plantation
belied the lack of technological innovation with regard to field operations.
In reality, the GM and Wall Street executives were invested in an industry
in which key labor tasks were virtually the same as on seventeenth-century
plantations. Especially during the harvest season, an adequate labor force
was critical for the success of a sugar plantation. Timely harvesting took
on special urgency in subtropical south Florida, where frosts, though infre-
quent, were always a midwinter threat.

The lack of mechanization was not for want of trying. The simpler field
operations had been mechanized. Tractors furrowed the fields, and railroads
transported cut cane to the mill. Planting, however, was done by hand, with

workers laying individual lengths of seed cane end to end in overlapping rows the length of the furrowed field. USSC and its predecessor, Southern Sugar, had used fourteen Australian-designed Falkiner harvesters during the 1930–33 seasons but mothballed them because of technical factors and because wages were greatly depressed, making them economically uncompetitive. In 1942, anticipating wartime labor shortages, USSC paid $60,000 for the continental U.S. patent rights to the Falkiner harvester, although Bitting estimated that not more "than 10 percent of the field force c[ould] be replaced by machinery during the [1942–43] harvest" (quoted in Manuel 1942a, 12965). Wartime materials shortages limited the usefulness of the balky machines, which frequently needed repairs. The Florida environment made mechanization especially challenging. The muck soils, the fact that Florida cane laid nearly flat, and periodic freezing presented significant technological barriers to mechanization well into the 1980s (Burrows and Shlomowitz 1992). For the time being, USSC would not be able to solve its labor crisis through mechanization.

A second way to ensure a successful harvest would have been to attract a sufficient number of workers by paying prevailing wages for agricultural labor, or somewhat higher, given the arduousness of cutting cane. The local labor market, comprising both migrants and residents, was price-sensitive: "Cane cutters imported from Southern States have discovered that they are able to earn much more picking beans and other vegetables, and that such work is less arduous than cane cutting. Beans are picked only in good weather and the hours are short; cane may be cut in damp weather and the workers put in a full day" (Pascal and Tipton 1942, 12940).

However, USSC and French, its primary federal labor recruiter, resisted raising wages to remain competitive with the local labor market. In the 1940s, French headed the regional field office of the USES in West Palm Beach and was thus a key actor in regulating the agricultural labor market for USSC and other growers. In his own reports, he explained his role vis-à-vis the "situation" of the local labor market, which had "resolved itself into one of trying to make labor work which won't work; redistributing labor here on FSA contracts; and trying to make intra-area distribution of labor not now working."[20] In direct contradiction to economic theories of labor markets but in line with the racism of the day, he argued for keeping wages for black agricultural workers low and repeatedly recommended a ceiling wage for piece work. French's reasoning—and its racist underpinnings—are revealed in one of his typical assessments of the local black labor force:

It has been clearly demonstrated year after year that production is in inverse ratio to piece-work wage scales. A very large number of negroes (a majority of them, in the opinion of the growers), are not interested in making more money, but they are interested in making the same money quicker. . . . The ones who will work all day are few and far between, and those who will work every day are practically non-existent.[21]

Similarly, USSC's personnel director M. E. Von Mach testified at a 1937 federal hearing that "if you were to give the 'nigger' more money than he gets now he would leave 2 months sooner because he has too much money to spend" (quoted in Jones 1992, 185–86).

Opposition to wage increases went along with paternalistic claims that the company provided for workers' housing, health care, and entertainment. Within the local agricultural labor market, however, cutting cane was not a favored option. For example, under far less physically arduous conditions and in less time, bean pickers on farms surrounding the plantations could earn three times as much per day as cane cutters.[22] Thus in December 1942 French reported from "400 to 1000 idle negroes are seen daily in the Glades area who will not work because they have made so much money under the insane price-bidding war which has been current in that area."[23] The problem of "idle negroes" was thus attributed to irrationally high wages, and the solution was to import agricultural workers from outside the local labor market in order to "induce or force idle labor to work."[24] Sounding like an agro-industry spokesperson, French explained,

It is believed by a great many growers that if all of the laborers in the area could be put to work and made to put in full days and full weeks, that no shortage would be apparent this winter. This is one of the strong reasons why growers favor importation of Bahama negroes since they would be subject to control. With a sufficient number of Bahamans in here, it is believed resulting conditions would force idle domestic labor to work also.[25]

The growers' thinking as reported by French reveals their two primary concerns about agricultural labor—mobilizing an adequate seasonal workforce and controlling workers once they are at the worksite. These two problems were inextricably linked on the sugar plantations and their ultimate resolution depended on finding a new method of labor recruitment.

Even in the face of rising agricultural wages in Florida, as well as throughout the country, USSC representatives insisted that the company did not need to raise wages to secure a labor force. They were right. An alternative to mechanization or higher wages emerged mid-war out of the conflicting agendas of various federal and state agencies and the combined interests of growers' organizations, farmers, and USSC. As the depression-induced oversupply of desperate migrant labor diminished, the role of the federal government vis-à-vis migrant workers shifted from New Deal paternalism to war-time *padrone* (Hahamovitch 1997). World War II farm labor programs echoed World War I policy responses to perceived labor shortages, but the geography of east coast agriculture had changed in the interim, as McWilliams explained:

> A fully developed cycle of migratory labor could not be organized on the Atlantic seaboard until the Florida muck lands were brought into production. Before 1920 the undeveloped swamp and palmetto scrub lands of south-east Florida had no agricultural importance. . . . Mainly within the last ten years, a large-scale industrialized type of agriculture has developed in the region which, to-day, involves the employment of some 50,000 migratory workers. (McWilliams 1945, 170)

To FSA leaders concerned with regulating migratory labor, World War II seemed to present the opportunity "to help rationalize and stabilize the farm labor supply" and "to do more than attack the symptoms of the farm labor problem" (Baldwin 1968, 223).

In the midst of USSC's labor crisis, French claimed that there were "probably from one to three thousand idle negro farm hands in the Glades and the Coastal areas," who would not work.[26] USSC officials, after expanding sugar production on the backs of southern black labor for over a decade, curiously argued, "Black men in America simply lack this skill [of cutting], just plain don't have it" (quoted in Jones 1992, 195). Underlying these efforts to characterize suitable sugarcane harvest labor were the desire for an easily controllable workforce and the need to avoid raising wages in order to recruit workers. As French explained, "It is hoped that some methods may be evolved for preventing the competitive bidding for agricultural labor, for compelling the vast amount of domestic idle labor to work, and for importation of labor to augment the insufficient supply of willing labor."[27]

With American blacks now labeled as unsuitable, the industry turned to the so-called offshore workers—blacks who lived in the Caribbean islands. Growers first looked to the nearest islands, favoring the "importation of Bahama negroes since they would be subject to control."[28] The key to controlling Bahamian labor was to bring them into the United States "allocated strictly to agricultural work, with provisions that they be returned in case they would not work for the growers to whom they are assigned."[29]

House Joint Resolution 96, written by the American Farm Bureau Federation with input from various farm organizations, gave U.S. agricultural employers in general, and Florida growers in particular, what they needed to control a racialized work force (Hahamovitch 1997). The resulting bill, Public Law 45, gave the Farm Bureau practically every concession it demanded from the state's program for regulating wartime migrant labor allocation, including the discontinuance of various restrictions related to minimum wage, housing conditions, and unionization activities (Grubbs 1961). Farm worker advocates, among them Eleanor Roosevelt, had urged President Roosevelt to veto the bill, but, having promised his support to House and Senate leaders, he signed it on April 29, 1943. Public Law 45, otherwise known as the "Peonage Law," removed FSA oversight of migratory labor and shifted control to the locally controlled Extension Service (Baldwin 1968).

Two aspects of the bill altered the geography of agricultural labor markets. One was the provision authorizing the U.S. government to temporarily admit "native-born residents of North America, South America, and Central America, and the islands adjacent thereto, desiring to perform agricultural labor in the United States." The other was the provision that no farm workers could be moved out of a county without the prior consent of the county agricultural agent. Taken together, these two provisions further marginalized domestic workers by constructing barriers to their movement and shifting the funding for transportation toward offshore workers. In addition, the Office of Labor within the newly created War Food Administration took over administration of the migratory labor camps, renamed "farm labor supply centers" with priority of use assigned to foreign workers (Hahamovitch 1997).

The first agreement to bring offshore workers to Florida was made between the USDA and the Bahamian government shortly before Public Law 45 was signed. The first shipment of Bahamian workers arrived on April 13, 1943, in an operation coordinated among several federal agencies. The FSA recruited on the islands, U.S. Navy doctors performed the physicals, and U.S. Army cargo planes flew the workers to Miami. There they were "met by FSA personnel, loaded on chartered buses and taken to points of destination, where the USES takes over."[30] These workers were brought in at the behest of sugar growers exclusively to harvest cane, and were sent back to the Bahamas when the work was completed. Each worker was processed from point of origin to field and back under tight government and company control and subject to summary deportation. Contracts stipulated that a worker would be immediately returned "to his point of recruitment" for any "act of misconduct or indiscipline."[31] Such procedures and terms, French happily noted, "gives us our first control of farm labor which may be used as an entering wedge toward stabilization" of wages and labor supply.[32]

Encouraged by U.S. officials, numerous other agreements were made during the war with the governments of Jamaica, Barbados, and British Honduras for offshore workers. Arrangements for bringing in offshore workers were made government-to-government until late 1947, when, with the demise of the War Food Administration, the U.S. government ended direct participation in the program. At that time, USSC personnel director, F. C. Sikes, traveled with a "nationwide group of agricultural employers" to the headquarters of the Immigration and Naturalization Service. There they "discovered a clause of the 1917 Immigration Act under which we could petition the U.S. government to employ offshore workers in agriculture" and from that came the British West Indies program (Sikes, quoted in Kramer 1966, 3). Beginning in 1947, the program was conducted under a private enterprise-to-government arrangement involving a tripartite contract between the companies, workers, and governments of the countries of origin, with federal oversight concerning immigration and naturalization laws.

In March 1951, the President's Commission on Migratory Labor, appointed by President Harry Truman, issued its report. The section of the report dealing with the British West Indies (BWI)/Bahamian program criticized the lack of "official vigilance for the protection of living and working standards of alien farm laborers" (U.S. Senate 1978, 10). Because BWI/Bahamian workers' contracts included provisions for withholding forced savings, the commission concluded that the "greater vulnerability of the British West Indian workers to financial discipline" was a reason why the "British West Indians deserted from their contracts much less frequently than the Mexicans" (11). Following the commission's report, Public Law 78, enacted July 12, 1951, established the basic framework for the Mexican *bracero* program. According to U.S. Senator Holland of Florida, the bill as originally drafted included "agricultural workers within the Western Hemisphere" but at the request of "the agricultural interests of Florida" who "much prefer not to have a subsidy from the Government in this connection" the BWI program was excluded (13). Enacted the following year, Public Law 414, the Immigration and Nationality Act of 1952, set the terms under which temporary workers, other than *braceros*, would be admitted, defining the category of temporary alien workers in section H-ii, or H-2. Through the H-2 worker program, the Florida sugar industry was able to secure for decades a steady supply of black field labor from the former slave plantation economies of the Caribbean.

Engineering the Landscape for Water Control

USSC's third and final point of interest in lobbying for direct federal support was water control, on which Bitting also wrote extensively. In February
1943 he submitted his "Report on the Everglades and Contiguous Areas" to
Florida Governor Spessard Holland, weighing in on the pressing question
of water management and land development in south Florida (Bitting 1943).
The political, economic, and ecological context sheds light on the timing
of his report. The Everglades Drainage District had been near bankruptcy
from 1932 until 1942, having defaulted on its bonds in 1931 (Dovell 1947b;
Manuel 1942a). In 1941 Governor Holland signed a bill enabling the legislature to restructure the district's debt, which defined seven zones in the
district and set taxes according to levels of drainage benefits received. Back
taxes were lowered, and defaulters were required to pay a maximum of two
years' taxes. After two years, the district could claim title to land on which
more than a years' taxes were owed (Blake 1980). One of the major beneficiaries of tax restructuring was USSC, which had "had issue with bondholders of the Everglades drainage district over unpaid taxes on lands owned
by the corporation" since its incorporation (Manuel 1942b, 12961). In 1940
a bondholders' protective committee sued USSC and other landowners; a
$5,660,000 Reconstruction Finance Corporation loan enabled the district
to settle with the bondholders' committee, while USSC was able to clear its
back taxes by paying a maximum of two years' taxes at significantly reduced
rates (Manuel 1942b). Regardless of the questionable nature of its recovery,
by 1943 the district was financially able to address questions of regional concern regarding land development and water control (Blake 1980).

These were vital issues due to the "rapid deterioration of the resource
base during the 1930's and early 1940's." The dry years of 1938, 1942, and 1943
had exacerbated the problems caused by careless drainage, such as soil subsidence, shrinkage, and oxidation, as well as muck fires (Ford 1956, 29). In
1939 the Soil Science Society of Florida was established, with R. V. Allison
of the Everglades Agricultural Experiment Station as its secretary-treasurer
(Carter 1974). By 1940 this group of soil and drainage experts had published
a series of studies of the problems of haphazard development in the region.
The society was influential with farm and other local leaders, who respected
its members for their previous work, such as Allison's discovery that the
muck soils were deficient in trace elements, the application of which vastly
improved the agricultural possibilities of the region (Blake 1980; Carter
1974). In 1942 and 1943 there was a spate of publications from various agencies dealing with issues of Everglades soils and water management, includ-

ing studies by the U.S. Soil Conservation Service indicating that in substantial portions of the Everglades soils were too shallow for cultivation.

Bitting's report began by noting that the financial recovery of the district paved the way for addressing the physical problems. Most generally, he made a twofold argument: (1) that the state should prevent private ownership and development of submarginal lands and should manage them so as to conserve water and soil, to increase hydrostatic pressure in urban aquifers, and to maintain wildlife; and (2) that existing agricultural land should be fully utilized: "It has been demonstrated . . . that the presently developable portion of The Everglades has, agriculturally, agro-biologically, agro-industrially, and chemurgically, definite possibilities and potentialities for the immediate future far beyond the dreams of the past" (Bitting 1943, 7). Bitting presented a grand vision of the agro-industrial development of the Everglades (fig. 4.7). His vision underscored the profound transformation that had occurred regarding knowledge of the environment and of its potential products, in contrast to Governor Broward's vision of several million acres devoted to a single crop. Bitting detailed dozens of crops and hundreds of products produced from an agro-industrial region more limited by physical geography than Broward imagined, but also more agriculturally diversified because of the restrictions created by sugar-production quotas. In Bitting's imagined economic geography, the Everglades would play a role similar to the tropical colonial possessions of other industrial nations, supplying the key components of an emerging, durable food complex as well as inputs for synthesizing nonfood products, such as "lastics and synthetic rubber" (Bitting 1943, 40). Along with the numerous benefits of intensive agro-development, rewatering of interior lands would also provide environmental protection and renewal: "A thorough and continuing rewatering of the open undeveloped 'Glades will exert an ameliorating influence on the temperature of the lower peninsula . . . eliminate soil fires, re-establish food supplies for native wild life, bring to an end the present dry-season emigration of native wild life, and restore the natural and unique beauty of such country" (21).

Bitting thought it fortunate that state-owned lands and tax delinquencies were concentrated in agriculturally marginal areas. He argued that all "State-owned or State-controlled lands other than those immediately contiguous to presently existing developments" should be withdrawn from sale, pending completion of a comprehensive regional study. We should not mistake his well-argued position on the use and conservation of the Everglades for an environmentalist stance as we understand it today. Bitting was primarily interested in defending and consolidating USSC's dominance of

Figure 4.7. By the 1940s, the upper Everglades had been transformed into a highly industrialized landscape. This United States Sugar Corporation refining facility produced a
variety of products from cane, including alcohols and synthetic resins. In the foreground
is a new "starch house" for processing sweet potatoes for starch for the U.S. war effort.
Courtesy of Historical Museum of Southern Florida, *Miami News* Collection.

sugarcane production in Florida. Nonetheless, his arguments for the retirement and state ownership of marginal lands for the purposes of regional environmental management foreshadowed those made decades later in Everglades' restoration planning.

The report is remarkable in its synthesis of contemporary knowledge,
defining the physical problems in terms of scale and interconnectedness in
ways that seem quite familiar today. The "five segments" that comprise the
"one great hydrological problem" as Bitting defined them included (1) the
watershed areas, most importantly, the Kissimmee Valley; (2) Lake Okeechobee; (3) the developed and (4) the undeveloped portions of the Everglades; and (5) the areas contiguous to the Everglades. In an accompanying
abstract, Bitting emphasized the "menace to the well-being of the entire
peninsula" posed by haphazard development and drainage. Overdrainage

not only led to soil loss, oxidation, muck fires, and salt-water intrusion, but also, because dried out soils "do not retain the heat of day," local climate change. "All of these adverse conditions will be eliminated when and as the undeveloped portions of The Everglades are restored to their original condition and nature will then begin its healing processes" (Bitting 1943, n.p.).

The report was notable for its political recommendations as well; not only that submarginal lands be withheld from sale but that undeveloped lands "no matter where situated, should be restored to their natural condition until ready for immediate productive use thereof" (Bitting 1943, n.p.). In this report, Bitting seemed to have anticipated the present-day concern for restoration by demonstrating that the relationship between agricultural land reclamation and the ecological "health" of Everglades National Park has been long recognized:

> The restoration of natural conditions will re-establish the native wild-life and thus create an allure of inestimable value in the attraction of additional tourists. Not only will such restoration be of value in and of itself but will render more valuable the proposed National Park and thus create indirect earning value for much of the land not presently adapted to agricultural development. The natural condition of the lower 'Glades will be automatically restored when and as the physical characteristics of the region have been corrected. (n.p.)

Ideas such as those expressed by Bitting were gaining political momentum at the time. Meeting with the trustees of the IIF in August 1943, Everglades Drainage District commissioners requested that publicly held lands too shallow for cultivation "be withheld from sale and designated 'a water holding and conservation area'" (Blake 1980, 175). Agreement between the president of USSC and the commissioners was not entirely coincidental, since J. E. Beardsley, land agent for the corporation, was also one of the drainage district commissioners and, in at least five of the seven subdrainage districts in which USSC held land, company employees were district presidents (Manuel 1942b). The IIF trustees placed a temporary ban on sales, requesting further study, for which the district hired an engineering team. Their report of May 1944 recommended improving canals for existing agricultural land but that drainage of all wild lands should cease. The district lacked the capital for even this modest plan, so instead plans were drawn to designate publicly held land as reservoirs. In 1946 the district unveiled "an elaborate map designating three water conservation areas—one in Palm Beach County, a second mostly in Broward County, and a third mostly in Dade County" (Blake 1980, 175).[33] With IIF approval of the plan, the district requested that the

legislature designate water conservation areas and donate land to the proposed national park.

Rainfall during the summer of 1947 was heavy, and after two hurricanes
hit the state that fall, flooding was extensive. The first flood affected the upper agricultural area of the Everglades, and the second affected the urban
areas of southeast Florida. Flooded urbanites imagined that a great levee,
longer than Hoover Dike, would protect them from Everglades waters. Desiring federal aid for this project, voters in all three counties overwhelm-
ingly approved their referenda on respective water conservation areas. Former governor Holland, by fall 1947 a U.S. Senator newly appointed to the
Public Works Committee, expressed his commitment to "Congressional
consideration of a permanent flood control project" (quoted in Blake 1980,
178). A report—already in progress before the hurricanes—by the Army
Corps of Engineers was hurried along and made public in February 1948.

The years 1947 and 1948 witnessed several events critical to the Everglades region; the formal designation of Everglades National Park, the publication of Marjory Stoneman Douglas's *The Everglades, River of Grass,* and
the completion of the report by the Army Corps. Douglas's poetic environmental history of the Everglades was a sustained lament for what had been
lost to date, especially during the dry years of that decade: "The whole Everglades were burning. What had been a river of grass and sweet water that
had given meaning and life and uniqueness to this whole enormous geography through centuries in which man had no place here was made, in one
chaotic gesture of greed and ignorance and folly, a river of fire" (Douglas
1988, 375).

The key to this phrase is *chaos;* the one point on which the environmentalist Douglas and big-sugar spokesperson Bitting seemed to be in agreement was the need for water *control.* Remarking on a set of interagency documents produced during the war, Douglas wrote:

> The most important single recommendation of the Everglades Project Reports
> was for a single plan of the development and water control for the whole area,
> under the direction of a single engineer and his board. Only in that way could
> the conflicting demands of local areas be equalized, so that the soils fit for
> high cultivation could be used and maintained without detriment to the wa-
> ter supply of the lower areas. . . . A well-planned system of canals that would
> discharge excess lake water into the open Glades would permit the river of
> grass to flow again with sweet water. (1988, 383)

The Corps recommended an apparently "well-planned system" comprising
780 miles of new or improved levees and 492 miles of canals that would

transect the landscape of south Florida so as to provide water control. The plan, the Central and Southern Florida Flood Control Project (C&SF Project), named and defined an Everglades Agricultural Area (EAA) comprising seven hundred thousand acres to the south of Lake Okeechobee (see fig. 1.1). East, southeast, and south of the EAA were the three water conservation areas. The southeast urban areas would be protected by a levee paralleling the coastal ridge. The federal government would pay 85 percent of the construction costs, estimated to be $208 million (Blake 1980). Ultimately the project would cost much more and take much longer than anticipated. When the Corps finally completed the project in 1962, they had created a system of 1,000 miles of canals and 720 miles of levees, which collectively diverted 70 percent of the water flow out of the lower Everglades, including the national park. The Corps substituted human management for the natural dynamic between landscape and water and in doing so thoroughly politicized that interrelationship (fig. 4.8).

The muck fires of the 1930s and 1940s were Florida's equivalent of the dust bowl, interpreted at the time as a senseless and wasteful tragedy brought on by "haphazard development" (Bitting 1943) and "greed and ig-

Figure 4.8. The grid of drainage canals and ditches define the spatiality of the "second nature" created in the Everglades Agricultural Area. Courtesy of SFWMD.

norance" (Douglas 1988). Both Bitting and Douglas saw the problem as inherently political and the solution to be water control under a centralized authority. The Corps' proposed C&SF Project was on balance more an engineering than a political solution. At this critical historical juncture, when scientific knowledge of the landscape was increasing through studies of soils, hydrology, agriculture, and wildlife, the Corps' plan was based on an oversimplified interpretation of this ecosystem. It was not uncritically accepted at the time. Edward Menninger, publisher of the *Stuart* [Florida] *News,* wrote to Senator Holland: "Some hard-shelled conservationist needs to arise in Congress and awaken his associates to the fact that we are not interested in getting rid of the water. The engineers think only in ditches. . . . The longer I live here, the more I am impressed with the necessity of stopping this infernal ditch digging."[34] What Bitting thought of the project is unknown and irrelevant, as USSC directors had relieved him of his duties in May 1946 (*New York Times* 1946a and 1946b). His successor supported the C&SF Project. Senator "Holland received a timely boost from R. Y. Patterson, president of the United States Sugar Corporation, who sent supporting telegrams to Congressional leaders" (Blake, 1980, 180).[35] When the project was completed, the Everglades would, indeed "be truly like a great factory creating new wealth,"[36] not only for agro-industrialists but for the engineers, contractors, and constructors who would build it.

CHAPTER FIVE

The Cold War Heats up the Nation's Sugar Bowl

For three decades following its establishment, Florida's sugar-producing region struggled to achieve steady but slow growth. Then, between 1960 and 1965, the sugar region grew spectacularly, increasing planted acreage nearly fivefold (fig. 5.1). During this period the elements necessary for expansion coalesced for Florida sugar: labor control, water control, and, most important of all, a geopolitical situation in which its chief foreign rival for a share of the U.S. market was disqualified from the competition. When, following the 1959 revolution, Cuba quite suddenly stopped selling sugar to the United States, the Florida industry was well-positioned to gain ground among its domestic competitors. Whereas in 1960 Florida's production was only a small fraction of Louisiana's, five years later they were nearly equal. These changes were the result of bold and financially risky actions by growers and investors. The scramble for domestic quotas in the aftermath of the Cuban Revolution happened not only through Congressional lobbying, but quite literally also on the ground in Florida, where planting continued feverishly around the clock in a race to outpace expansion in other sourcing regions. As old and new investors expanded sugarcane acreage in Florida, they also furthered the transformation of the already drained Everglades landscape into the more intensively managed system of leveled fields, dikes, canals, and pumps necessary for sugar cultivation.

This chapter addresses the competing interests among sugar-producing regions in the Cold War years leading up to the Cuban Revolution and then examines the consequent growth of Florida's sugar industry. While scholars commonly recognize the significance of the Cuban Revolution to the

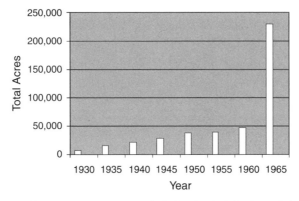

Figure 5.1. The acreage in sugarcane exploded in Florida following Cuba's Castro-led revolution.

Florida industry, many accounts miss the importance of the long-standing political-economic relationship among U.S. sugar-producing regions, domestic and foreign. Though the change in the U.S. sugar system seemed abrupt, it was the culmination of nearly a century of rivalry among U.S. producing regions. In the context of the Cold War, and, most dramatically, in the aftermath of the Cuban Revolution, the meaning of the sugar question shifted once again. Now it revolved around the tensions inherent in balancing domestic interests and U.S. imperial interests in the political and economic stability of its overseas client states.

Because of the complex geography of U.S. sugar production and refining, sugar was a commodity that did not fall neatly within national borders. Because refined sugar could be obtained from two very different crops, cane and beets, grown in different regions of the United States, national alliances among commodity producers sometimes cohered and sometimes dissolved in conflict. As we look at the conflicts engendered in the process of restructuring the U.S. sugar system after the Cuban Revolution, we see differences in interests at every level. We find domestic producers vying with foreign producers. Among domestic producers, we see that beet growers' interests diverged from those of mainland cane growers, who were themselves a heterogeneous group comprising the quite disparate agricultural regions of Louisiana and Florida. Furthermore, even within each region, critical differences of interest existed between old, established growers and newcomers looking to profit from the bonanza. Finally, within the U.S. government differences arose between the executive branch and Congress, and within each of these, between the State Department and the USDA and between the Senate and the House of Representatives, respectively. What becomes apparent as we consider the struggles among these factions is that the tremendous expansion of the Florida sugar region and the concomitant environmental transformation of the Everglades wetlands was in no way preordained. To better grasp the geopolitics of mid-twentieth century sugar, we need to return to World War II, when, as one contemporary observer wrote, "the sugar bowl [had] become the symbol for one of the difficult and complex problems which the United States face[d] in its domestic politics; in its administration of its island territories and possessions; in its relations with Cuba and, indirectly, with all of Latin America; and in its present concern with national defense" (Baldwin 1941, 102).

Sugar on a War Footing

As was the case during World War I, World War II reconfigured the geography of sugar sourcing on a global scale. Again, shipping lines were disrupted, and commercial ties were severed. And once again, in the United States and elsewhere, access to sugar supplies became an issue of national security, this time with even greater intensity. Advances in sucrochemistry during the interwar years had made sugar an essential ingredient in all manner of wartime necessities: "Sugar, as acetone, is in every normal bomb and bullet. It goes into the making of synthetic rubber. It's in practically every military and naval ration and it adds 5,000 feet to a fighter pilot's ceiling. The stuff is, therefore, a basic raw material of war" (Gervasi 1945, 20). Sugar had become a strategic commodity—essential for the production of explosives, synthetic rubber, and other products—and therefore once again Cuba was asked to "turn all its sugar production over to the U.S. government, its 'ally'—naturally, at 'reasonable' prices'" (Zanetti and Garcia 1998, 348).

In October 1941 the United States negotiated to buy the entire Cuban crop (minus two hundred thousand tons for Cuban consumption) for 2.9 cents per pound. The price, which was higher than any since 1927, did not please Cuban sugar industry representatives at a time when world market prices were expected to rise. Following the attack on Pearl Harbor, when Cuba declared war on Japan, Germany, and Italy, the United States and Cuba concluded a more extensive trade agreement. On January 28, 1942, Cuban sugar harvests were pledged to the United States through 1947 at a price of less than 3 cents per pound for the first three years. At first the United States required Cuba to restrict its sugar output, fearing overproduction, so the 1943 harvest was limited to 2.8 million tons. However, within the year it became evident that the United States had underestimated sugar needs and shortages, so Cuba was asked to increase production. Subsequent production levels climbed to more than 4 million tons in 1944 and 5.8 million tons by 1948.

As in the previous war, the United States entered, somewhat reluctantly, into wartime food planning with the United Kingdom. In June 1942 the U.S. President and U.K. Prime Minister announced the formation of the Combined Food Board, "the supreme Allied body for the planning of food supplies during the four years 1942 to 1946" (Roll 1956, 3), which Canada joined in 1943. The Combined Food Board Sugar Committee was one of the first commodity-specific units to be formed, taking over administration of a purchasing and allocation agreement that had been signed earlier in the year. In general, there were major differences between the United States and the United Kingdom with regard to food planning. First, food distribution was highly centralized in the United Kingdom, with 100 percent of imported food handled through the Ministry of Foodstuffs, while no single, comparable agency existed in the United States. Second, in contrast to the United Kingdom, the United States had a large rural farming population and was virtually self-sufficient in most staples, with the exception of sugar (Roll 1956). Third, for a variety of reasons, people in the United Kingdom accepted the need for rationing, whereas it met with tremendous skepticism in the United States. As Lord Brand, the head of the British Food Mission in Washington, D.C., said to one of his American counterparts, "Our difficulties, I think, arise from the fact that your country has never considered food as a problem at all" (51). These differences were especially salient with regard to certain commodities such as meat and especially sugar (Levenstein 1993).

In the United States, sugar was the first food commodity to be rationed during the war, and the last to be released from rationing. Representatives of the sugar industry were skeptical of the need for rationing, suggesting that the primary purpose was "to bring the Nation to a complete consciousness of total war" and that the government recognized that "sugar is a psychological medium" to do so (Bourg 1942). The renowned anthropologist Margaret Mead, who headed the Committee on Food Habits of the National Research Council, suggested that for cultural and psychological reasons, wartime conditions increased the desire for sugar: "Sugar in our culture at present is a symbol by which parents can be indulgent to children. It is also a way of indulging yourself. We are not certain, but we think that consumption of sugar has gone up in so called mushroom defense towns where people are living under deprivation conditions."[1]

However, sugar industry representatives feared quite the opposite and held Mead and her ilk responsible. She and the nutritionists with whom she worked provided support for the U.S. government's position that sugar consumption could be reduced without hardship and that doing so would, in fact, be healthful. Fear that postwar American habits would shift permanently toward reduced sugar consumption galvanized sugar capitalists. In March 1943, cane-sugar refiners, beet-sugar processors, and raw sugar producers from Louisiana, Hawaii, and Puerto Rico, led by Ellsworth Bunker, president of National Sugar Refining Company, and Louis V. Placé Jr., president of McCahan Sugar Refining and Molasses Company, incorporated as the Sugar Research Foundation (later renamed the Sugar Association).[2] Their stated purpose was to support "research relating to sugar, and any and all uses [of sugar, and to] disseminate . . . information as to [the] uses, purposes, utility and effects of sugar."[3] That is, they were dedicated to spreading propaganda to counter U.S. government nutritional information and dietary advice by funding and publicizing research that found, for example, no positive correlation between levels of sugar consumption and the incidence of dental caries or rates of diabetes. With the aid of a public relations firm, the foundation identified the "primary source of the anti-sugar infection" to be nutritionists, home economists, and dieticians. Its strategy therefore was to develop teachers' manuals, teaching materials, educational films and recordings, as well as speaking tours, radio programming, and motion pictures (Hollander 2003). A typical advertisement emphasized the strategic role of sugar in the war effort: "Saved by Sugar" read the caption beneath an illustration of two servicemen about to be airlifted from a lifeboat: "For nearly a week, sugar—contributing 95% of the food value in their emergency rations—has kept them alive."

The *Sugar Situation Report,* published in 1945 by the Special Committee to Investigate Food Shortages for the U.S. House of Representatives, expressed great frustration with regard to bureaucratic management of sugar and identified various factors that contributed to shortages, beginning with the limits placed on Cuban production in 1942 and 1943. Second were the incentive payments and production goals that stimulated production of other crops at the expense of domestic beets, while material and labor shortages restricted expansion of sugarcane. In addition, military sugar rations were relatively high; 1945 military requirements increased by 26 percent to two hundred and twenty pounds per capita per year, while the supply estimated to be available for civilians had declined by 19 percent. Other factors included the diversion of nine hundred thousand tons of the 1944 sugar supply to the synthetic rubber program and the lack of coordination among the twenty or so agencies that had a voice in U.S. sugar policies. In much of this criticism the Special Committee echoed British and Canadian sentiments and frustrations. In 1944 the Combined Food Board, reflecting the U.S. government's overly optimistic outlook, reported that food supplies were generally adequate with one exception: "Only the sugar position was characterized as 'extremely difficult.'" In reality an international food crisis to which the United States—for domestic political reasons—did not want to admit was looming (Roll 1956, 181).

In contrast to the situation during World War I, by the time of World War II the Florida sugar industry had become well established. At that time, USSC operated the world's largest single-tandem raw sugar mill, moved cane via its own railroad system, owned twelve locomotives and five hundred cars, maintained 960 miles of roads and canals, and had an agricultural research laboratory. Between 1936 and 1945, Florida produced an average of three tons of sugar per acre, almost double Louisiana's (Sitterson 1953). Because of production limits set by sugar quotas, USSC had diversified, with 150,000 acres devoted either to pasture or lemon grass, the latter processed in Clewiston into an oil that was utilized in varnish, soap, and other products (Dovell 1947b). Sugar rationing and shortages gave ammunition to Florida boosters such as Senators Ernest Graham (who came to Florida to manage Pennsuco) and Claude Pepper and the directors of the State Chamber of Commerce of Florida, who claimed the state "could grow almost enough sugar to feed the nation" and petitioned Congress to lift production quotas (*New York Times* 1942, 30; see fig. 5.2). At this time, booster rhetoric began to criticize the monopolistic structure of the Florida industry. Thus when W. D. Outman, Director of the Florida Economic Advancement Council, asked the War Production Board to approve the application filed by the Okeelanta Growers and Processors Cooperative to establish a mill and refinery near South Bay, Florida, he argued his case on several points. First, "there is a matter of good faith involved. Our great Navy Department took over the sugar lands and equipment of certain owners in Puerto Rico" with the understanding that the machinery and equipment would be utilized in the United States. Second, "The State of Florida, if it had been allowed to produce sugar, would have served as the Sugar Bowl of the World." And, finally, "this operation is urged as a check against monopolistic control. It must be borne in mind that sugar production in Florida is rapidly approaching the status of monopoly. Granting production facilities to a genuine cooperative would be a very potent factor in restoring true competition."[4]

The request was granted, not surprisingly, given the circumstances—that the U.S. Navy was expropriating more than twenty-five thousand acres from large U.S. and Puerto Rican landowners (Ayala 2001). The Okeelanta Growers and Processors Cooperative bought the Playa Grande mill and moved it

Figure 5.2. Cartoon from the *Miami Daily News* expressing Senator Ernest Graham's lobbying of the federal government to recognize the Florida sugarcane industry's potential for providing strategic raw materials. Courtesy of Historical Museum of Southern Florida.

from Vieques Island to the small town of South Bay in Palm Beach County. In 1947 it produced its first sugar, bringing to three the total number of mills operating in Florida. The Okeelanta Growers and Processors Cooperative included "men of considerable experience" in the sugar and rum industries, among them members of some of the oldest sugar-producing families in Puerto Rico. The experience of this group of "exiles" with "homes in Miami, where they have established their own Pan-American Bank," foreshadows the sugar diaspora that would further transform the Florida Everglades (Hanna and Hanna 1948, 313).

1930s Redux: Cuba and Florida Compete for Sugar Quotas

By world standards, U.S. per capita sugar consumption remained high throughout the war, with 1945 consumption only 18 percent below prewar levels, compared to that of the United Kingdom, for example, where consumption had fallen by a third, or even Canada, where it had dropped by one-quarter (Roll 1956). At the close of the war, when it became evident that there were tremendous food shortages in Asia and Europe, other nations, including Canada, continued to ration food in order to divert surpluses to areas of need. However, by November 1945 the U.S. government abandoned rationing for all food commodities with the exception of sugar, which continued to be rationed through 1947 (Ballinger 1975). At that time, Cuba, having switched from forced underproduction before the war to unfettered maximum production, provided at least half of the U.S. sugar supply. In 1946, when world sugar prices reached six cents per pound, the United States contracted to purchase the entire Cuban harvest for two years at 3.675 cents per pound, with a slightly higher price in the second half of each year if necessary to offset potential price increases of U.S. imports to Cuba (Thomas 1971). In popular media, Cuba was still portrayed as the "American Sugar Bowl," despite Florida growers' best efforts to claim this designation for the Everglades. Melville Bell Grosvenor, soon to succeed his father as editor of the *National Geographic,* published this personal domestic vignette:

> "We lost two sugar stamps today," my wife lamented. "They lapsed because I couldn't find any sugar to buy." Here was a tragedy. Our family talked about it for days. "Where does most of our sugar come from?" my young son asked. "From Cuba. This year Cuba will send to this country more than 30 pounds of sugar for every man, woman, and child." "Cuba is really our sugar bowl then," he said. (Grosvenor 1947)

Domestic production quotas had been suspended in April 1942 and remained so until 1948. In 1946 the USDA asked the sugar industry to develop guidelines for a new sugar act. "Led by C. J. Bourg of the cane growers, the producers tried to resolve their differences. As usual, every group desired a larger quota" (Heston 1975, 141). Bourg suggested a quota of 525,000 tons for Louisiana and Florida; when a bill was introduced that allotted only 453,915 tons to southern cane growers, they found it unacceptable, especially Florida producers, who wanted an allotment more than twice the size of their prewar share (141). Throughout 1946 the federal government pressured the industry to draft a bill, but it was deemed too controversial for either the executive branch or Congress to attempt to influence. Attendance at industry meetings was critical to the specific regional interests competing for quota shares. In April 1947 Florida's Commissioner of Agriculture, Nathan Mayo, telegraphed Economic Advancement Council Director Outman in Washington, D.C.: "Please attend conference and hearing on trade agreement and report. I think we are at a disadvantage in the sugar deal."[5] However, Outman was out of town, and his secretary replied that "no one from Florida representing the sugar producers had testified or was scheduled to testify at these hearings."[6]

Cuba had reason to expect fair or even generous treatment in the 1948 Sugar Act. As a reward for supplying most of the sugar consumed by the Allies during the war, "the State Department assured Havana it would do all it could to improve the island's position in the postwar market" (Heston 1975, 398). Cuba requested a larger allotment than it had in the Sugar Act of 1937, when it was given 28.6 percent of the U.S. market. Now Cubans were interested in 40 to 53 percent of the market, reflecting their wartime expansion. This was complicated by the desire of some in the State Department to use quota levels for leverage with the Cuban government in protecting U.S. business interests in Cuba, "threatening Havana with decreased sugar sales unless it acted more effectively to advance the interests of U.S. capital on the island" (Benjamin 1990, 113).

The Sugar Act of 1948 returned to the fixed-tonnage quotas of the 1934 Jones-Costigan Act. The act guaranteed Cuba only 28.6 percent of the market, allowed it to fill 98.64 percent of demand over 5,220,000 tons, and gave it 95 percent of the Philippine deficit (a quota of 952,000 tons at a time when the Philippines required sugar imports). Thus, the act was temporarily generous but did not hold guarantees of stability nor extend that generosity beyond the time when other producing regions had recovered from the war. "Moreover, the U.S. Congress—influenced by domestic sugar producers desirous of keeping inexpensive Cuban sugar off the U.S. market—tightened the noose by tying sugar purchases to fair treatment for U.S. investments" (Benjamin 1990, 113). The United States Cuban Sugar Council argued that the act slighted the "real interests of [the] U.S." by permitting an increase in domestic quotas as early as 1952, "something to which their record does not entitle them, and which United States consumers should not tolerate" (U.S. Cuban Sugar Council 1947). The council also presented its arguments cartographically, with a map of shipping distances to offshore U.S. sugar sources to depict Cuba's geographic advantage and a U.S. map filled with pictorial commodity symbols to demonstrate that "Cuba's purchases benefit every section of the United States" (U.S. Cuban Sugar Council 1948). Indeed, Cuba was the sixth largest trading partner of the United States, reflecting on the one hand the dependence of the Cuban economy on sugar and its consequent lack of diversification, and on the other, the reciprocal concessions Cuba was forced to make to American exporters in order to have access to the U.S. sugar market (Mahler 1986). From a quarter to a third of Cuban national income derived from sugar, and the Cuban economy's sensitivity to sugar prices and production was magnified by laws that tied nonsugar wages and prices to the price of sugar "in the interests of humanity" (Thomas 1971, 1152).

The Sugar Act of 1948 was similar in form to the Jones-Costigan Act and the Sugar Act of 1937 in its return to a system of fixed quotas, but it differed fundamentally in the underlying logic that determined how those quotas were assigned. The intention of the previous acts was to maintain stability and supply, to which end quotas "reflected the restraints of the historical principle," whereas with the Sugar Act of 1948, "they were now arbitrarily established so as to accomplish the short-range goals of the industry, the Administration and Congress" (Gerber 1976, 114). This shift has been attributed to "the mood of economic nationalism in . . . the first Republican Congress in 16 years, [which] brought with them a 'determination to do everything possible to encourage the domestic industry'" (115). Their protectionism was tempered only slightly by a sense of moral obligation "to repay Cuba for her sacrifices during the war," which led to the provision of an initial bonus for Cuba that was designed to "minimize any immediate effects on the domestic industry and to provide for a phasing down of the bonus over time" (115). For Florida producers, the Sugar Act of 1948 offered opportunity, provided they could successfully compete with other domestic producers for a share of the U.S. market. Although relatively small, they were economically competitive, as shown in the 1947 crop analysis appearing in a "comparative study of plantations" undertaken for Bernardo Braga, a grandnephew of Manuel Rionda. Among the plantations listed, primarily in Cuba but also in Hawaii and Puerto Rico—with profits ranging from $1.42 to $4.88 per one-hundred pound bag—the "Clewiston, Fla." plantation's profit of $3.41 was quite respectable.[7]

The Cold War Geopolitics of Sugar

One of the venues for U.S. government discussions regarding the strategic importance of sugar during the Cold War was under the auspices of the National Security Resource Board (NSRB), a creation of the Truman Administration that was part of the National Security Council. The NSRB Sugar Industry Task Group, the Advisory Committee for Mobilization Plan for the Sugar Industry, was composed of representatives of the U.S. sugar industry with interests in domestic and offshore supplies, among them George Atkinson Braga, then vice president at Czarnikow-Rionda. Their report was framed by the geopolitics of the Cold War, predicting that "by, or before, 1955, Russia will be in complete control of all Asia, and will dominate the Middle East, Turkey, and even Greece" and emphasizing how Russia's real and imagined territorial expansionism "implies threats to and unrest in all of the Pacific Islands, including Japan, the Philippines, and Australia." In this bipolar world, a "necessary assumption is that the United States will do all in her power to see that her Allies are fed." The report noted that U.S. domestic producers failed to increase production during the war and highlighted Cuba's critical role in past wars and expected role in future wars: "In the last two wars Cuba was the only supplier which materially increased its sugar production. Cuba's labour was not required for other purposes and its proximity to the United States prevented any serious interference in transporting sugars. In the light of the experience of the past two wars, we find that we must look towards Cuba for any real increase in sugar production."[8] Thus, from the perspective of the NSRB, the relationship of the United States to Cuba in the Cold War was critical, and the key to confirming that relationship was sugar.

The national security discourse for sugar during the Cold War cut two ways, however. Domestic sugar interests and their political supporters, loathe to stand by as the U.S. government provided incentives to foreign producers, renewed their argument that self-sufficiency in sugar was vital to national security. For a brief period, the Korean War kept sugar prices high and postponed the inevitable confrontation in Congress over marketing quotas. In 1952 Cuba set records for sugar production, with more than 7 million tons produced. Sugar had to be stored in warehouses, and Cuba had to reinstate a system of restriction on future harvests (Zanetti and Garcia 1998). By 1953 the United States was also experiencing production surpluses. U.S. producers, holding unusually large inventories, applied "extreme political pressure for higher domestic quotas" to be obtained at Cuba's expense (Gerber 1976, 119). However, the State Department and National Security Council (NSC) warned against reducing Cuba's quota. In 1954 Secretary of State John Foster Dulles wrote,

> The proposal of the domestic sugar producers would be a bitter repayment for Cuba's effective efforts to stabilize the sugar market. It would seem to ignore the fact that Cuba has been a reliable expansible source of sugar in both war and peace, a strategic concept which should not be jeopardized. [It] might easily tip the scales to cause revolution in Cuba and would certainly

increase instability and promote anti-American feeling and communist activity in an area of great strategic and economic importance to the United States. (Dulles 1954, 901)

Thus various interests in the United States framed the question of Cuba's sugar quota as a security issue in terms of, on the one hand, U.S. national supplies of sugar and, on the other, Cuba's political stability.

Meanwhile, elected officials representing the "sugar states" began lobbying President Eisenhower to support the rewriting of the Sugar Act before it was due to expire in December 1956, with the specific proviso that Cuba's quota would be reduced. In January 1955 Eisenhower met with these representatives; his memorandum of the meeting mentions by name only Senator Holland of Florida, who had explained to the president that the timing of legislation was critical to his state "for the reason that cane is not an annual crop" (Eisenhower 1987a, 782). When Eisenhower maintained that he would not vitiate present agreements by signing into law a retroactive measure, USDA undersecretary McConnell "remarked on the great political strength of the [domestic] sugar interests" and suggested that "the President should carefully weigh the political difficulties which would result" from opposing them (McConnell 1987, 791–92).

As the question of sugar legislation heated up, an internal State Department memorandum entitled "Political Aspects of Cuban Sugar Problem" warned that a reduction in Cuba's participation in the U.S. sugar market would "weaken the Batista government, render it more vulnerable to revolution, strengthen the hand of the 25,000 active communists in Cuba," and perhaps "strengthen revolutionary elements to an extent which might enable them to overthrow the Government" (Newbegin 1987, 797). Even more to the point, a memorandum of April 7, 1955, noted that Henry Holland, assistant secretary of state for inter-American affairs, reported that the U.S. Ambassador in Havana believed any change in sugar legislation before the expiry date would mean the fall of the present government.

But domestic sugar producers wanted immediate relief from their accumulating surplus, and in 1955 fourteen sugar bills were introduced into the House. It was the sugar question redux, as the hearings to amend the Sugar Act reprised arguments recycled from the turn of the century that had then been further developed at the Smoot-Hawley Hearings. In 1955, 53.7 percent of the U.S. sugar supply came from domestic areas (including twenty-two states producing sugar beets and Louisiana, Florida, Hawaii, Puerto Rico, and the Virgin Islands producing sugarcane), 33.1 percent from Cuba, 11.8 percent from the Philippines, and the remaining 1.4 percent from other foreign countries such as the Dominican Republic, Mexico, and Peru (U.S. House 1955). Not only were domestic interests hoping to secure larger quotas, but foreign countries other than Cuba and the Philippines sought new or increased quotas, which "had one particularly significant effect; it caused the extensive hiring of domestic lobbyists." As the politics of sugar sourcing intensified, what emerged came to be known as the "sugar subgovernment; . . . the system of power relationships between the congressional committees dealing with sugar, the administrators charged with responsibility for sugar programs and the domestic industry representatives [that] had acquired almost total hegemony over sugar regulation" (Gerber 1976, 120).

The principal focus of sugar lobbyists was the House Agricultural Committee, which had primary responsibility for setting foreign quotas for several reasons. First, because sugar legislation contained an excise tax and was therefore considered a revenue measure, it had to originate in the House. Second, once the legislation was written, it was moved to the Senate, where it was handled by the Finance Committee, which, lacking expertise in agriculture, was "in a clearly subordinate position [that] enhanced the relative power of the House committee" (Gerber 1976, 121). Thus the chairman of the House Agricultural Committee, Harold D. Cooley, was in a position to wield tremendous power, which he relished. Cooley, a North Carolina Democrat who served as chairman from 1934 to 1966, became known as the "Sugar Czar" due to his "zest" for the "princely power" over sugar fortunes that allocating quotas gave him. Cooley was also known to be "tightly knit" with the sugar refiners, who depended primarily on foreign supplies of raw sugar and were also invested in offshore production (Markel 1975).

With key exceptions, the hearings pitted representatives of the sugar-producing states against officials with the State Department and representatives of the Cuban industry. One exception was Senator George Smathers, a Florida Democrat, who did not adhere to the rhetoric of domestic protectionism. Senator Smathers represented a state that not only produced sugar, but was also closely linked to the Caribbean, home to many Cubans and Cuban Americans and the recipient of much Cuban investment. In Senate testimony, he chose to stress that during the war Cuba had supplied the United States with large quantities of sugar "at a price lower than they could have gotten on the world market" (U.S. Senate 1956). Smathers, a close personal friend of John F. Kennedy but also "a conservative friend of Eisenhower" (Benjamin 1990, 192), apparently saw the political-economic issues at stake in a different light than did many of his colleagues. Indeed, Smathers was known as both an internationalist and as a strong advocate for Latin American development, which earned him the sobriquet, "the senator from Latin

America." Since Florida "benefited enormously from trade and tourism" with Latin America, Smathers's efforts on behalf of the region were generally interpreted as conforming to "the part of a good home-state senator" (Crispell 1999, 105).

Exacerbating producers' dismay over the sugar surplus was a discernable downward trend in U.S. sugar consumption per capita. As *U.S. News and World Report* noted, "U.S. sugar consumption seems to be sliding off. So a tug of war is developing between U.S. and Cuban growers for a shrinking market. There's not much doubt who will win out. The Cubans will get the short end of the stick."[9] However, Cubans from all ranks of the industry had joined forces with the Sugar Research Foundation in a propaganda campaign against the common enemy, artificial sweeteners. Correspondence between the New York and Havana offices of Czarnikow-Rionda showed widespread support for the effort. From the secretary general of the National Federation of Sugar Workers, who declared "that sugar workers were willing and anxious to collaborate in a campaign to combat artificial sweeteners in the United States," to the General Assembly of the *Asociación de Colonos*, which "agreed to contribute $100,000 to a fund for propaganda in defense of the sugar industry," the Havana Rotary Club, which "planned to create a technical committee for the defense of the sugar industry, composed of all professional groups in Cuba," and, finally, the *Asociación de Hacendados*, which "declared that the campaign would be [on a] national scale in the United States and would not only defend sugar from the attacks of synthetic sweeteners but also promote higher sugar consumption," all segments of the Cuban industry united with U.S. sugar interests in an attempt to maintain and expand market share for sugar in the diversifying U.S. sweetener market.[10] Thus, in an ironic historical twist, rural Cuban sugar workers were among those tithed to convince U.S. women that consumption of sugar—"only eighteen calories per teaspoon"—was slimming. They paid for advertisements in U.S. magazines that claimed, "Research findings show how raising your blood sugar level helps keep your appetite—and weight—under control."[11] As reported in Havana, after meeting in the United States with the directors of the Sugar Research Foundation, the vice president of the *Asociación de Hacendados* declared, "The greater help Cuba grants to the Sugar Research Bureau will not only promote consumption but also eventually increase Cuba's quota to the U.S.A."[12]

Despite Cuba's best efforts and the strongly worded opinions emanating from the State Department, in May Congress passed the Sugar Act of 1956, and President Eisenhower signed it. The act, which extended the Sugar Act of 1948 until December 1960, was retroactive to January 1, 1956. It reapportioned Cuba's share of consumption growth, reducing it from 98.64 percent to 43.2 percent in 1956 and 29.59 percent thereafter. Domestic producers received a share of consumption growth, beyond their fixed quotas and deficit share. The law divided a percentage of growth between beet and cane areas, and gave the Puerto Rican deficit to domestic producers. These adjustments meant an increased quota for Florida. The loser in the 1956 amendment provisions was Cuba; increases in other foreign quotas and in both domestic and foreign share of growth came at the expense of Cuba's quota and share of consumption growth.

The response from Cuba was immediate. A telegram sent May 14 from Havana to the New York office of Czarnikow-Rionda recounted the statement of Cuban Secretary of Agriculture Amadeo Lopez Castro published in the newspaper *Alerta* to the effect that the law had not done justice to Cuba. Lopez was reported to have said that President Batista, his government, *hacendados, colonos,* sugar workers, and all other groups of industry had exhausted their efforts to defend Cuba's participation in the U.S. sugar quota, but domestic producers and other sugar producing regions apparently held greater influence within the U.S. Congress.[13] Laurence A. Crosby, chairman of the United States Cuban Sugar Council, analyzed the outcome as due to the combined power of domestic and cane beet producers along with the full-duty countries. "Thus Cuba, the largest and most dependable supplier, already suffering from drastic cut-down of her sugar crops, was the intended victim of attacks on two fronts." Cuba, he said, found the quota reduction "a strange 'reward' for her wartime cooperation with the United States."[14] Crosby noted that the administration ended up supporting full-duty countries at the expense of Cuba, which could be attributed, in part, to the influence of Cooley, the "Sugar Czar." Cooley found that he could maintain more perfect control if his fellow congressional representatives remained so confused as to accept "on faith" the bills that he put before them.

In July Presidents Eisenhower and Batista met in Panama City at the ambassador's residence, where President Batista expressed concern regarding the sugar situation, reminding President Eisenhower that sugar accounted for 87 percent of Cuba's exports. He noted that his attempt at alleviating the grave economic situation in Cuba through public works projects was not succeeding. In response, Eisenhower "commented that from his observation everyone is mad about the sugar situation" and that "the American form of government is such that it is impossible to make promises as to what

we can or cannot do." He then suggested that they step outside to be photo-
graphed by the waiting press, to which "Batista stated it would be a pleasure
for him" (Eisenhower 1987b, 833).

Photo opportunities were not sufficient salve for the Cuban economy. In
April 1957 Dr. Joaquin Meyer, the Washington representative of the Cuban
Sugar Stabilization Institute, called the State Department to inform them of
a pending sale of a hundred and fifty thousand tons of Cuban sugar to the
Soviet Union. The State Department official "voiced his surprise" that Cuba
would sell to the U.S.S.R. when "a ready market existed for sugar amongst
the consuming member countries of the World Sugar Agreement" (Leonhardy 1987, 842).[15] Dr. Meyer attributed the sale to three factors; first, that
the world market price exceeded the U.S. price for sugar; second, that Cuba
had "undergone many lean years"; and third, that "the U.S.S.R. was willing
to place a large, firm order" when other countries were buying limited quantities. Suggesting that Cuba might therefore have difficulty filling its U.S.
quota, a USDA official warned Dr. Meyer that "cognizance would be taken
of this [sale] when our present sugar legislation is amended" (843).

From the U.S. perspective, Cuba was vitiating the bipolar structure of
the world sugar market. From the Cuban perspective, the political maneuverings of domestic U.S. producers since 1954 had created the need for an
alternative approach to the world sugar system. In 1955 the *New York Times*
had reported that "the Soviet Union is wooing Cuba as part of a plan to
'neutralize' Latin America and isolate the United States. Appealing to Cuba's greatest weakness—her mercurial sugar industry—now threatened
through legislation pending at Washington with reduction quota [*sic*] in
the United States market, the Soviet Union began to buy Cuban sugar last
year" (Phillips 1955, 30). As the political and economic tides shifted in a
world once again glutted with sugar, the *Times* noted, "While the Communist party has been outlawed and communism is being wiped out in Cuba,
Soviet commercial agents are received with pleasure by the Cuban Institute
for the Stabilization of Sugar" (30).

The Sugar Act of 1956 reflected in great measure the strength of domestic sugar interests in Congress. To some extent, it also reflected, on the one
hand, the confidence of the U.S. government in general and of the Eisenhower administration in particular that relations with Cuba were robust
and, on the other hand, their ambivalence toward the increasingly fragile
Batista regime (Thomas 1971; Benjamin 1990). With all evidence suggesting that the act would be received very badly there, it does not seem that
the administration expended much political capital to get legislation more
favorable to Cuba. Rather, in December 1958 Eisenhower gave his assent to

another of his conservative friends, William Pawley, to use his influence to convince Batista to step down to make way for a U.S.-engineered junta.

William Pawley was a quintessential "cold warrior," a key figure in U.S. foreign operations and a central player in the unfolding Cuba-Florida drama. His family moved to Cuba when he was a child, where he grew up speaking both English and Spanish. After graduating from the Gordon Military Academy, he spent the next twenty years developing businesses throughout the hemisphere, among them the Cuban national airline and, after purchasing and dismantling the streetcar system, the Havana bus system. In the 1930s he expanded the geographic range of his investments to become president of China's national airline, subsequently organizing the elite Flying Tigers unit during World War II. In 1945 President Truman appointed him U.S. ambassador to Peru and reassigned him the following year to Brazil, where he stayed until 1947. Despite his "reputation as demanding, opinionated, and intemperate," in 1951 Truman asked him to return to the State Department, where he served briefly as an advisor on East Asia during the Korean War (Holland 2005, 44). In 1952 Pawley, "heretofore an ardent internationalist Democrat," switched parties, raising hundreds of thousands of dollars for Dwight Eisenhower's presidential campaign (Holland 2005, 46). Pawley maintained major investments in Cuba as well as Florida, was a close friend of Central Intelligence Agency (CIA) Director Allen W. Dulles, and participated in overthrowing the Guatemalan government of Jacobo Arbenz in 1954. With this background, Pawley seemed well chosen for the task of negotiating on behalf of the United States to convince Batista to step down. Nevertheless, Batista rejected Pawley's proposal, not certain that the proposition that he retire to his estate in Florida had official U.S. government approval. Nonetheless, at the end of the month Batista prepared to leave Cuba, and by January 8, 1959 Fidel Castro was taking power in Havana. Pawley, having failed to avert the communist revolution, would reappear in Florida's sugarcane region, where he helped ease Cuban exiles' entry into U.S. agro-industry.

The Sugar Question Sans Cuba

In the months after the revolution, as the U.S. administration attempted to assess the direction that the new Cuban administration was taking, the sugar quota was central to negotiations. The revolutionary government had contradictory ideas about the future role of sugar in the island's economy. On the one hand, Castro and his closest advisors aimed "to escape from the monoculture," and therefore "to avoid the production of sugar on the existing scale" (Thomas 1971, 1154–55). On the other hand, in May 1959 Castro talked of producing as much sugar as possible and flooding the world market, after which, in response, world prices temporarily fell. In June 1959 Castro requested that the United States increase its sugar purchase from three million to eight million tons that year, which it did not. In October, as relations between the countries deteriorated, the Eisenhower administration was "beginning to wonder whether to continue the sugar quota" (1249). By early 1960 the U.S. "Departments of State, Commerce, and Agriculture were wrestling with the problem of how to use the tremendous North American economic power over Cuba. . . . The major weapon in that arsenal was sugar" (Benjamin 1990, 194). In June 1960 Congress passed a bill that gave the U.S. president the power to cut the Cuban sugar quota. The State Department preferred that the quota be kept as a tool of negotiation, and the U.S. Ambassador to Cuba, Philip Bonsal, was adamant that part of the quota should remain. However, on July 6 President Eisenhower reduced the quota by the amount remaining unfulfilled for the year, including what would have been purchased to make up for other countries' deficits. Thus the remainder of the 1960 Cuban sugar quota was cut to zero. Notified only hours in advance of Eisenhower's announcement, Bonsal was "deeply disturbed" and attempted to dissuade the president from making Cuba a "gift" to the Russians. In cutting the sugar quota, the United States had, Bonsal felt, "turned its back on thirty years of statesmanship in Latin America" (Bonsal 1971, 151).

The suspension of the Cuban quota threw the whole U.S. sugar system into chaos, and domestic growers saw their chance. In August 1960 Lawrence Myers, director of the Sugar Division, USDA Commodity Stabilization Service, wrote to True D. Morse, undersecretary of agriculture: "The demand for acreage by new growers, both in sugar beets and in Florida, is reaching the breaking point. Something must give. The most vociferous demands are from grower groups wishing to get new mills established in Texas and in Florida."[16] Myers was concerned regarding "the grave danger of delay in extending the Sugar Act. March 31, 1961 is a most unfortunate date for the Act to terminate." The new Congress would not have sufficient time to amend the act before it expired. He suggested that the objective of the USDA should be to "provide for Presidential discretion in determining foreign quotas and the extent to which domestic producers will be given a right to fill any deficits." On December 16, 1960, Eisenhower fixed at zero the quota for imports of Cuban sugar during the first quarter of 1961. "Since my proclamation of July 6 of this year the Government of Cuba has continued to follow a policy of deliberate hostility toward the United States and to commit steadily increasing amounts of its sugar crop to Communist countries."[17]

In September the House Committee on Agriculture requested the USDA to "conduct a study of sugar and its relationship to our national economy in its broadest sense" and to report "on the various alternatives which might be pursued."[18] The *Special Study on Sugar* (U.S. House 1961) provided a snapshot of the U.S. sugar situation in relation to the world market, historical background of the U.S. sugar program, and probable outcomes of alternative policy scenarios. At that time the United States was by far the world's largest importer of sugar, followed by the United Kingdom, Japan, and Canada. Annual per capita consumption in the United States was estimated to be 97 pounds. The largest sugar producer in 1960 was the U.S.S.R., where production had risen sharply in the previous six years. In general, world production was increasing faster than world consumption, with all producing countries except Indonesia experiencing growth in their sugar output. Thus the report anticipated a world market characterized by abundant sugar supplies at relatively stable prices.

In 1959 approximately 25 percent of U.S. sugar was supplied by domestic beets, 20 percent came from cane grown in Hawaii and Puerto Rico, and 7 percent from Louisiana and Florida sugarcane. Of the remainder, about 11 percent of U.S. supply was imported from the Philippines, while more than 33 percent came from Cuba, which was at the time the world's largest exporter. When the Cuban quota was cut, replacement sugar was obtained from other countries with existing U.S. sugar quotas (the Philippines, Dominican Republic, Mexico, and Peru) and from areas outside the U.S. Sugar Program (Brazil, West Indies, and British Guiana). While the United States found new suppliers, Cuba also found alternative markets. The U.S.S.R. doubled its previous order for sugar to two million tons for five years, purchased largely on a barter basis. In addition, China ordered five hundred thousand tons per year for five years, also on a barter basis. The report speculated on whether the world market had in reality been cleaved in two, or if instead Cuban sugar "shipped to the U.S.S.R. ha[d] reappeared in markets outside of the bloc" (U.S. House 1961, 2).

In the Cold War context, it was necessary to note "the virtues of free trade and a competitive world" (U.S. House 1961, 8), but neither was relevant because neither the U.S. nor the world sugar markets operated "freely." Since most sugar was consumed in the same country in which it was produced, only 30 percent of world sugar production entered into international trade, of which more than half was traded through various preferential trade arrangements. "The so-called free world market for sugar applies only to those residual quantities of sugar moving without the benefit of preferential arrangements" (26). Of that amount, signatories to the 1953 International Sugar Agreement, which determined members' export quotas to the "free market," exported 84 percent. The structure of the U.S. Sugar Program, which attempted to reconcile "the several interests of the domestic sugar industry, the foreign sugar suppliers, and the American consumer,"— all pulling "in diverse directions"—meant that the United States had "developed a thoroughly managed sugar economy" (5). The principle feature of the U.S. system, which was to limit supplies of sugar to maintain price stability, required the secretary of agriculture to determine each year what the domestic requirements at that price would be. Once that determination was made, quotas were assigned to various producing regions, domestic and foreign, calculated on the basis of the percentage of the U.S. market they had been allotted.

Of course there were always uncertainties to be reckoned with, such as the failure of various producers to fulfill their assigned quotas or the capacity of the market to absorb sugar at a particular price. With respect to these uncertainties, "the United States benefited because Cuba maintained sufficient reserves of sugar to meet increases in U.S. demand and to cover shortfalls in supply from other areas" (U.S. House 1961, 60). Until the latter half of 1960, Cuba was the swing producer that absorbed the risk of uncertainty in the U.S. sugar market and lent this otherwise very structured market the flexibility necessary to meet the vicissitudes of production shortfalls and unmet demand. "The economic break with Cuba created or helped to create three main economic problems: getting enough sugar, getting the sugar when it was needed, and getting it where it was needed—all three of which arose because the quota system had been structured around Cuba's role as a buffer" (Gerber 1976, 127).

The primary purpose of the USDA's 1961 report was to analyze future prospects for the U.S. sugar system in the absence of this primary and flexible supplier. Consumption growth was predicted to result from population increase rather than increasing demand per capita, which had stabilized. World supplies were deemed abundant, and foreign supplies available to the United States—primarily from countries holding quotas and from other Latin American and Caribbean countries—were considered adequate. The question of future domestic beet and cane production was examined in some detail, with projections for acreage and production of sugar crops in 1970 based on three different price situations: a price per ton that was 25 percent below, the same, or 25 percent higher than that prevailing in 1959. Based on this assessment, the report concluded that the domestic sugar industry had "substantial capacity to expand at present prices," with significant potential in beet areas and "mainland cane areas, particularly Florida" (U.S. House 1961, 3). "Florida has a considerable potential for increasing sugarcane production through drainage of land south of Lake Okeechobee (54). In addition, in 1960, "expansion in raw sugar operations was occurring in Florida despite a great deal of uncertainty with respect to future developments in sugar legislation. With some degree of assurance that increased production could be marketed, substantial investment in new processing facilities seems likely in that State" (56). With Cuba seemingly—even if only temporarily—out of the U.S. market, the lid was off of the Florida sugar bowl.

Several days after the USDA's *Special Study on Sugar* was issued, Secretary of Agriculture Orville Freeman circulated among close colleagues— very cautiously— a confidential memorandum on sugar that had been sent to him. The unsigned memo offered an alternative "analysis of the sugar problem" that differed from the official report. The central argument of the anonymous writer was that sugar beets were not by any definition or in any world region economically competitive. "If the pattern of production and consumption were determined by market forces virtually every country would rely on sugar from sugar cane grown in tropical countries."[19] The essence of the sugar problem in the United States was that "legislation about sugar beets has always involved good politics and bad economics." As a result, U.S. consumers paid a double penalty: higher prices for less sugar than would be available if beets were unprotected. The demand for larger domestic quotas could be justified neither in terms of agricultural policy nor for reasons of national security. "Any additional permanent quota would be pure sweet gravy—something that was unexpected until Castro." As it was, domestic beet and cane growers already enjoyed larger returns than they expected when they went into business. "The average farmer asking for a price support is on the defensive; the beet grower asking for a larger quota is on the offensive."

The memo writer dismissed the national security arguments often advanced in the name of protectionism: "An increase in domestic sugar production quota is sometimes advocated on the ground that it is vital to the national security: imports may not be able to reach our shores during waretime [*sic*]. In any limited war there would be no serious sea transport problems and if we have an unlimited war with thermonuclear bombs and warheads, a shortage of sugar will be the least of our worries." Moreover, in response to those who suggested that domestic quotas should be expanded because of "more Castro-like revolutions in Latin America," the answer was that "permanently larger quotas for domestic producers would damage our relations with our strategically located neighbors in the Caribbean. The national interest is clear. But what will the political realities allow?" The answer, the memo observed, lay in a "few hopeful elements in the political situation," such as a hesitancy to expand on the part of beet growers because of the uncertainty created by unstable international conditions and the Farm Bureau's "passionate belief in free markets." In addition, the "fluidity of the Carribean [sic] situation" might forestall a permanent domestic quota increase. "You and the President would want, I believe, to avoid a policy that would do great damage to our desperate position abroad, even if a certain political price had to be paid."

Indeed, over the course of the year, the Cold War geopolitics of U.S. sugar sourcing became increasingly heated and convoluted. A memo to Secretary Freeman from a USDA administrator, Robert Lewis, expressed alarm about the Byzantine methods being used to fulfill the suspended Cuban quota. Specifically, amendments to the Sugar Act had suggested that in reallocating the quota, special consideration should be given to countries purchasing U.S. agricultural commodities. "This language has resulted in a number of suggestions that the sugar quotas be utilized for the purpose of disposing of surplus agricultural commodities through barter or barter-like transactions."[20] Lewis saw this as highly problematic for several reasons. First, the foreign policy goal of the sugar program was to retain the Cuban quota "as an incentive to Cuba to reform," but complicated barter transactions would remove the quota from play. Second, barter complicates the process of securing supplies and could create resentment among countries that are forced to take surplus U.S. commodities rather than cash. Third, barter transactions violate the principles of multilateral trade: "In the eyes of the world, 'barter' trading is associated with Nazi and Communist trading procedures, and is resented because it involves coercive pressure by the strong nation upon the weak to accept goods which it may not necessarily desire in preference to others."

In an effort to simplify the politics of the U.S. relation to the international sugar market, the Kennedy Administration and the USDA began to consider a global quota rather than the country-by-country system in place. The proposed global quota would be filled on a competitive basis by sourcing from a designated group of countries, with an import fee that would recapture some or all of the excess of the United States over the world price. The administration found that the existing country quota system was "harmful to foreign relations and gave at least the appearance of possible corruption" and argued that "a global quota would increase the adequacy and reliability of supplies" (Gerber 1976, 131). When "Sugar Czar" Cooley caught a whiff of this suggestion, he responded with a sharp rebuke to Freeman in a memorandum listing three resolutions that had been adopted that day (September 7) by the Committee of Agriculture. First, the committee directed the USDA secretary to allocate domestic quotas to new mills to support development of sugar in new areas. Second, the committee "would regard with extreme disfavor any action or statement by any representative of the United States which would commit the United States to adopt any system of sugar importation other than fixed statutory quotas such as have heretofore operated so effectively." Third, the committee resolved that the president instruct those in charge of the sugar program "that in making any such purchases of sugar for the calendar year 1962 clear preference is to be given those countries which offer to buy a reasonable quantity of United States agricultural commodities in return for our purchase of their sugar."[21]

That the USDA was at loggerheads with Cooley and the committee is evident in the confidential administrative memo of September 20 on the topic "what the USDA should do with respect to new legislation and administrative action with regard to sugar."[22] The memo called into question the logic of the entire quota system, which had been based on conditions that no longer held. The system was developed when "it was in the interest of U.S. foreign policy to provide substantial assistance to Cuba by reserving a large quota for Cuba."[23] Without that key relationship, "the quota premium serves no useful purpose under today's conditions. We now have no generally acceptable basis, historic or otherwise, for allocating quotas to foreign countries." While Cooley relished the political maneuverings and the power that the quota system gave to the granter, the USDA found "the endless pressure from foreign countries" distasteful: "These pressures will also come from domestic exporters and their agents, as an effort is made to use quota reallocation as a device to increase exports of other U.S. products. We feel that the quota premium is an awkward, inefficient and perhaps dangerous method of building buying power for our farm exports, or for aiding foreign countries."[24] The overarching recommendation of the USDA, in consultation with the Foreign Agricultural Service and the Agricultural Stabilization and Conservation Service (ASCS), was to eliminate the quota premium and then, bowing to political realities, to increase the domestic market share slightly to 60 percent of U.S. consumption.

The Scramble for Quotas

While Congress and the administration battled over sugar legislation, the files of the USDA began to bulge with queries, requests, and protests emanating from all of the domestic sugar-producing states and various foreign regions regarding the allocation of quotas. A typical inquiry was that of U.S. Representative Charles Bennett of Florida: "Many people in Florida are asking me why Florida could not be allowed to produce all the sugar it can in view of the international situation with regard to Cuba. I wish to understand this situation and to be as helpful and constructive as I can in it. Can you please give me any advise [sic] and suggestions that you may be willing to give?"[25] Prior to the reduction of the Cuban quota to zero, several entities planned to expand existing enterprises or to develop new ones to take advantage of the allotment to Florida of a percentage of the Puerto Rican deficit. For example, before economic relations with Cuba were severed, USSC was planning for construction of its second mill. In addition, a group of 'Glades-area vegetable growers had formed a cooperative and was also planning to erect a mill. With Cuba's quota cut and the concomitant USDA announcement of unrestricted sugarcane acreage in 1960, interest and investment in Florida sugar production mushroomed far beyond what was being envisioned by local industrialists and growers. Farris Bryant, the Democratic gubernatorial nominee, announced at a press conference in New York that Florida could triple sugar production almost immediately and in "a longer period, it might be doubled again. The state could probably turn out ten times its present quota and more efficiently and cheaply than Cuba does" (New York Times 1960, 65).

Meanwhile, on the ground, there was "a moving day parade from Louisiana" of mills headed for Florida (Business Week 1961, 58). By the fall of 1961, contemporaneous news accounts echoed the booster rhetoric from earlier in the century when, in 1929, the "Florida sugar bowl" had been declared open. "Florida, the land of booms, has another sweet one going—in sugar. Dormant for years while Louisiana cane and northern beet producers passed Florida by, the sugar country around Clewiston, Lake Okeechobee and Belle Glade has suddenly come alive" (Birger 1961, 8A). Land prices in the Florida "sugar bowl" had reportedly doubled in one year, with the choicest land near Lake Okeechobee commanding the highest prices. The region was being transformed with an influx of capital estimated at $100 million, representing the investments of exiled Cubans and of Americans, many of them displaced from the Cuban sugar industry. Among them was Alfonso Fanjul Sr., grandnephew of Manuel Rionda; in 1959 the Rionda group was estimated to be one of the two greatest landholding companies in Cuba (McAvoy 2003). The players in this sugar diaspora were carefully inventoried in the pages of *Avance in Exile,* a Spanish-language newspaper published in Miami by exile groups, with support from the CIA (Pfeiffer c. 1970). *Avance* highlighted the key role of the Miami financial center in underwriting the expansion, listed the names of the investors and their Cuban *ingenios,* and noted that some were also or instead investing in Brazil and Costa Rica (*Avance* 1961).

As recounted in the pages of *Business Week,* Czarnikow-Rionda led the parade: "The company, which lost six mills to Castro, dismantled a 2,500 ton mill in Louisiana through a new subsidiary, Osceola Farms, Inc., and is rebuilding it near Canal Point, Fla." (1961, 59). Czarnikow-Rionda also provided 15 percent of the financing for the Sugar Cane Growers Cooperative mill and secured a management contract to run that mill as well as to provide the cooperative with a hundred and twenty thousand tons of cane from the company's newly planted Florida fields. Another prominent company in this parade was the Cuban American Sugar Corporation, established in 1899, which had close ties to U.S. sugar refiners, had a refinery in Louisiana, and, prior to Castro, vast holdings in Cuba (Ayala 1999). Headed by David M. Keiser—who was the president of the New York Philharmonic as well as the chairman of the board of Cuban American Sugar—the corporation established the Florida Sugar Corporation with a mill moved from Louisiana. For Keiser, as for Fanjul, this was a "family" business; his father, George E. Keiser, had been president of the Cuban American Sugar Company.

Several other mills were "on the horizon" at this time in south Florida. Talisman Sugar Corporation, located about twenty miles south of South Bay, was "created by the joint efforts of several Cuban expatriates with the backing of Henry Ford II" (Salley n.d., 27). Four major investors along with eleven others paid a total of 21.8 million dollars for 16,800 acres of "virgin muckland," a sale heralded by realtors as "the biggest ever in the state."[26] Fernando de la Riva, who was heading this group, had also invested in a central in Brazil (*Avance* 1961). Backers planned "to hire only Cuban exiles with experience in the sugar industry" in order to "alleviate the exile problem in South Florida" and give "help to the many Cuban sugar technicians forced out of their country by Castro."[27] Representatives from Allis-Chalmers Manufacturing Co. of Milwaukee led a tour to the site so that visitors could "watch tractors chew up the rugged terrain and Everglades vegetation and transform it into miles of flat, black, warm muck beds" ready for planting. News accounts described the "snake-infested wilderness" as it was "being broken by the noisy teamwork of draglines, dredging a vast and complicated secondary drainage system, complete with its own pumping stations" (Birt 1961). Talisman—named to bestow good luck on the "new life" of the exiles—would, within a year, need to be rescued from bankruptcy by William Pawley.

Second was the South Florida Sugar Corporation, begun by the owners of a large Puerto Rican mill and refinery, who moved a mill from Puerto Rico to Florida; by the close of the 1964 harvest, this company was purchased by Talisman Sugar Corporation, which absorbed its marketing allotment and acreage shares. Third, the Glades County Sugar Growers Cooperative, composed of about fifty members and spearheaded "by two big-time Cuban sugarmen" who held a 25 percent controlling interest, was located in the environs of Moore Haven, Florida. Fourth was the Atlantic Sugar Association, a farmers' cooperative with a mill located about fifteen miles east of Belle Glade, Florida. Finally, a small mill was erected near Clewiston by a Cuban investor, Adalberto Mesa, which ground cane for one season before going bankrupt (Salley n.d.).

Thus the sugar diaspora and the resultant expansion of the south Florida industry brought a deepening and thickening of its ties to the networks—corporate and family—that structured the sugar business in the American Sugar Kingdom. While the Florida industry was already part of an international circuit of expertise and labor, to this point it had been dominated by USSC. We might see the mill at Okeelanta as a precursor: moved from Puerto Rico, it was sold in 1952 to the Okeelanta Sugar Refinery, Inc., a corporation organized by the operator of five mills in Cuba, which in 1959 bought the mill at Fellsmere as well (Salley n.d.). This, suggested *Time* magazine, provided a good example for the "Cubans who are moving to the U.S.": "The Okeelanta Sugar Co. was started by two Cuban families in 1952 as a sideline to their island companies. Now their two Florida mills and vast acreage are worth almost as much as their $27 million worth of mills and land that were confiscated" (*Time* 1961, n.p.).[28]

The national media attention paid to Florida's sugar expansion elicited protest among competitors from other regions and commodity sectors. Representative Joseph M. Montoya of New Mexico wrote to Secretary Freeman:

> I have just read an article concerning the vast acreage in Florida which is presently being planted to sugar cane. I feel that the assignment of additional cane acreage at this time circumvents the intent of Congress. I am sure that you are aware of my strong efforts to obtain increased sugar beet acreage for the southwestern section of our country, but to little avail. I would certainly be inter-

ested in learning just how these extensive operations received approval to put vast acreages in the Everglades of Florida under sugar cane cultivation.[29]

Assistant Secretary of Agriculture James Ralph replied to Montoya, explaining that growth had occurred because acreage restrictions for cane sugar had been lifted at the close of 1959 and would remain so through 1962. Ralph admitted that "considerable expansion in sugarcane acreage and processing facilities has occurred in Florida" but noted that it was being done at some considerable risk. "The expansion of acreage, as well as the building of new sugar factories in Florida, is occurring with full awareness by those involved that substantial cut-backs in acreage may be necessary in future years."[30]

At the time, the mainland cane quota was 775,000 tons, of which 500,000 tons were allocated to Louisiana. The two extant Florida companies, USSC and Okeelanta, were then producing 160,000 tons, leaving just 115,000 available when the expansion already underway was estimated to produce additional sugar in excess of 180,000 tons within the year. But as *Business Week* suggested, "Capital such as is now pouring into Florida sugar is not attracted by the comparatively small immediate gain. The growers and mill operators who are doing the investing . . . are betting that Florida will become a more important supplier for good" (*Business Week* 1961, 58). To become an "important supplier," Florida needed favorable treatment in the soon-to-be-rewritten Sugar Act. The haste with which mills were moved and land put into production was not a matter of jumping the gun but of aiming it. By establishing an expanded territorial production complex with significant fixed capital, owners hoped the fact of their investment would secure a generous quota. Harry T. Vaughn, president of USSC, was "encouraging others to enter the business, feeling it [would] be a big help to Florida's chances of getting an increased quota when the new act [was] written" (Birger 1961, 8A).

Such strategies to increase the size of the sugar industry were not limited to "runaway expansion" (Lobo, quoted in Birger 1961, 8A); they also involved intensive political organization. Upon taking office, Governor Farris Bryant took steps to coordinate industry efforts, which the business community duly appreciated. The president of a local public relations firm wrote to Bryant offering the services of his firm free of charge: "Your promotion of Florida's sugar production, through the instrument of the appointment of a select committee to implement your objective of stimulating the economy of Florida, is exemplary leadership. We request that we be afforded the privilege to serve you to achieve those goals mutually desired as Floridians."[31] Bryant appointed Harry T. Vaughn, president of USSC to chair the Governor's Sugar Advisory Committee. In April Vaughn wrote to inform Bryant that Chairman Cooley had promised to begin hearings in May for a new sugar act. In preparation for negotiations at the national level, all of the Florida producers would begin meeting that week to "arrive at a program to present in meetings with representatives of other areas. I have had numerous conversations with representatives of other domestic sugar producing areas and I have found no opposition to our request for a substantial increase in the basic quota of the mainland cane area."[32]

The Florida Legislature voiced support at the national level for the Florida sugar industry through a report entitled "A Memorial to the Congress of the United States" asking that Vaughn and Charles Stewart Mott, who still served as chairman of the board at USSC, be recognized as "two pioneers in the United States Sugar Industry." To do this the State Legislature suggested that the U.S. Congress increase the quota for domestic sugarcane growers: "Whereas, there is a vast amount of rich and fertile soil in Florida suitable for the expansion of the sugar industry now being developed especially around the Lake Okeechobee area, the Florida Legislature respectfully requests the members of the United States Congress to provide additional sugar quotas for domestic growers."[33] More urgent and to the point was the telegram sent by Florida Commissioner of Agriculture Doyle Connor to Secretary Freeman:

> New mill construction and a sharp rise in planned sugar acreage indicate a progressive increase in Floridas [sic] sugar production in the next few years. Floridas annual production 10 years ago was about 80,000 tons. It is expected to be 300,000 tons in 1962 and 500,000 tons within a few years. I strongly urge your support for early legislative action to increase the mainland cane quota allotment sufficiently to protect Floridas rapidly growing sugar industry.[34]

Hearings on the Sugar Act had been forestalled throughout 1961 because neither the industry nor the administration was yet prepared to make recommendations. Secretary Freeman sent his staff a memo on "the touchy matter of sugar" with the intent "to clarify relationships in light of political developments so that we will know where we are and will be able to move in unison. We must coordinate and cooperate closely."[35] Freeman clarified the chain of command in the department with respect to sugar, identifying Undersecretary Charles Murphy as having "the number one responsibility for coordination and the making of recommendations." With respect to a pending 150-million-ton discretionary allotment, President Kennedy had expressed interest in favoring Colombia, while the department had Argentina in mind. Freeman counseled the president that "we do nothing about this quota for at least one month" in the hope of keeping domestic prices stable and in an effort to terminate barter agreements. There were three key political considerations to keep in mind with respect to sugar legislation. First, the sugar supply was considered a matter of national security. Second, USDA staff must remain cognizant of "the political situation with many new producers desiring to come into the field and substantial investments for refineries in the offing." On the other hand, the United States was urging the European Common Market not to encourage uneconomic production: "Taking this position internationally, we find ourselves in a very strange position if we take a counter-position domestically."

In meetings held in December, the domestic industry as a whole agreed on some issues but failed to do so on others. Vaughn, in his capacity as chairman of the Governor's Sugar Committee, wrote to "All of the Florida Sugar Companies" to inform them of the status of the industry's proposals for new sugar legislation.[36] Domestic producers had agreed that they should receive between 60 percent and 62 percent of the domestic market and that mainland growers would receive two-thirds of market growth, which was a hundred thousand tons. But they had not yet agreed upon how to divide the market between beet and cane growers. Thus, early spring saw a period of heavy lobbying: industry representatives, growers and local boosters throughout the country flooded the USDA and the administration with appeals for regional favoritism. Florida was no exception. Some appeals were addressed to the state administration, in an effort to bolster the governor's demands at the national level. For example, the president of the local Allis-Chalmers equipment dealer wrote to Governor Bryant to request "that your office use all of its influence with the United States Congress . . . for 150,000 tons to be released to the State of Florida[, which] would completely change the economical picture in the South Florida area."[37]

As a tentative agreement among industry representatives emerged, Florida sugar boosters made it clear that that they were eager for the new Sugar Act—which they felt represented "an acceptable compromise"—to be passed. The governor's Florida Sugar Advisory Committee sent a telegram to each of the members of Congress from Florida, urging its passage: "It is most important to Florida to secure a new Act, as the present law and regulations imperil the tremendous capital invested in field plantings and processing facilities."[38] The Florida Council of 100, "composed of leading business men and educators," made a similar request "for favorable consideration by the administration of the new proposed sugar legislation" via a telegram sent to Secretary Freeman, reminding him of the "private capital expenditures approximating one hundred million dollars in capital outlay alone for new sugar producing facilities in Florida."[39] Governor Bryant transmitted a resolution "adopted unanimously by the elected Cabinet of the State of Florida" to President Kennedy, urging the passage of the Sugar Act: "Your personal familiarity with Florida and with the sugar industry make additional explanation unnecessary. May I merely point out that we consider the question of sugar quotas to be of vital concern to our state, its economy and its future."[40]

Indeed, President Kennedy was quite familiar with south Florida, counting Senator Smathers among his closest personal friends. They became friends as freshmen Congressmen in 1946; a measure of their closeness was that Smathers ushered at Kennedy's wedding and spoke for the groom's side at the rehearsal dinner. Now Smathers, who was known for his advocacy on behalf of Latin America, turned his attention to the support of the Florida sugar industry: "Switching adeptly from Cold War globalist to home-state senator, Smathers worked to ensure greater sugar quotas were allocated for mainland producers" (Crispell 1999, 175). Thus Harry Vaughn suggested the following strategy to Florida's Commissioner of Agriculture Doyle Conner, as the latter sought to convey to the Kennedy administration the importance of the new sugar legislation to the state:

> If Senator Smathers approves, we believe the following language will accomplish our purpose: "My friend George Smathers has reminded me of the necessity for new sugar legislation to replace the existing sugar act which expires June 30th this year. Upon recommendation of Senator Smathers, this administration favors the new sugar legislation, which will go a long ways towards alleviating Florida's sugar problems and permit orderly growth of this important industry."[41]

While it is not certain whether Conners followed Vaughn's advice, the Florida booster network was undoubtedly aware of Smathers's important connections. Smathers obviously had the ear of the president, and both men — Smathers and Kennedy — clearly felt some responsibility toward the south Florida industry, in part because of the relocation there of sugar interests that had been displaced from Cuba.

As Florida waited for the administration to reveal its intentions, Louisiana asserted its claims to an increased share of the quota. In February 1962 the Sugar Division of the USDA held an informal meeting on acreage controls with mainland sugarcane growers in New Orleans. As reported in the *Sugar Bulletin,* primarily a Louisiana growers' journal, the division within the sugarcane industry was not based on geography but on history. In contrast to the *Times Picayune* report of a clash between Louisiana and Florida growers, the *Bulletin* suggested that "a more accurate report of the proceedings would have been to say that 'old established sugar cane growers of both Louisiana and Florida clashed with new Florida growers over the development of a 1963 cane acreage control program.' It was obvious that the established Florida growers were as determined as the established Louisiana growers to keep the 'Johnny come lately' Florida people from muscling in on their acreage allotments" (Dykers 1962, 106).

In April, Senator Allen J. Ellender of Louisiana wrote to Undersecretary Charles Murphy to protest a statement by a representative of the Sugar Division of the USDA to the effect that new producers would receive weightings from 85 to 100 percent of new acreage, which, he argued, was not a compromise position between old and new growers. At the heart of his complaint was the relationship between allocations for Louisiana and Florida, and, within Florida, between old and new growers. He began his letter by noting that the February meetings of the mainland cane growers had established two facts: "First, an unexpectedly large acreage of sugar cane had already been planted in the State of Florida, and, secondly, and most important, planting was continuing in Florida at that time, some of it on a 24-hour a day basis, with plans being made by numerous individuals to plant as much cane as possible in the year 1962." Ellender noted that the extent of the plantings "were a surprise not only to the established producers of Louisiana and Florida, but also to representatives of the Department." He cautioned, "These disclosures had serious repercussions in the State of Florida. New producers were tempted to redouble their efforts and old producers, both large and small, openly threatened to resume planting immediately. Their reasoning was that the larger they made their 1962 crop acreage, the larger would be their 1963 proportionate share acreage when controls went into effect. A foolish and useless planting race was in the making."[42]

At a second meeting, Louisiana and Florida growers, according to Ellender, "recognized the threat" posed by a planting race and saw that the USDA "would not take action in time to prevent such a thing from happening." Consequently, producers of both states voluntarily limited their own planting plans and agreed that the proper ratio between Florida and Louisiana for 1962 plantings and for history purposes would be 127,248 acres and 320,000 acres, respectively. However, at a March 13 hearing "to the complete amazement of the Louisiana producers, and presumably to the Florida producers, a representative of the Sugar Division at the time of the hearing voiced grave doubt as to whether or not this recommendation would be accepted by the Department." This reopened the cut-off date, and "the way was once again open for the Florida producers to plant purely for history purposes."[43]

Perhaps reflecting the "surprise" factor of Florida planting, in May Secretary Freeman sent a correction to Cooley to revise upward the department's estimate of mainland sugarcane acreage, from 445,000 acres to 465,000 acres, representing an increase of 105,000 acres from the 1961 crop.[44] In June Cooley was finally able to bring a bill to the House, which was revised in the Senate and then signed in July. The new Sugar Act raised the mainland share of consumption growth from 55 percent to 65 percent and added a special provision for new beet growers, reserving 26,000 acres specifically for them with the implication that new processing facilities would be built to accommodate their crop. The number of foreign countries receiving quotas grew from eight to twenty-three, and the size of the Cuban quota that was being held in reserve was significantly reduced and assigned as a global quota. In resisting a more ambitious application of the global quota, Chairman Cooley suggested that the existing system provided necessary political camouflage for domestic growers: "If we do away with quotas, the only thing you have left will be payments to domestic producers [which] would stand up like a sore thumb and will be difficult to justify" (quoted in Gerber 1976, 131).

When the Sugar Act had passed, mainland cane producers turned their attention to one another, and the competition heated between Louisiana and Florida producers for their shares of marketing allotments for 1963. In hearings held in November, each group argued its case.[45] At stake was the basis on which the USDA would calculate production histories, which would in turn determine allotments. Louisiana producers contended that during the preceding few years when acreage and marketing controls had been lifted, expansion in their state had been "reasonable," while in Florida it had been "explosive." Between 1958 and 1963, acreage in Louisiana had climbed from 240,083 to 327,000; the corresponding figures for Florida were 35,848 and 139,000. Using the 1962 crop — the first for the new Florida producers and the largest for the old ones — as the measure would discriminate against the Louisiana processors and farms, "historically long established," unless they were permitted an alternative measure. Florida producers argued that the USDA allotments were in fact discriminatory because they placed the entire and "unconscionable" burden of storing the excess 67,902 tons of sugar for the year on Florida.[46]

From Surplus to Shortages

By the spring of 1963, Louisiana and Florida producers had agreed to differ. That is, each state was given a goal in terms of allotted acreage and permitted to reach that goal according to "procedures and provisions applicable" to that state alone.[47] More significant, the situation with respect to the United States and the world sugar market had changed as well. For a variety of reasons, the problem in spring 1963 was one of sugar shortages rather than a surplus, a circumstance that the USDA had not anticipated in its *Special Study on Sugar* completed two years earlier (U.S. House 1961), which forecast abundant supplies. A December freeze had reduced Florida's production by 30 percent from previous estimates. At the same time, a poor European beet harvest coincided with a reduction in Cuban production, as well as the diversion of most of the Cuban crop to the communist bloc. However, the USDA believed that Florida output would increase due to expansion: "The retention of good cane stubble that would otherwise be plowed-out and minor additional plantings in Florida prior to a closing date of April 16, 1963, should result in about 15,000 additional acres." Therefore they continued to advocate production controls to avoid "another spring planting race in Florida" and increased proportionate crop shares by only 10 percent to meet demand.[48]

However, U.S. sugar prices were high enough to be a political liability for the administration as sugar users—from "housewives" to industrial food processors—addressed their pleas and protests to the first lady. Director Lawrence Myers of the Sugar Policy Staff, ASCS, wrote to Murphy, "In accordance with Mrs. Kennedy's request, I met with Mr. Mike Feldman and the two representatives of the Coca-Cola Company. They called attention to the high sugar price situation and pointed out that it had serious political as well as economic potentialities."[49] The *Washington Star* noted that the price of sugar had "shot through the ceiling" and that "as far as industrial users are concerned, the White House deserves the blame." On April 22 the price was $7.70 per hundred pounds and only two weeks later it was $8.50, its highest price in 43 years. "Responsibility for the situation is being attached to the Kennedy administration."[50] The White House and the USDA had fought hard and won major amendments to the Sugar Act, which, for the first time in thirty years, tied the U.S. price to the world price. In addition, the 1.6-million-ton quota that was still being reserved for Cuba was filled by a global quota. This aspect of the act was secured at the request of the USDA and the administration against the wishes of the industry and the "Sugar Czar." The aim of the global quota was to recapture the difference between the world price and the U.S. price, the latter usually being higher. Now that the world price was higher than the U.S. price, the recapture device was useless, and U.S. power in the world sugar market diminished.

The White House was sensitive to the political heat generated by rising sugar prices. Myer Feldmen, the deputy special counsel to the president, sent a copy of the *Star* article to the USDA, highlighting within it a list of alternatives for bringing down the price of sugar put forth by "Members of the Industrial Sugar Users Group." Secretary of Agriculture Freeman responded, offering a pessimistic forecast with respect to world sugar supplies: "Our sugar specialists foresee no relief in the supply situation until this year's beet harvest, at the earliest, and do not anticipate a really easy balance between supply and demand for 3 or 4 years—assuming the Cuban supply is not recovered. It will probably be much longer before the world sugar price again reaches the low levels of last year." At the global scale, cane-sugar production could meet demand, but it would take two to three years to come on-line and improve world supplies. "No immediate large-scale increase appears to be underway anywhere."[51]

Acutely aware of the political costs and the "sensitivity of the sugar situation," the USDA's director of information launched a public relations campaign in June called "Information on Sugar." Undersecretary Murphy and Lawrence Myers each taped interviews for radio and television broadcast on all the major U.S. networks, as well as for the Voice of America. In addition, the USDA press service gave "much wider than normal distribution" to statements concerning sugar, disseminating press releases to numerous outlets, among them the Associated Press and UPI, and hundreds of magazines, trade journals, farm publications, and agricultural radio programs. By late fall, after the beet harvests, the USDA found that the world sugar situation had not improved, largely due to a drought in the Soviet Union and hurricane damage in Cuba. Thus the expectation that the "free world" would obtain 3.5 million tons from "Bloc countries" was unlikely to be fulfilled, unless "they need foreign exchange bad[ly] enough to restrict internal consumption."[52] Therefore, consumption at the global scale continued to exceed production; thus, prices were expected to remain high and world stocks to decrease for the third year in a row. Balancing production and consumption would "require another sizable increase in production next year." Part of the remedy—already enacted earlier in the year and welcomed by Florida producers—was the lifting of production restrictions on mainland sugarcane growers for the 1963 and the 1964 crops.

A "Bitter War" within the U.S. Sugar System

With the foreign-quota provisions of the Sugar Act due to expire at the end of 1964, Congress was expected to pass legislation that year to continue foreign quotas in some form so as to maintain the overarching structure of the Sugar Program. Because the U.S. market was entirely managed—and managed as an entirety—failure to legislate with regard to foreign quotas was expected to cause chaos in the whole sugar system, including both foreign and domestic sources. Uncertainty was exacerbating tensions within the system among various factions of producers, processors, and growers. A USDA administrator described the problem: "The almost irreconcilable dilemma of reserving a substantial place in our market for Cuba while encouraging other foreign countries and domestic areas to provide that sugar in the meantime, has after three years built up tremendous pressure by a number of parties with diverse interests."[53] In March 1964 the *New York Times* headlined its financial pages, "Beet and Cane Sugar Men Are Waging a Bitter War Over U.S. Market" (Maidenberg 1964, F1).

The battle lines in this "war" were numerous; the most hardened divided the beet-sugar industry—growers and processors—from the urban refining industry, which processed cane, primarily from foreign sources. While the beet industry was located predominately in rural areas of the U.S. Midwest and West, and cane growers in the rural South, the cane-refining industry was mostly urban and coastal. Representing Massachusetts, Congressman Thomas ("Tip") O'Neill wrote to President Lyndon Johnson and Secretary Freeman to protest the expansion of "the highly subsidized beet sugar industry."[54] O'Neill complained that by withholding the global quota from foreign suppliers and encouraging the growth of the beet industry, Secretary Freeman was depriving domestic cane refiners of their raw material. He asserted that this had "a direct and injurious impact on employment and business in my State, on the Port of Boston, and on the sugar refining and sugar using industries of my State." He argued that refinery employees worked year round and were well paid in comparison to seasonal beet-processing employees, and further, that U.S. exports to sugar-producing Latin American countries would be curtailed if sugar imports were reduced. Finally, he reminded the president and Freeman that only eighteen months earlier domestic quotas—to which beet growers had agreed—had been fixed for a period of four and one-half years, to expire December 31, 1966. Yet now, with the allocation of the global quota in flux, "leading elements of the beet sugar industry have already announced that they will seek to capitalize on this situation to augment their own quotas." A similar protest was lodged by the Harris County AFL-CIO on behalf of longshoremen, warehousemen, and refinery workers in Sugarland, Texas, and the surrounding Gulf Coast, where an estimated fifteen to twenty thousand people were employed in handling imported sugar, "seventy five per cent being organized." Cutting sugar imports "would produce an immediate and detrimental effect on labor in this city and our ports."[55]

The *New York Times* confirmed the basis of refiners' fears, reporting that "the beet sugar industry is sinking vast sums of money into processing plants across the nation." With respect to shares of the U.S. market, "processors are asking for 750,000 more tons for 1965. Since this is an election year, many believe that the domestic beet as well as cane growers may be given a larger share of the sugar market at the expense of foreign suppliers." Because the mainland supply was relatively small, most refiners had traditionally processed offshore cane, but they were "not particularly opposed to the increase in cane sugar acreage in Florida or Louisiana or anywhere else for that matter"(Maidenberg 1964, F1). However, mainland cane growers were not subject to the restriction on producing "direct consumption" sugar that offshore cane growers faced and which the refiners preferred. Even if mainland sugarcane growers' interests did not conflict with those of the refiners, they faced another problem identified earlier in the century: expansion depended on the propagation of seed cane, which required "making haste slowly" (Dacy 1929, 8). Although beet sugar was considered less "economic" than cane sugar with respect to production costs, the differences in propagation methods meant that beet growers could respond more quickly to market signals and increased allotments. Beet growers could purchase and plant seed, responding rapidly to expanding acreage and marketing quotas within a single season. Cane growers, in contrast, required a season to produce a "seed cane" crop, which then had to be cut into sections and buried. The new stalks that sprouted from the bud ("eye") on each section required another season to mature. Thus, beet growers could beat cane growers in the race to establish the "historical" production acreages on which the government based its system of allocation.

For example, consider the Immokalee Sugar Growers Cooperative Association of Florida. In May 1964, Congressman Claude Pepper of Florida, a former Senator, wrote to Secretary Freeman on behalf of the members of the association, who were protesting the allotment of thirty-three thousand acres to beet growers in Maine.[56] Although sugar had been recently in short supply, mainland cane growers anticipated that if Congress did not rewrite the Sugar Act, they would be subject to acreage restrictions in 1965 based on 1964 production history. In his reply to Congressman Pepper, Secretary Freeman summarized the situation:

> We understand that members of the Association have planted 2,000 acres of 1964-crop sugarcane to be harvested as seed this fall. They plan to use this seed in the planting of about 33,000 acres of cane for sugar for 1965-crop harvest. They urge the Department not to restrict plantings earlier than December 31, 1964, so that their acreage objective can be met.[57]

However, the USDA had already announced May 1, 1964, as the cut-off date for determining the 1965 proportionate shares that would be used in the event of acreage restrictions. Whether acreage would be restricted was a different question, yet to be determined.

Officials in the USDA were frustrated with Cooley, and both the administration and Cooley were stymied by competing sugar interests. An internal USDA memorandum regarding sugar meetings held in June noted that "initially Cooley refused to hold hearings" because he "was deeply concerned at the volatility and emotion and political difficulty of a sugar bill at this time." At the meeting, Cooley proposed that the existing Sugar Act be extended by one year, but Freeman thought that this would not be acceptable to "the beet people," who were in a strong bargaining position with respect to increased quotas. Both the administration and Cooley knew that the sugar question would be used for partisan purposes, which made it difficult to advocate domestic production controls: "It was generally agreed that a solid Republican group in the Committee plus a half dozen Democrats subject to local sugar pressures meant that it was doubtful what the Committee would finally bring out."[58]

As hope for sugar legislation dimmed, the secretary of state and the secretary of agriculture prepared a memorandum to inform the president of the impasse. Foreign quotas were due to expire at the end of the year, and cane-sugar refiners and the beet-sugar industry had "taken opposing positions as to the division of the U.S. market between domestic and foreign producers."[59] Unable to reconcile their differences, they appeared "to have decided that they would prefer no legislation rather than the kind of legislation the opposing faction will acquiesce in." Each was capable of blocking legislation, leading to deadlock. In August Cooley conveyed his pessimism to Freeman, noting that "it appears that representatives of the different segments of the industry and the agencies of the executive branch of the government, have not been able to compromise. The situation at the moment looks rather hopeless."[60] He noted that domestic cane producers were willing to cooperate by reducing production but beet producers were not. By September USDA officials were worried that failure to enact legislation would lead to chaos in U.S. markets, which would be vulnerable because the world market price had fallen rapidly: "This impaired demand together with its implications [for] domestic sugar prices would occur at the time when the sugarcane producers in Louisiana and Florida are harvesting their crops and endeavoring to market them. It would be difficult, if not impossible, to hold domestic prices high enough to avoid financial ruin for many farmers and processors, especially for the relatively new ones."[61] Amazingly, less than a year had passed since high sugar prices had been deemed a political liability for the Kennedy administration.

Risky Business in Florida

Though sugarcane acreage increased substantially in Louisiana while production controls were lifted, it did so at a much faster rate in Florida. From 1961 to 1962 sugarcane acreage in Florida more than doubled, while Louisiana acreage increased by 10 percent.[62] Acreage allotments, initially established for the 1963 crop, were increased by 10 percent because of a damaging freeze, and in May crop restrictions were lifted for that year and the next in an effort to encourage maximum production in view of the world shortage of sugar and high sugar prices. The unrestricted 1964 crop in Florida represented a 45 percent increase from the previous year, while Louisiana's crop grew by only 10 percent. The combined growth in production in the two states led growers to anticipate that restrictions might be placed on the 1965 mainland cane acreage. While planters in the two states shared some of the difficulties of cane propagation, they faced dissimilar cultivation conditions because of differences in the growing season. A key distinction was that Florida growers were able to replant damaged cane within the same growing season, whereas Louisiana planters could not.

Though Florida was the relative newcomer, it was now in the firm embrace of the transnational networks that once included the Cuban industry and continued to include Puerto Rico, the Dominican Republic, and others. Central to this was Czarnikow-Rionda, which until 1961 was primarily engaged in importing raw cane sugar and molasses to U.S. refineries; in 1958 it ranked second in size as a sugar trader after Galban, Lobo & Company (*Business Week* 1961). While profitable capital investment was the raison d'être of these corporations and their interlocking directorates, vital personal and political issues were at stake as well for the people involved. For example, Michael J. P. Malone, the vice president of Czarnikow-Rionda who was most closely linked to its Cuban operations, was—after the Cuban Revolution and the expropriation of the company's assets there—closely engaged with the transition of former Cuban employees to life in exile and with the development of new sugar companies in south Florida.[63]

Malone, who had been responsible for the company's diversification into cattle in Cuba, now applied this expertise to south Florida by establishing the New Tuinucú Ranch near Belle Glade, Florida. Named for the Tuinucú Sugar Company, one of the Rionda family's first plantations, incorporated in 1891, the ranch was intended "to keep operating gainfully in order to obtain income which would be non-taxable due to our Cuban losses; and to maintain a financial solvency for our return to Cuba."[64] Malone was also in close touch with James Monahan, a senior editor of *Reader's Digest* and co-author of *The Great Deception: The Inside Story of How the Kremlin Took Over Cuba* (1963), also excerpted in *Reader's Digest*. Their correspondence suggests that prior to publication Malone reviewed the "rough, uncorrected copy of what is still going through the mill" and that they later collaborated in the "hope [that] we may be able eventually to do something with the Guantanamo exile government idea" as well as "the Cuban training schools."[65]

So it was Malone, committed to the exile community and involved in anti-Castro activities, who represented the Wall Street offices of Czarnikow-Rionda in developing vertical links between the expanding Florida sugar industry and downstream users of molasses. Without his efforts, it would have been difficult to expand so quickly and successfully. For example, with Malone's help, the Sugar Cane Growers Cooperative was able to find several outlets for its molasses, including Cargill, Inc. in Minneapolis. Malone's memoranda to co-op representatives contained detailed instructions regarding the workings of molasses markets. Whereas four to five million gallons of molasses had been shipped from the EAA in previous years, Malone estimated that nearly nine million gallons would be marketed in 1961 and predicted that in the near future this figure would be much higher.[66] He explained that Florida molasses would be priced in relation to New Orleans molasses, with which it would compete; in 1962 Florida molasses was discounted 2.5 cents per gallon to reflect the difference in transportation costs between the two.[67] With visions of a Florida molasses glut, he noted "the great need for an orderly market" and counseled producers to look for some alternative uses. To that end, co-op members queried a Brazilian sugar firm as to whether sulphured molasses could be used as animal feed and were presumably pleased with the response that it was fed without harm to cattle, horses and mules; soon thereafter, Malone reported that local ranchers wanted "to utilize some amount of molasses."[68]

In addition to the competition between Louisiana and Florida, other rivalries arose as well, such as that between Czarnikow-Rionda and Galban Lobo, which played out in the south Florida landscape. As Alfonso Fanjul Sr., reported to Malone, "I notice Mr. Falcon is constantly in Talisman office; so I imagine he is trying to compete with us not only in sugar but also in molasses in representation of Lobo. I hope that our endeavors will crystalize [*sic*] not only in being able to handle the molasses of Osceola and the Coop., but also of Talisman, which I anticipate is going to be difficult."[69]

Although acreage and marketing restrictions were suspended for 1963 and 1964, Florida investors were anxious about the scale of the operations being developed and the extended length of time between land preparation and first harvest, during which circumstances could change. Uncertainty mounted as new sugar legislation failed to materialize from the House Agricultural Committee. As they made haste to plant, Florida's sugar growers sought to gain assurance from the USDA that their efforts would not be in vain. With the possibility of acreage restrictions looming, some operations appeared especially vulnerable. Unlike the situation in the beet industry, in which a portion of the quota was reserved for new producers, recently established cane growers were not guaranteed a share of the quota. Several examples serve to illustrate the difficulties faced by new producers in Florida.

The first example, Talisman Sugar Corporation, established by Cuban expatriates with the backing of Henry Ford II, as noted earlier, was "on the horizon" in 1961. In January 1962 the plantation was taking shape under the direction of Fernando de la Riva, an exile who had been the second largest sugar producer in Cuba in 1960. At that time thirteen thousand acres were being planted, "probably the largest cane-planting operation ever carried out at one time in the United States" (Phillips 1962, 57). By late 1963 Talisman was already struggling. One of the key figures in the effort to salvage Talisman from bankruptcy was William Pawley, who had served the Eisenhower administration by asking Batista to step down in 1959. After the revolution, Pawley kept his hand in political activities with respect to Cuba. He was instrumental in the planning stages of the Bay of Pigs and has been credited with providing critical intelligence during the Cuban Missile Crisis (Pfeiffer c. 1970). He also had a hand in assisting the economic transition of Cuban investors to the south Florida sugar industry, stepping in to help reorganize the Talisman Sugar Corporation by using his significant political capital on behalf of the company. Not coincidentally, at least ten of the Cubans working at Talisman had taken part in the Bay of Pigs invasion (Phillips 1962).

In December 1963 Senator Spessard Holland wrote to Secretary Freeman on Pawley's behalf to ask how future quotas would be calculated, noting that if the three-year history formula were used, as a newcomer Talisman "would be crippled before it actually [had] time to get underway." Holland also forwarded to Freeman a memorandum from Pawley that detailed the brief history of the operation: "Canals were dug, roads built, water control established, and about 11,600 acres of land was planted in cane. The whole program was done on a crash basis at very high cost, and the management spent more than was budgeted for the mill." Even so, the mill was completed two months late, with disastrous consequences, because in 1962 "the worst freeze in sixty years hit the Glades area December 8th, and as there was no mill to grind the cane, it deteriorated and was lost." At this point the company had "over $10,000,000 of pressing obligations" so "personnel could not be held [and] the cane stood frozen in the fields." For most of 1963 Ford failed to find sufficient backing to rescue the company from bankruptcy until, as Pawley explained, "after considerable time spent with the principal sugar producers of the area, I decided to try to reorganize Talisman." Pawley intended to advance $4 million in equity capital and to secure another $8.5 million equity loan to save Talisman from foreclosure. This investment would be futile, however, if "quota restrictions were placed against this property and we were unable to plant the 25,000 acres now proposed."[70] Pawley predicted that 5,000 acres would be planted by the end of April, and the rest by the end of 1964.

Responding in early January, Freeman merely reminded Senator Holland that the Sugar Act required the USDA to restrict acreage on the basis of past production history if necessary for the purposes of managing the U.S. sugar market.[71] At that time it was too soon to know whether the 1965 crop would be restricted. Not until April 3, 1964, did the department announce that it was considering restrictions and that, for purposes of production history, only the acreage planted by April 15, 1964, would be counted. A USDA administrator explained that the previous year's crop of mainland cane exceeded the marketing allotment by three hundred thousand tons and that "acreage to be harvested in one of the two States, Florida, is expected to be 60,000 acres or 40 percent larger than in 1963." As a result, even if Congress increased marketing allotments for mainland cane as requested by the USDA, restrictions would be needed. The April 15 planting deadline "was recommended by both the American Sugar Cane League (Louisiana growers), and with certain minor variations by a number of the established Florida producers." Thus, with less than two weeks warning, Talisman and several other new sugarcane enterprises in both Florida and Louisiana were caught in an economic squeeze, with too much capital sunk but too little cane planted to receive adequate proportionate acreage allotments.

Senators Smathers and Holland sought a meeting with Secretary Freeman, for which he in turn requested preparatory briefing: "It appears that what Senators Smathers and Holland want to talk to you about is the number of acres Florida can plant to sugar cane."[72] Since the April 3 announcement, the USDA had received two communications, one from Pawley "urging that the April 15 deadline be extended" and another from the Immokalee Sugar Cane Growers Cooperative Association, "one of five other groups that would like to build sugar cane factories and plant extensive acreage of sugar cane in Florida." All told, if there were no restrictions, it was expected that one hundred thousand additional acres would be planted, which would require a severe cutback of acreage for both old and new mainland cane growers the following year. "I have been unable to find out whether Smathers and Holland have any gimmicks in connection with this." Freeman's jottings scrawled across the memorandum identified the key players and their positions: "La. sugar cane league [sic] strong for cut of [sic]—Five largest strong for it—U.S. Sugar Corp, Okeelanta, Belle Glade Coop, Moorehaven Coop, Osceola Sugar Co. favor cut of [sic]." His notes indicated that Pawley had requested at the very least a two-week reprieve to plant an additional 750 acres by May 1, 1964.

Two days later Freeman reported to his staff on his meeting with Congressmen Smathers, Holland, and Rogers, all from Florida, who "urged on me that we proceed with some care so that a heavy investor will not find his investment destroyed." Holland's primary emphasis was on Pawley and Talisman, "which had been restored with great advantage to the community at a $9,000,000 investment and which must have more acreage to be efficiently operated." The six other operations, including the Immokalee Sugar Cane Growers coop, for which the senators were advocates, were "another thing. Here they feel is the first real success in breaking the almost complete dominance of the U.S. Sugar Corporation and the large operations which have completely dominated Florida sugar. Therefore they are sympathetic to the prospect of extending the cutoff date so far as possible so these groups can also produce."[73]

These were not small or poor producers but "young men who moved into this field which previously had been tightly held." Freeman explained to the senators that new acreage "granted from here on out would have to be taken away from old growers. Thus, the matter was left up in the air." Meanwhile a staff member answered Freeman's request to investigate the availability of seed cane in Florida. "I don't think I left any tracks; I got the information from the best informed sources in Fla. He agrees with what Mr. Pawley told the Secretary that there is no seed cane—that Mr. Pawley probably has very little too." The implication of this intelligence was that without sufficient supplies of seed cane, not much additional acreage could actually be planted in sugarcane in Florida that year.

The second example, Big B Sugar Corporation, was a case in point. By April 1964 more than seven million dollars had been invested to convert part of the thirty-thousand-acre Big B Ranch to sugar production, including both a cane plantation and mill: "Bridges and pumping stations have been built. 12 miles of roads have been completed as have 6 miles of main canals."[74] However, at that time only fourteen hundred acres of sugarcane had been planted, to be used the following fall for seed, with no economic return until the harvest of fall 1965. Big B's investors appealed to Undersecretary Murphy, asking him to "keep in mind" that reimposing quotas would destroy the capital that had been sunk into Florida sugar by numerous companies and cooperatives. They proposed the creation of a tonnage or acreage reserve for new sugarcane producers similar to that created for sugar beet producers in the Sugar Act of 1962.

The third example, Atlantic Sugar Association, was one of the cooperatives that was struggling that summer. It included forty farms with approximately seventy investors and was backed by the Columbia Bank for Cooperatives and Allis-Chalmers. In September the president of the association, Sam Knight, met with Secretary Freeman to ask his consideration with respect to marketing allotments and quotas. Afterward he wrote to Freeman that he felt "very much relieved having the opportunity to explain our case to you and feeling how you understood the problem we are facing."[75] Had Knight been able to read the handwritten note that Freeman scrawled across the copy of his letter before forwarding it to Undersecretary Murphy, he might have been even more reassured: "This is the man I talked to you about re sugar. It seems that these guys need help. I couldn't care less about the fat cats."[76] In short, the secretary thought "little sugar" more deserving of political support, a common populist sentiment that "big sugar" manipulated in its public relations and Congressional lobbying activities for decades.

USDA staffers determined that the situation for Atlantic was better than for most of the newcomers, as it was "the only new sugar factory in Florida which got started fast enough to get covered in." That is, Atlantic had established sufficient cane acreage to be viable; however several of the new processors, including Atlantic and two other cooperatives, would not have marketing rights until January 1965. A possibility for the Atlantic mill, and

any other in the same situation, would be to proceed that fall in delivering sugar to a refiner, who would then charge it against the 1965 allotment.
In October Freeman wrote to Knight to make that suggestion and to offer
words of encouragement: "I know that you have a fine new sugar storehouse
and the Department will make every effort to have a preliminary 1965 marketing allotment available to your Association by New Year's Day."[77] At the
bottom of the page he appended a postscript, "We will do all we can to help
you. OLF."

The Sugar Act of 1965: Florida Secures Its New Share

Without a revision of existing legislation, USDA administrators had to enforce the terms of the 1962 Sugar Act, so in the fall of 1964 marketing allotments were reinstated, with cane acreage reduced by 13 percent and beet
acreage by 5 percent for 1965 (Heston 1975, 157). This was the prelude to the
political maneuverings leading to the Sugar Act of 1965. By March of that
year, the domestic industry was able to reach an agreement that called for
an immediate quota increase of 375,000 tons and 205,000 tons for domestic
beet and cane, respectively (*New York Times* 1965a). Reassured, Alfonso Fanjul Sr., wrote to George A. Braga,

> In reference to our Florida investments, this undoubtedly with two hurricanes
> and a severe freeze is not a good year, but on the other hand I feel, especially af-
> ter my talks this morning with Harry Vaughn in Washington, that if we get, as
> he seems to be confident, the Sugar Act to increase the Mainland Cane Quota
> by 205,000 tons, the increases in our business will be consolidated by defi-
> nite quota. From now on we should strive to have the factory operate more
> efficiently and the agricultural part to do the most efficient agricultural job,
> leaving as you have to in any agricultural venture only to the uncontrollable
> elements the risks that in agriculture we will always have.[78]

The administration entered the fray somewhat less reform-minded than
it had been in 1962. Two years of high sugar prices and the public perception
that the global quota was partly to blame meant that the USDA and the executive branch were ready to take on neither the House Agricultural Committee nor the Sugar Czar. Meanwhile Cooley was not in a hurry to proceed
with sugar hearings in committee. The central problem with respect to stabilizing the U.S. market, according to USDA administrators, was "how to
support the domestic price while assuring adequate supplies of sugar in the
absence of both a) the Cuban warehouse and b) a variable import level."[79]
As the world price of sugar declined relative to the U.S. price, foreign ex-

porters would seek to fulfill their already-promised U.S. quotas. Previously Cuba had provided flexibility to the system. Now, since the entire foreign market was allocated, flexibility would have to come from domestic areas. By the terms of the Sugar Act, the government could require domestic areas to carry inventories at a high level to maintain price.

In addition to the decline in the world sugar price, administrators identified another impediment to stability in the sugar system: "Recent contacts with sugar users suggest that the inroads made by non-sucrose sweeteners may be greater than current statistics reveal."[80] The overall outlook was not improved by "the specter of a sugar price war that Premier Fidel Castro raised" by "referring to the possible sharp rise in the production of this vital commodity." By pressing Cuba's "comparative advantage" in sugar production, Castro was ironically challenging the "free market" economies: "Under competitive conditions with any capitalistic nation we can ruin them by producing sugar" (Maidenberg 1965, F1). With increasing uncertainty for domestic producers, rumors abounded. In December the marketing manager of the Sugar Cane Growers Cooperative of Florida wrote to Malone, "We have received reports that Cuba is making an effort to be nice to Uncle Sam so that she can also edge back into the trade. This, on top of present marketing conditions I am sure you will agree, would be disastrous [sic]."[81] In January a reporter from *Time Magazine* called USDA officials to inquire about the possible resumption of Cuban sugar imports. It had been reported that, in asking Florida sugar growers to reduce their acreage, the Florida Commission of Agriculture had given the possibility of Cuban imports as the reason.

By midsummer, USDA administrators were attempting to outline the "chronology of events if there is no sugar legislation."[82] According to this scenario, certain mainland cane producers were especially vulnerable. The first "event" would be a 20 percent restriction of crop acreages, which would mean that "new producers and processors in Florida and Louisiana [would] be especially distressed." Though the established producers in Florida were the most profitable of the domestic cane industry, they too would be adversely affected because—as vertically integrated growers and processors—they would be required to hold large inventories of raw sugar.[83] If the Sugar Act were to expire, the USDA would have difficulty supporting domestic prices through 1966, with disastrous results for the domestic industry. Only two regions—one producing cane and one beets—would remain viable. "Some farmers in Florida and possibly some in the Red River Valley might be able to compete but on a belt-tightening basis."[84]

Vulnerable cane producers appealed to Secretary Freeman for special legislation to protect new growers and processors. Reverend George Speidel, a member of the Atlantic Sugar Association who had 172 acres planted in cane, wrote to ask that the protection guaranteed new sugar beet growers be extended to cane farmers. "I am like many others in Atlantic Sugar Association mill, a little person, financially. Now with a loss of $28,892.00, I am truly a hardship, disaster, struck case, and so is Atlantic Sugar Association."[85] Another member of the association, near bankruptcy with 415 acres planted in cane, explained that after two hurricanes and two freezes "our first year's experience teaches us that Florida weather is undependable."[86] However, after industrywide meetings, USDA personnel concluded that new producers did not have the support of the established mainland cane growers, who "believe that the Department has been too considerate of the new growers with respect to both the 1965 sugarcane acreage program and the 1965 sugar marketing allotment program. Their spokesman stated that the industry would oppose any special legislative consideration for the new growers."[87]

Through September Cooley continued to delay committee action on the Sugar Bill. Finally, in October, Cooley and the Agricultural Committee moved the bill to the floor for consideration, "the final controversial measure on the House agenda this year" (*New York Times* 1965b, 62). One controversial provision in the bill as written was a ten-thousand-ton quota for the Bahamas, which did not at the time grow sugar cane. This was to benefit the Owens-Illinois Glass Company, which had an exclusive arrangement with the Bahamian government to go into the sugar business if the quota were secured. When the bill passed the House and moved to the Senate, one point of debate was the difference between the U.S. price and the world price of sugar, 6.8 and 3.4 cents per pound, respectively. With 35 percent of the U.S. market allocated to foreign suppliers who would receive the U.S. price, several senators protested the fact that U.S. consumers "would pay a $1 billion subsidy to foreign countries" (*New York Times* 1965c, 63). The Senate's version of the bill was similar to that of the House in raising domestic quotas by 580,000 tons through 1971, but differed in that it only extended foreign quotas through 1967 rather than through 1971. This was done in an effort to split consideration of foreign and domestic quotas. "For years, however, Mr. Cooley has used the domestic quotas as a lever to win approval of his version of foreign quotas. He was not considered likely to give up this weapon without a fight" (*New York Times* 1965d, 43). Cooley's strategy meant that domestic growers—cane and beet—had to work in concert even though their interests differed.

After reconciling the differences between the House and Senate versions in committee, Congress forwarded the bill, which was signed into law in

November. The Sugar Act of 1965 provided "an immediate outlet for excess continental beet and cane supplies" by substantially increasing fixed quotas, but "to compensate for this increase those areas were not permitted to share in the growth of the consumption estimate for approximately four years" (Gerber 1976, 138). The latter provision was included at the behest of domestic refiners to protect their supply of foreign raw cane sugar. In addition to the crop-specific quotas, special provisions answered the concerns of vulnerable growers in the form of an "amendment to the Act which authorize[d] the [USDA] Secretary to allocate acreage to new producers for the purpose of relieving hardship."[88] In addition, Section 205(a) of the Sugar Act was amended to include an increase of 16,000 tons in the marketing allotment for mainland cane sugar, to be allocated under the provisions of the amendment and by the recommendation of the USDA. Of the 16,000 tons, 6,200 tons were to be allotted to a particular Louisiana processor on the basis of congressional intent as stipulated in the amendment, and "the balance of such reserves should be used to alleviate hardships and inequities in other areas."[89] Nearly 10,000 tons remained to be disbursed, so hearings were convened in late November to determine which processors would benefit.

The USDA gave priority for increased allotments to the newest processors, of which three had just started producing sugar in 1964. It granted two Louisiana companies and the Atlantic Sugar Association increased marketing allotments equal to their total excess inventory. Next on the list were those that had started production in 1963, South Florida Sugar Corporation and Talisman Sugar Corporation, both of which also received marketing allotments equal to their excess inventory. Finally, the three new processors that had started before 1963 and were therefore more established—Glades County Sugar Growers Cooperative, Osceola Farms, and Sugar Cane Growers Cooperative of Florida—each received a marketing allotment equal to a fraction of their excess inventory, approximately one-sixth of their excess tonnage. In contrast to the situation a mere five years before, as 1965 came to a close there were nine Florida processors instead of two. Total mainland cane-sugar production had increased from about 700,000 to 1,100,000 tons. Of that, Louisiana produced 557,000, and Florida—which in 1960 was a distant second—now produced 542,000 tons.

For the Florida industry, the most important element leading to this spectacular growth was the shift in geopolitical relations between the United States and Cuba. Arguably for Cuba, a major consequence of this relationship was political and economic instability, for it "was dependent on the unpredictable outcome of sugar tariff and quota contests in the U.S. Congress" (Benjamin 1990, 66). Without contending that the 1956 Sugar

Act triggered the Cuban Revolution, it is important to acknowledge the widely held *perception* within factions of the U.S. government that cutting the Cuban quota would be economically and politically destabilizing and, by extension, a national security threat. The State Department was clear on that, while the USDA remained solely concerned with the "great political strength of [domestic] sugar interests." The Sugar Acts reflected the contradictions embedded in the relationships between the two countries, between Congress and the executive branch, and between the State Department and the USDA, which regional agricultural interests were able to exploit to their advantage.

Ironically, the agricultural interests in Florida included the Cuban producers who were able to make the transition to the Everglades Agricultural Area. In December 1961, one of them, Alfonso Fanjul Sr., wrote to the head of the Czarnikow Ltd. Office of London: "The potentials that Florida has in sugar are such that we in Cuba never could have realized them until we came here and saw them and I hope to look back in the midst of our tragedy and say how fortunate we were to decide to go into it."[90] Many of his exiled compatriots were far less enthusiastic about Florida's potential. Thinking ahead to the day of Fidel Castro's deposition, which they assumed would be soon, they felt that any investment in Florida sugar production was "a treason to the fatherland" (*Avance* 1961, 49). Writing from exile in Miami in 1962, Jose Miró Cardona—who had been the first prime minister of postrevolutionary Cuba—published a booklet entitled *In Defense of the Position of Cuba as a Supplier of Sugar to the United States Market*. As president of the Revolutionary Council of Cuba, which was akin to a government in exile, Miró Cardona pleaded against the reduction of Cuba's quota, which "Would be an Insurmountable Obstacle to the Reconstruction of Cuba After its Liberation" (Miró Cardona 1962, 21; uppercase in original). Taking a contrary position, Fernando de la Riva, president of Talisman, stated "The sugar we produce here has nothing to do with the quota for foreign countries and we hope that the United States will restore to Cuba her former quota as soon as the Castro Communist regime is overthrown" (Phillips 1962 57).

An article in *Avance* entitled "Cuban Investors Guide Sugarbowl Expansion in Florida" recounted numerous ironies. For example, it noted that the proprietors of Talisman made a "sensational purchase" when they acquired

16,000 acres of virgin swampy land at the fantastic price of $21,800,000, considered the largest sale that has been realized in the history of this state, for which in 1819 North America paid the sum of $6,674,057 to the Spanish empire. The operation . . . paid three times what was paid for all of Florida, for a relatively small strip of land that has never been cultivated and will have to be drained in order to develop a plantation. (*Avance* 1961, 8)

In light of the seven *centrales* then under construction, a Miami public official, quoted without attribution, said "the state of Florida should give thanks to Fidel Castro for its internal sugarbowl expansion" (*Avance* 1961, 9). The article emphasized the factors that had previously inhibited the Florida industry. "The most astute minds of the State of Florida always have cherished the dream of raising a gigantic sugarbowl in this region. They have viewed with disgust that Louisiana would be the region preferred by government planners" (47). As to whether it was treasonous to invest in Florida now, the debate was not just between those who feared that Florida's expansion would "harm Cuba as soon as it reestablishe[d] a democratic state" and those who argued that it had no bearing; an emerging viewpoint argued "that the participation of Cubans in producing the Florida sugarbowl will allow Cuba and Florida in the future to form political organizations to jointly address Congress and the White House." That is, some thought the sugar diaspora could foster a transnational commodity lobby that would engage the U.S. government on behalf of producers on both sides of the Florida Straits. However, "a large part of the Cuban exiles strongly criticize[d] the money invested in Florida sugar"; concerned about "not only the capitalists but the Cuban sugar technicians, many of them thinking that they are not going to return to Cuba," who might unintentionally swing the balance in Florida's favor (49).

CHAPTER SIX

A Restructured Industry

While the Sugar Act of 1965 afforded some relief to domestic producers, the elimination of Cuban sugar from the U.S. market was not the cure-all for the Florida industry's sugar-quota blues. Whereas the Florida industry expanded rapidly in the immediate post-revolutionary period, 1960–64, the latter half of the decade found producers engaged, not always successfully, in a political struggle to maintain their gains in acreage allotments and marketing quotas. After 1965, the industry and the federal government had to come to terms with the latest permutation of the sugar question, which included a greatly expanded cane industry in Florida as well as the context in which it came into being—political revolution in the Caribbean. At the same time, many more countries were now supplying sugar to the United States, all making strong moral claims for access to the U.S. market in the context of the Cold War, a situation over which Cuba cast a long shadow. Florida producers also made moral claims regarding their right to larger quotas, arguing that they had answered the national call for increased production at a time of worldwide shortages. To complicate matters further, many workers in the Florida industry were recent immigrants from Cuba, who argued that the livelihood and well-being of the nascent Cuban-Floridian community depended greatly on the health of Florida's sugar industry. The twenty-year period from 1965 through 1985 that is the focus of this chapter saw a profound restructuring of the sugar market, including the rise of high-fructose corn sweeteners and the decline of the Caribbean sugar agro-industry.

Florida Sees the Future

In May 1964, a few weeks after new sugarcane growers finished their feverish planting to establish maximum acreage on which their 1965 proportionate shares would be based, the state of Florida, through the auspices of the Institute of Food and Agricultural Sciences at the University of Florida, convened the DARE meetings. The 1964 meetings, named for the statewide initiative, Developing Agricultural Resources Effectively, included a two-day session devoted entirely to the cane-sugar industry. This session offered a chance to take stock of an agro-industry recently transformed in numerous ways—geographically, economically, politically, and socially—and served as a venue to contemplate the near future, specifically "an estimation of sugar production in 1975" that had been called for by "Operation DARE" (DARE 1964, 7). The session was organized by the state's Sugarcane Committee, composed of agricultural economists and agronomists from the Florida Agricultural Experiment Station in Gainesville, the Everglades Experiment Station in Belle Glade, and the USDA Sugarcane Field Station in Canal Point, as well as John B. Boy, who was at that time vice president of USSC, and a few growers representing several EAA sugarcane cooperatives. Yet despite the assembled expertise, the task of peering into the future proved daunting, for the simple reason identified at the top of their final report: "The future opportunities for the sugar industry in Florida depend primarily on what happens in other states and in other countries of the world" (1).

In various ways and at several geographic levels, the political-economic terrain facing the industry in 1964 was much more complex than it had been five years earlier. First, a dozen or more countries, primarily Brazil, the Dominican Republic, Mexico, and Peru, were now filling the large sugar quota previously allocated to Cuba. Second, whereas USSC had previously dominated the EAA in terms of acreage and output, now the industry was much more diverse, in terms of structure, scale of enterprise, and geography. For example, the location of much of the land that had recently been brought into production differed in terms of microclimate and soils from that of the older, established farms and plantations. Third, a major shift was taking place in the domestic sugar market. Between 1955 and 1959, U.S. sugar use was evenly balanced between industrial and nonindustrial users; that is, at the national level, sugar purchased for direct household consumption was equal to that sold to industrial food processors. By 1964 the balance had tipped toward industrial users, who, because they dealt in large volumes, were especially sensitive to small differences in price; the balance would shift even farther toward industrial food manufacturers by 1975.

The DARE report considered three possible scenarios for 1975 quotas. Even in the most optimistic of these, Florida's "assumed allotted production" in 1975 was predicted to be 710,000 tons, while by 1968 the state would have production capacity of more than a million tons "if companies now in formation should become a reality." Brushing aside interstate diplomacy in foretelling the future, Operation DARE suggested that "on this basis Florida alone could produce in 1968 almost the total amount of sugar allotted to mainland cane producers in 1975. In such case, Louisiana, and possibly California which has recently asked for a share in the mainland cane allotment, would have to discontinue their sugar production" (DARE 1964, 8). However, contributors to the report identified numerous problems limiting the realization of Florida's potential for sugar production. During the recent period of expansion, 1961–64, cold damage had substantially reduced yields per acre. Expansion had occurred on land farther from Lake Okeechobee, which was subject to earlier and more severe cold. J. W. Beardsley, an independent grower who had been in the business since 1938, argued that there was a dearth of research directed toward the needs of smaller farmers, who instead relied on experience. As a result, for each cultivation practice, there were numerous methods: "In planting techniques, if you ask ten farmers you get at least five answers. In the harvesting of cane, if you ask five growers you will get more than five answers because one man may not use the same techniques more than two weeks in a row" (34).

Two concerns loomed large: adequate quotas and sufficient harvest labor. John B. Boy of USSC addressed the first topic, arguing that domestic growers in general should be given a larger share of the quota, and, more specifically, that the beet producers' insistence on a three-to-one split between beet and cane had to be modified in favor of cane. "We have already expanded in Florida and we are already in trouble. We are not asking for quotas to help us expand in the future, but are asking for help to get us out of the situation which we are already in due to the encouragement that was given us to expand" (DARE 1964, 44). Boy contended that Florida producers deserved federal support because their expansion had occurred in response to "propaganda put out recently concerning possible sugar shortages, which propaganda was put out in order to drive the price of sugar lower. The price is down now and we are presently worried when it will stop" (42). F. C. Sikes, personnel director for USSC, addressed the second topic, identifying "the Achilles heel of sugarcane production in Florida" as "the failure to have an adequate labor supply, particularly for harvest"

(38). He argued that complacency regarding the offshore labor supply was problematic. The problem of maintaining an offshore labor program was made difficult by the poor image of agricultural employers in the public mind, which could be remedied by expanded public relations: "Although you cannot influence organized labor you can work through your church or other community organizations. Once you have done that you can get next to the do-gooders" (39).

With respect to quotas and labor supply, the DARE report emphasized the need for the local industry to unite and coordinate efforts to attain legislation favorable to the Florida industry. Deep cleavages remained between the established and the newer enterprises, but to the extent that they were able to overcome differences, they faced formidable opposition fueled by ideas and ideologies concerning the proper role and functioning of markets and about the U.S. role in the world economy in general and vis-à-vis the developing world in particular. Over the next two decades, the sugar question would gain prominence in successive administrations, as presidents from Johnson through Carter sought to balance U.S. sugar policy between foreign policy initiatives and domestic political realities. Moreover, not only did each administration seek to balance U.S. sugar policy, but they also saw in sugar quotas the means to build and maintain circles of influence in foreign affairs that extended beyond the realm of commodity interests per se.

The Cuban Shadow Over U.S. Sugar Policy

When John B. Boy of USSC mentioned government "propaganda" regarding sugar prices, he was referring to the attempt at deliberate manipulation of sugar markets, both national and international, undertaken by the U.S. government after severing economic relations with Cuba. As noted above, the 1959 Cuban Revolution and its aftermath led successive presidential administrations to view sugar policy and markets as matters of great significance with respect to national security and the conduct of foreign policy. On the one hand, to the extent that the 1956 reduction of Cuba's sugar quota might have played a role in the revolution, there was the concern not to repeat that the experience. On the other hand, now that direct economic ties to Cuba had been severed, the U.S. government aimed to reduce Cuba's income from sugar by attempting to influence the world price. For example, a "top secret" memorandum listing "Possible Additional Measures Against Cuba" included "low level flights, unleashing of exile groups, stepped up sabotage" and "manipulate sugar market."[1] Senator Smathers strongly advocated undermining the Cuban government by devaluing sugar, estimating that if the United States would cut consumption by 50 percent, Castro would fall within days (Crispell 1999). However, the preferred way to manipulate the market was to stimulate production elsewhere, including within the United States.

Thus, in May 1963 National Security Advisor McGeorge Bundy sent a Confidential Security Action Memorandum to Secretary Freeman to tell him that "the current situation in the world market for sugar is a matter of considerable concern to the Standing Group of the National Security Council. The unusually high price of sugar is significant both in its relation to the economic prospects of Cuba and in its relation to restrictions on the production of sugar in free countries, and not least in the United States."[2]

Bundy asked the USDA to prepare a report on the situation and possible alternatives, "bearing in mind the particular interest which we have in preventing any long continuation of unjustifiably high prices for Cuban sugar on the world market."[3] Within ten days he received a "secret report" entitled "The World Price of Sugar," which predicted that prices would remain high for several years and explained that quotas prohibited the marketing of any further increase in domestic cane sugar.[4] However, by July 1963, when the NSC met to discuss the "Contingent Plan to Reduce Price of Sugar," the feeling was that world production had increased sufficiently to ensure a drop in price.[5] Nonetheless, again in April 1964 the State Department warned that "new firmness in the world sugar market can help Castro." Castro's refusal to predict crop levels and his promise to sell more to the USSR "suggest that the Cuban leader may want the world to believe the crop is short." Thus, Castro could "capitalize on the stronger market and even encourage an upswing in sugar prices by stimulating rumors of a short crop."[6]

A second way to manipulate markets was through counter-rumors. Anticipating that "Castro might make an additional fifty to seventy-five million dollars,"[7] by "doing things to give the impression that the Cuban sugar crop is short this year," the State Department and the CIA were enlisted to "combat this effort."[8] In late April, the *New York Times* reported, "Cuba was accused today of deliberate maneuvers to increase world sugar prices to reap major benefits later this year" (*New York Times* 1964, 18). Cuba's purchase of twenty thousand tons of sugar to meet promised delivery in Bulgaria and China was labeled "transparently phony" by these specialists, who also estimated that the Cuban harvest would be 4 million tons, not the 3.7 million forecast by Cuban officials. A confidential memorandum sent the same day to the White House explained that the news story was "a bit premature" but "should help in preventing a rise in sugar prices."[9] Thus, although Cuba no longer played the critical role of swing producer in the U.S. sugar market, the specter of Cuban sugar still strongly influenced U.S. sugar policy, and, conversely, sugar played a key role in foreign policy strategies.

The problem for the U.S. government was that, having "lost" Cuba, it was important not to appear fickle in allocating foreign sugar quotas, yet somehow it was necessary to maintain a measure of flexibility in supply. The case of the Dominican Republic illustrates the sensitive nature of post-Castro sugar politics. In July 1960, when Congress authorized President Eisenhower to cut the Cuban quota, it mandated an allocation of a hundred and twenty thousand tons to the Dominican Republic against Eisenhower's wishes. Until then, most Dominican sugar was sold to European buyers (Tucker 2000). At that time the Dominican dictator Rafael Leonidas Trujillo dominated the Dominican economy, including the sugar industry, owning twelve of sixteen *ingenios* in the country and at least 1.5 million acres of sugar land. Thus Trujillo, who had been in power since 1929, a period of tyranny "which ranks among the most ruthless and efficient in the history of the entire world" (Williams 1984, 465), stood to profit from the U.S. quota. In February 1960, Senator Smathers and William Pawley paid Trujillo a visit, with Pawley reprising the role he played with Batista, again asking a dictator to relinquish power in an orderly fashion, again to no avail.[10]

Eisenhower sought a legal remedy to overturn the mandatory quota allocated to the Dominican Republic. In early fall, before the Cuban quota was cut to zero, he received a memorandum pleading on behalf of the Dominican resistance movement, which sought to defeat Cooley's attempt to assist Trujillo: "Trujillo is causing Dominican people to believe he is more powerful than the government of the U.S. because he can control its Congress through certain elements in it who serve his interests and if the importation of Dominican sugar should be permitted that would be viewed as confirmation of his political propaganda which would be fatal for our democratic future." The accompanying State Department memorandum agreed that "granting this benefit to the Dominican Republic would diminish the prestige of the United States and hinder the conduct of our foreign relations in Latin America."[11] However, in December the president was informed it would be "legally necessary to continue purchases of sugar from the Dominican Republic."[12]

The situation changed abruptly with Trujillo's assassination in May 1961, after which "the Americans feared another Cuba" (Williams 1984, 466). Now the concern was that the 1962 sugar legislation, in reducing the Dominican quota, threatened "to damage seriously our relations with the Dominican Republic" and had "given an important propaganda advantage to the Castro/Communist elements."[13] The fact that the Trujillo government received a larger quota than its elected successor was embarrassing and harmful. In 1963 the elected president, Juan Bosch, was ousted by a military coup and replaced by a military-backed civilian council. After eighteen months the council relinquished power to "a conservative Dominican politician" whose regime collapsed in late April 1965 (Richardson 1992, 95). By May 1965 the United States had twenty-three thousand troops on the island, the beginning of a U.S. military occupation that lasted for a year (Richardson 1992).

But in certain ways the United States did find in the Dominican Republic "another Cuba," as the economic relationship between the two countries tightened through sugar trade. Although there had been substantial U.S. investment in the Dominican sugar industry dating to the 1920s, sugar imports to the United States were small, only 15,000 tons in 1950, climbing to "a late-1950s average of 92,000 tons" (Tucker 2000, 54), spiking to 900,000 tons when the Cuban quota was cut, and settling at around 600,000 tons through the 1960s. More important, that figure became critical to each successive Dominican government, whose economic and moral claims echoed those of Cuba from prior decades. Thus in August 1966, following the inauguration of President Joaquin Balaguer, Secretary of State Dean Rusk advised President Johnson to show support by giving the Dominican Republic the combined shortfalls of Panama and the Philippines—about 118,000 tons—which Johnson did.[14] Again in May 1967, President Johnson agreed to Secretary Rusk's request that he authorize a special deficit allocation of 105,000 tons. Johnson's economic advisor, Walter Rostow, laid out for him the pros and cons: "The special allocation translates itself into U.S. political support which is a stabilizing influence in the DR." However, "other Latin American sugar producers may protest," but "they will receive slightly larger quotas than they did last year" and "there was no hue and cry last year when you gave the DR a special allocation."[15] Although other Latin American countries acquiesced, a "hue and cry" would soon arise from the domestic cane sugar industry.

On March 17, 1968, President Balaguer wrote to President Johnson to remind him of "the singular importance to the Dominican Republic's economic and political stability of the sugar quota," with 70 percent of the country's foreign exchange earnings dependent on sugar. His depiction of "the Dominican Republic as the nearest and most reliable source of sugar supply for the United States market and the only important sugar producing country in the Western Hemisphere that has to depend entirely upon the export of sugar to maintain a balanced domestic and international economy" was striking for the resemblance it bore to Cuba in earlier decades.[16] In accord with U.S. emphasis on development, he stressed not only the country's dependence on sugar but also progress made with respect to agricultural modernization and diversification that had been funded by sugar income. Balaguer's purpose in writing was to request for his country the Puerto Rican deficit of approximately two hundred thousand tons.

However, domestic cane growers, who had produced sugar in excess of their marketable quota for 1967 and 1968, also coveted the Puerto Rican deficit. Without a significant reallocation to them, mainland growers were facing extensive acreage cutbacks. On March 20, 1968, Senators Holland and Smathers wrote to Secretary Freeman, "It is our understanding that efforts are being made to have the Department of Agriculture reallocate the Puerto Rican sugar deficit. In view of the large surplus of Mainland Cane, we earnestly request that the Department take no action until such time as the Congress may consider possible legislation to have such deficit reallocated to Mainland Cane."[17]

Mainland cane growers had senators, congressmen, bankers, and various other industry representatives writing on their behalf for the Puerto Rican deficit, such as the vice president of the Exchange National Bank of Tampa, who wrote,

> In the early 1960's, at the urging of the Department of Agriculture, Mainland Producers responded to a shortage of sugar for the American consumer by building seven new milling facilities and investing an estimated $200 million in land, processing facilities, agricultural equipment and improved drainage. Unfortunately, before these producers could recover their costs, the acreage was cut by 18.1 percent and then an additional cut of 4.76 percent was imposed. As a major funding agency in Florida our bank has been adversely affected by these acreage cuts.[18]

Meanwhile, Dominican Republic representatives used another tactic. In May 1968 the U.S. ambassador in Santo Domingo telegraphed Washington expressing alarm about the behavior of Ambassador Garcia Godoy. Godoy had publicly proclaimed that he "was very optimistic" that the Dominican Republic would receive a quota of 700,000 tons, though "he also expressed belief it would be delayed some time because of Louisiana-Florida complication. I am frankly disturbed that Garcia Godoy is thinking of even the possibility of a total quota of up to 680,000 tons." The problem was that Godoy's public statements had helped to "greatly embed [the] magic figure of 700,000 tons in Dominican consciousness."[19] The Johnson administration had the authority to address Dominican expectations, but could not alter domestic marketing quotas. Palm Beach County commissioners wrote directly to President Johnson to transmit a lengthy resolution in support of the Florida sugar industry, invoking moral claims about the responsibility of the USDA in provoking expansion: "[The] refusal to permit the sale of a substantial part of the effective inventory will drastically hurt the economy of Palm Beach County, and in all likelihood will cause the economic demise of some of the mainland cane sugar mills and farmers heavily invested in producing sugar because the Department of Agriculture threw open the controls in 1963."[20] The resolution formally solicited the assistance of the president, the secretary of agriculture, the governor of Florida and the Florida Congressional Delegation in securing the Puerto Rican deficit on behalf of mainland cane producers.

That same week another voice from Palm Beach County was heard in the struggle to gain quotas for Florida, namely, that of the Cuban exile community residing there. Representatives of five groups—the Cuban Revolutionary Nationalist Front, II Front Alpha 66, Students Revolutionary Directory, Christian Democratic Movement of Cuba, and the Cuban Liberation Army—sent telegrams addressed to President Johnson appealing for his assistance: "The Cuban community in Florida depends greatly on the sugar cane industry for its livelihood. 3,000 families of Cuban workers and technicians of the sugar industry will suffer great hardship if further cut backs in proportionate share acreage and reduction in marketing allotments of domestic cane sugar is [sic] imposed."[21] Merely by reminding the president of their presence in southeast Florida, these groups evoked moral claims regarding U.S. foreign policy and domestic responsibility to Cuban exiles. The transnational identity of the restructured Florida industry muddied the line between foreign and domestic policy and gave industry supporters a compelling argument for supporting sugar as a domestic regional development strategy.

Mainland Cane Gets Cut

By June 1968 the U.S. sugar supply situation was characterized as "tight" by Horace Godfrey, national administrator of the ASCS. The problem was that as the Puerto Rican industry downsized, it was unable to fill its quota, but reallocating the quota was impossible "until final action by domestic growers." Meanwhile, Godfrey complained, "We are being criticized in the press and are probably being criticized in the Congress for not taking action to provide an ample supply of sugar."[22] While sugar prices rose, mainland cane growers held surpluses that they could not sell because of quota restrictions. With severe acreage cutbacks looming, the USDA, the industry, and state representatives considered three remedies. The most logical, from the point of view of Louisiana and Florida producers, would be to provide them two hundred thousand tons of Puerto Rico's deficit, but the rest of the industry opposed this. Therefore, Senator Holland, with the help of USDA staff, began to pursue two alternatives. One was that the federal government would purchase mainland cane sugar under the auspices of Public Law 480, to distribute abroad as part of the U.S. government's program of food aid. The problem with this "solution" was that most of the countries receiving aid were seeking export markets for sugar themselves. The second was that the federal government would purchase mainland cane for the domestic Food Assistance Program. Consumer and Marketing Services estimated that up to two hundred thousand tons of sugar could be used "in the program for feeding the needy at a cost of around $15,000,000."[23]

Over the course of the summer, these proposed solutions came to naught. No countries were willing to take surplus U.S. sugar as part of a food aid package. Beet growers and processors and cane refiners adamantly opposed an amendment to increase the mainland cane quota, but did not protest Senator Holland's proposed amendment to an appropriations bill that would allow purchase of sugar "to feed hungry people and to relieve the cane sugar surplus."[24] Nevertheless, that remedy failed as well, not over any ethical concerns, but because it was found in clear violation of the Sugar Act. In early August, Godfrey informed Freeman that the politically sensitive declaration concerning cutbacks for sugarcane was immanent: "We have been requested to withhold the announcement until after the Louisiana primaries, which will be held August 17."[25] Senators Holland and Ellender were to be informed prior to public notice, which occurred on August 20 and called for an acreage reduction of 20 percent, with an exemption for farms of less than fifty acres. Protest missives from Louisiana and Florida flew into Washington, addressed to the White House and to the USDA. Within several weeks, President Johnson held a meeting with Secretary Freeman and the Louisiana congressmen, and soon thereafter the USDA began to reconsider the size of the cutback. This too elicited protest, in a telegram that "all segments of the American sugar industry, other than mainland cane" sent to President Johnson arguing "that there is no justification for a change. The integrity of the Sugar Act requires that production in any area be held to the quota level plus inventory requirements."[26]

Memoranda internal to the USDA continued to hash over the sugar dilemma, with one agricultural economist advocating restrictions on the 1969 sugar beet crop as well. The ASCS evaluated Freeman's proposal for a 15 percent reduction in cane acreage rather than the 20 percent already announced. One problem was that new varieties of cane, more suited to cold lands, had been released to Florida growers, so yields per acre were expected to rise substantially. A second problem was that this would be "quite disappointing" to the rest of the industry.[27] Among the numerous letters Freeman received on this topic was one from Doyle Connor, Florida's agricultural commissioner, asking for "any special consideration you can give to the hardship appeal made to you by Mr. Sam Knight of the Atlantic Sugar Association of Belle Glade."[28] Freeman, who had found "very painful" the decision to cut cane acreage, and who had expressed fondness for the smaller growers represented by Atlantic Sugar, wrote Connor of the decision to modify the reduction in acreage from 20 percent to 15 percent.[29] Meanwhile, Horace Godfrey, who had worked for the USDA since 1934, was about to leave the employ of the U.S. government to begin a private agricultural consulting firm and would soon reappear on the Washington sugar stage representing the Florida Sugar Cane League (FSCL) and the Sugar Cane League of the United States. In the intense lobbying prior to the 1971 Sugar Act, he would be joined by fellow North Carolinian Harold D. Cooley, former Sugar Czar, whose defeat in the 1966 Congressional campaign was attributed to his dealings with foreign agents over the 1965 Sugar Act and who would be lobbying on behalf of a first-time quota for Liberia and an increase in Thailand's quota (Blair 1971, 53).

The "Usual Conflict" Over Sugar Quotas

With respect to U.S. sugar policy, the administration of President Richard Nixon seems to have been even more schizophrenic than most. On the one hand, Nixon's appointed secretary of agriculture, Clifford Hardin, agreed with the General Accounting Office (GAO) draft report of May 1969 sent to him for comment, which read, "We recommend that the Secretary of Agriculture request the Congress to consider legislation to modify the Sugar Act of 1948, as amended, so as to enable the Secretary to allocate continuing, long-term domestic marketing deficits to other domestic producing areas rather than to foreign countries." In reply, Hardin concurred, suggesting "that consideration should be given to enabling the domestic areas to market a substantially larger proportion."[30] On the other hand, Nixon's own views were much more internationalist, less protectionist, and certainly concerned with maintaining the image of the United States as a free market economy open to the developing and nonaligned countries of the world. To that end, he created the Council on International Economic Policy (CIEP) in the Executive Office of the President by a presidential memorandum of January 19, 1971, later authorized by the International Economic Policy Act of 1972. As a consequence, the sugar question would be hotly contested within the administration.

Hardin and his staff frequently received letters pleading for relief from restrictive quotas, and even if the letters were not originally addressed to them—but instead, for example, to the president—someone at the USDA was usually called on to respond. For instance, S. N. Knight, president of Atlantic Sugar Association, wrote to Hardin in April 1969 requesting the return of the 15 percent acreage cut for the 1969 and 1970 crop; Florida weather had not cooperated, with freezing weather in both the spring and fall, so that sugar cane tonnage dropped from 427,000 tons in 1967 to 292,000 tons in 1968.[31] This was disastrous for the overcapitalized enterprise. As a result, during the spring of 1969 there was substantial correspondence regarding the Atlantic Sugar Association. Reverend Speidel, a member of the association who had written to Secretary Freeman several years earlier, now wrote to Secretary Hardin, Undersecretary Phillip Campbell, Assistant Secretary Clarence Palmby, as well as directly to President Nixon, whom he reminded of "his campaign promise to be mindful of the needs of the forgotten man."[32] Eventually it was Undersecretary Campbell who replied to Speidel, sending his department's "regrets" over their "inability to be helpful."[33]

Two Democratic congressmen, Robert Giaimo of Connecticut and Jamie Whitten of Mississippi, serving on the House Budget and Appropriations Committees, respectively, met with several of the smaller, independent farmers of the Atlantic Sugar Association and took up their cause. Giaimo wrote to Nixon "concerning the impossible situation in which some of our Florida cane sugar growers find themselves as a result of quota restrictions and reductions. The case of Atlantic Sugar Association and others is tragic. Involved in this predicament are not only the growers but also hundreds of Negro, Puerto Rican and displaced Cubans who depend upon these groups for employment." Giaimo included with his letter another, written by a representative of Atlantic Sugar Association, requesting that the secretary of agriculture sponsor "an 'An Act for the Relief of John Doe' in behalf of the farmers supplying sugar cane to Atlantic Sugar Association."[34] A meeting to discuss the situation of Atlantic Sugar was attended by USDA staffers and various Florida interests, including Reverend Speidel, D. T. Redfearn, chairman of the board of the Columbia Bank of Cooperatives, and David Angevine and Kenneth Samuels of the Farmer Cooperative Service. After the meeting, USDA staff members discussed special relief measures for Atlantic and Talisman, but, anticipating strong protests from producers in Louisiana, concluded that the USDA should not attempt to aid Florida growers.[35]

Talisman Sugar was in much the same straitened circumstances as Atlantic, which William Pawley made abundantly clear, as he called in his chips with the Republican Party in general and the Nixon Administration in particular in an effort to save Talisman. Secretary Hardin's correspondence file bulged with memoranda and letters from, to, and about Pawley. For his part, Pawley addressed most of his correspondence directly to his Cold War comrade, Nixon. In March 1969 Pawley sent a packet to Nixon that contained a letter addressed to "Mr. President" and another to Nixon's personal secretary, Rose Mary Woods, asking her to direct the president's attention to his letter and enclosures, which were three letters sent previously to Nixon. The earliest, dated January 15, 1965, was addressed to Nixon at his law practice and began "Dear Dick." In it, Pawley recited statistics on U.S. sugar consumption and distribution and on world beet- and cane-sugar production, as well as maps showing the Talisman property, all conveyed so that Nixon would understand two important points. First, that many other countries were given quotas larger than Florida's and, second, that "the Administration invited Americans to plant cane and beets and establish mills, with a promise that no restrictions of acreage quota would be put into effect in the year 1964, and it was based on this that many Americans in Florida and Louisiana invested large sums of money in developing their acreage, only to find by April of last year, quotas were reestablished on a basis that could conceivably bankrupt a number of these enterprises."[36]

The second letter, sent during the presidential campaign of 1968, began "Dear Dick: All of us here in Florida interested in your campaign are working with one specific goal in mind and that is to reduce the Wallace vote as much as possible in order to capture the state if it is at all possible." Making an explicit link between partisan politics and Florida sugar, he continued: "You will recall that in 1964 President Johnson and Secretary of Agriculture Orville Freeman were making speeches . . . urging Americans to go into the sugar business and help their country by producing more sugar." Since then the government had imposed cutbacks totaling 43 percent, with the result that "many individuals in Florida and Louisiana will be bankrupt." He concluded by providing "some remarks that . . . would be of tremendous value to our campaign both in Louisiana and Florida."[37] Pawley's letter of November 19, 1968, addressed "To Honorable Richard M. Nixon, The President-Elect," again recounted the history of his venture into the Florida sugar business, noting that the "only area in the world where there is any restriction is the mainland sugar cane area of Florida and Louisiana" and informing Nixon that he expected to sell Talisman within weeks.[38]

However, by the following spring, Talisman had not been sold because, Pawley explained in his March 18 letter to Nixon, the prospective buyer found the plantation's proportionate share acreage insufficient. Pawley contended that by rights Talisman's base acreage should be substantially larger than it was, and suggested three possible remedies to salvage the failing company; that Secretary Hardin be permitted to recognize retroactively a larger share for Talisman; that "some agency of the Government" offer a crop loan of $3 million; or, finally, that the law be amended to allow the secretary to "distribute solely to hardship cases" part of the Puerto Rican deficit. He closed with a personal appeal:

> My company is facing bankruptcy. I have hypothecated my entire personal assets of $13,000,000 for loans totaling $8,500,000 all of which has been put into the company and if the company is lost, everything I have in the world will go with it. I have hesitated, Mr. President, to write this because I realize the tremendous burdens with which you are faced; however, my problem is so acute and so urgent that I am taking the liberty of bringing it to your attention. [39]

On March 27, John D. Ehrlichman, counsel to the president and his close advisor, sent two missives. One was a letter to Pawley: "The President has handed me for attention your letter of March 18 and enclosures. Since this subject matter and your requests involve highly technical questions, I am forwarding your inquiries to Secretary Hardin."[40] The second was a confidential memorandum to Hardin: "Herewith is Pawley's letter and my reply together with his exhibits. My inquiry indicates that Mr. Pawley's factual allegations should always be checked. The President has stated that he wants nothing to be done for Mr. Pawley that would not be done for any other citizen in the same or similar circumstances. Likewise, he wants nothing less for him than anyone else would be entitled to."[41]

In reply, Hardin sent Ehrlichman a letter outlining why none of Pawley's "remedies" was feasible and a five-page memorandum that provided background concerning sugar legislation and an alternative view to Pawley's of the history of the Talisman operation, noting that "it was located on cold land" and initially was "not successful and its sugar operations were abandoned in 1963." His point was that investors, including Pawley in 1964, had taken a risk on a marginal operation. Hardin also pointed out that it was Congress, not President Johnson, who refused to lift restrictions from mainland cane marketing quotas. Finally, he noted that no acreage restrictions were imposed on the 1964 crop. The Department merely announced that if proportionate shares should be established for the 1965 crop, sugar planted after the normal time for planting the 1964 crop "would not be regarded as 1964 crop history for the purposes of computing the 1965 allotment." Hardin sent for Ehrlichman's approval his proposed reply to

Pawley, which concluded, "I regret my inability to be helpful to the Talisman Corporation."[42]

It is not difficult to imagine Pawley's reaction to Hardin's letter, which was approved and sent, nor was this the last the administration would hear from Pawley. However, his next correspondence on this matter was not addressed to the president but was sent through an alternative political channel. In May 1970, Pawley wrote to I. Lee Potter, executive director of the Republican Congressional Boosters Club located in Washington, D.C. Most of the letter concerned Pawley's sugar problems, but it segued to his political support of the president. He explained that his company had been profitable one year out of seven due to acreage restrictions, because out of 38,000 acres, he could only farm 17,000. He cited the same GAO report with which Hardin had agreed, claiming that $89,000,000 would have been saved in the previous year alone if the United States had been permitted to buy sugar from Florida and Louisiana rather than from other countries. Asking Potter's help with future sugar legislation, he turned abruptly to the subject of the president: "I am continuing my efforts to help the President in these really troubled times. I think great injustice is being done to him by some of our senators and congressmen. Our enemies have infiltrated every segment of our society and are misleading our youth, our teachers, our judges and large segments of the population in general."[43]

Cold warrior that he was, Pawley expected the administration to treat him as the covert hero he saw himself to be. He had been privy to the most intimate aspects of the break between the United States and Cuba, including the push to sabotage Cuban sugar either physically or economically through increased production. During the critical years following the Bay of Pigs invasion, he "supported Richard Nixon's presidential bid and hosted meetings between the intelligence agency and U.S. business" (Rosenberg 2005). He was credited as one of the financiers of the Students Revolutionary Directory, which infiltrated Cuba to return with the first information that Soviet missiles had been stationed on Cuba (Holland 1999). Indeed, it was partly because of his close ties to Pawley that Nixon was thought to have played "a major role in the Bay of Pigs activity" (Pfeiffer 1970, 250). A key planning meeting of CIA personnel, presidential advisors, and industrialists took place at Pawley's Miami residence on April 1, 1960, where "Mr. Pawley was told that the time had arrived for careful coordination of all activities; that permission had been granted for an all-out operation; a government in exile will be formed post haste" (250). To defray the cost to the U.S. government, Pawley arranged to float a bond issue in the name of the Cuban government in exile. Nonetheless, the CIA soon became uncomfortable with

Pawley, who fashioned a rival group from among much more conservative exiles than those supported by the agency and who circulated in Havana the rumor that his group had "entre to Veep" (256). The CIA wanted to cut ties with Pawley.[44] However, the "Veep" (Nixon) commissioned his national security aide, "to keep Mr. William Pawley happy and, in connection with this, he has also been instructed to keep Mr. Pawley briefed on how things are moving." At that time, Pawley was "a big fat political cat" that the vice president could not ignore (263).

Thus Pawley expected political support for his sugar enterprise from the Republican Party, and Potter, as chair of the Republican lobbying group, took him quite seriously. Potter forwarded Pawley's letter to Bryce Harlow, counselor to the president, to which Harlow replied "Many thanks for sending me Bill Pawley's letter about his sugar quota. I am having the problem examined and will get off a reply just as soon as possible. I continue to admire the agility and dedication with which you perform your critically important task." Harlow then forwarded the letter to Hardin with this note: "The enclosed copy of a letter from former Ambassador Pawley obviously requires consideration only by those quite expert in dealing with the sugar problem. Could an appropriate member of your Department suggest a suitable reply for me? It *has* to be right!" (emphasis in original).[45] On June 11 Assistant Secretary Clarence Palmby replied to Harlow, enclosing "a suggested reply to former Ambassador Pawley": "We recognize that sugar production in the mainland cane area has been stringently restricted in all years under the Sugar Act Amendments of 1965. We will keep this fact in mind when the time comes to recommend to the Congress amendments to the Sugar Act. Let me thank you on behalf of the President for your support of his policies."[46]

In the meantime, the world sugar market and the U.S. sugar market began to swing in favor of mainland cane producers, creating the economic conditions for making a plausible political argument to increase acreage allotments. In August 1969 the USDA announced an 11.3 percent increase for the 1970 cane crop, restoring by two-thirds the amount by which the 1968 crop had been reduced (*New York Times* 1969). By July 1970, as supplies failed to keep up with demand, quotas were raised three times in quick succession for mainland cane and beets as well as foreign supplies. That same month, proportionate shares for the 1971 crop of mainland cane were increased to 205,988 and 330,016 acres respectively for Florida and Louisiana, which was within 7 percent of the record high crop of 1964.[47] The USDA may have hoped that Florida producers would be placated, but now that the ASCS's Horace Godfrey had taken his expertise to the side of industry, their demands became more specific with regard to the technical aspect of sugar policy. Now, for example, letters and telegrams from the FSCL demanded that the USDA adhere more strictly to the close relationship that was supposed to exist between the spot price and the guide price.

All sides were gearing up for the coming battle over the 1971 amendment to the Sugar Act, which would involve lobbyists for foreign suppliers— whose number had reached 38 countries—in addition to all segments of the expanded domestic industry. The administration's position was to be shaped in the CIEP, since the purpose of the council was to coordinate the numerous agencies and groups involved in foreign economic affairs and to achieve a consistent foreign and domestic economic policy. It also advised the president on the whole range of international economic policy and assisted him in the preparation of his International Economic Report. In its first year, the CIEP was composed of President Nixon as chairman; Peter G. Peterson as executive director; the secretaries of state, treasury, agriculture, commerce, labor, and defense; the director of the Office of Management and Budget; the chairman of the Council of Economic Advisors; the assistant to the president for the National Security Agency; the executive director of the Domestic Council; the special representative for trade negotiations; and Ambassador-at-Large David M. Kennedy.[48] By creating the CIEP, Nixon attempted to integrate within the council the competing visions and diametrically opposed views of people in various agencies regarding the relation between foreign and domestic policy. On trade, his views were pragmatic and relatively moderate. Nixon supported a liberal trade policy but "did not want doctrinaire free traders in his administration," stating he would reject Peterson as executive director of the CIEP if he fit that description (Kunz 1997, 301).

The CIEP, having just been formed in January 1971, was playing catch-up on sugar policy. In March 1971 Congressman Belcher met with Peterson and Assistant Secretary Palmby to feel out the administration's position on the upcoming sugar legislation, specifically on the subjects of limiting the size of payments and the length of time the act would be extended. Peterson promised Belcher an answer within a week, alarming Deane Hinton, who was responsible for developing the administration's position on the sugar question, which was difficult because Secretary Hardin was stalling. Unable to draft a coherent policy position, Hinton wrote to Peterson: "Rightly or wrongly, on March 8 you told Congressman Page Belcher that you would try to get him an answer within a week. Subsequently I have been gently prodding Agriculture, bearing in mind your indication to me that Secretary Hardin was sensitive about his prerogatives concerning sugar and that we should move with care." Although limits on subsidies to a single producing unit had been enacted for other commodities, Hardin was thought to oppose payment limitations on sugar. Hinton recommended that the administration position should be neither to propose nor oppose payment limitations, pleading, "The politics on all this in Congress exceeds my present understanding."[49] On April 15 Hinton contacted all of the members of the CIEP, including those in the White House, to announce an April 19 meeting on sugar legislation and to distribute the USDA's recommendations, which included a three-hundred-thousand-ton increase in the mainland cane quota. "Mr. Peterson has asked that we develop an Administration position as an urgent matter."[50]

Following that meeting, Hinton prepared a lengthy, confidential report on sugar legislation for Peterson, which he sent with a memorandum asking him to read it before another meeting on sugar legislation, warning: "Tomorrow's meeting is rigged. There is neither OMB nor State, nor NSC representation. We should either get them there or use their absence to make certain no final decisions are taken." Hinton's report began by noting that for "at least four months the Administration has sporadically and futilely tried to reach a position on renewal of the Sugar Act." Meanwhile, the industry and Congress had reached substantial agreement on most issues so, the report continued, "it must be recognized how exceedingly difficult and politically hazardous it is for the Administration to make major new policy proposals at this late date."[51] However, in Hinton's analysis the emerging bill posed serious domestic and foreign policy issues, which would compound the problems of the existing Sugar Act, as it was the antithesis of a free market system, being highly protectionist and involving intricate government intervention. The changes proposed by the industry would shift benefits from foreign to domestic producers and foreshadowed the opening of new sugar areas within the United States. At that time, the bill had great political support in Congress, where key figures were "either from sugar producing states or from the southern conservative camp," leaving little "room for maneuver to limit the damage to consumer and foreign interests." Within the administration, views ranged from Dr. Houthakker's, of the Council of Economic Advisors, that the United States would be better off without a Sugar Act to the USDA's, that the matter should be left in the hands of Congress, "reacting only when points which clearly required Administration opposition came to the fore."[52]

From all of this, Hinton deduced and developed a "proposed Administration Position." He recommended that the administration support the renewal of the Sugar Act with as few fundamental changes as possible, but insist on a short renewal period of preferably two or at most three years (compared to the industry's proposed six years). In addition, he urged "a decision, with or without Congressional authorization, to carry out an independent Presidential commission . . . for a fundamental review of this highly protectionist program prior to the next go-around with Congress." Hinton explained that while the USDA recommended that the administration support the industry's proposals, the State Department and the NSC saw "serious foreign policy complications in this approach. In particular, they believe[d] severe damage would be done to the President's Latin American policy by shifting 300,000 tons of the Puerto Rican and Virgin Island quotas to domestic cane production." However, "the more realistic among State Department officials" saw "the handwriting on the wall," and were willing to accept the quota transfer if something else were promised to foreign producers.[53]

The April 27 meeting did not serve to "coordinate" agencies' positions on the sugar question. Instead, as Hinton reported to Peterson, "Acting Secretary Irwin overruled Katz this afternoon and decided State should fight the 300,000 tons switch to domestic cane production."[54] This was the context in which Hardin wrote directly to the president, urging him to "favorably consider a recommendation to Congress that 300,000 tons annually of marketing quota presently unused by Puerto Rico and the Virgin Islands be transferred to another domestic area—the mainland sugarcane area of Louisiana and Florida." Hardin explained that acreage had been severely restricted since 1965, so that the proposed increase, "which amounts to about 26% of the area's current quota, would permit the use of idle acres and factory capacity but would not entail commitment of additional resources." Because Puerto Rican production had fallen so drastically, the share of the U.S. sugar market supplied by domestic areas had declined from 62 to 55 percent, while imports rose from 4.3 million tons in 1966 to 5.2 million tons in 1970. "Farmers in Louisiana and Florida as American citizens resent the fact that their acreage has been restricted more severely than in any foreign country which markets sugar here."[55]

Although the views within the administration diverged widely, room to maneuver was limited. Just before administration representatives were to testify on sugar legislation, Senator Bennett gave the White House a list of the states affected by sugar legislation, outlining the political implications. For example, Louisiana Senators Long and Ellender chaired the Senate Finance and Appropriations Committees, respectively. Thus a White House aide wrote to Peterson, "For us now, two days before the Administration appears as a witness in the House hearings, to upset the delicate agreements reached by all phases of the industry which is united for the first time in 38 years, would be politically damaging to us and to the President's legislative program."[56]

Even though the State Department and the USDA were scheduled to testify on May 4, 1971, as late as May 1 they had not reconciled differences, and Nixon had not yet finalized the administration's position. Sardonically noting "the usual conflict between foreign policy considerations and domestic economic and political considerations," Peterson presented the president with a policy menu, outlining five possible positions on sugar legislation.[57] He explained to Nixon that the State Department and the USDA were far apart on the allocation of quotas between foreign and domestic suppliers, but unless the administration had a unified position, "Congress [would] proceed as it [saw] fit." The five options ranged from supporting in its entirety the domestic industry plan to opposing it completely, most notably the quota transfer of three hundred thousand tons to mainland cane. The White House advisors and the USDA supported the industry plan, position number one, while the State Department chose the position furthest from it, number five.[58] Nixon failed to meet the deadline of May 4, but on May 6 he communicated his support for a middle position, a two-year extension which included the 300,000-ton transfer to mainland cane, but he also specified two modifications: a shift of 300,000 tons of the Cuban reserve from temporary to permanent allocation to foreigners and a reduction of growth reserved for foreigners from 195,000 to 150,000 tons.[59] Because he did not simply choose one of the five options but instead personally tailored one of them, Nixon was evidently engaged at the finest level of detail in determining the administration's answer to the sugar question.

In June the House passed a three-year extension of the Sugar Act. In late July the Senate passed its version of the sugar bill, and on October 6 it was sent to the president for signature. Even as he was signing the bill into law, a CIEP memorandum to the president expressed displeasure over his reallocation of the Cuban quota, which "undermines the purpose of the reserve, which is to assure the availability to the Cuban people of an adequate sugar quota at such future date as diplomatic relations are restored."[60] Of course, by insisting on a relatively short extension of three years, administration officials expected to be able to deal in the near future in subsequent legislation with such contingencies as a change in U.S. relations with Cuba. As the *New York Times* reported, the administration chose a three-year extension because it "sees a changing foreign picture by 1974" (Blair 1971, 53). Indeed, much would change by 1974, but not in ways that the CIEP anticipated as it led the effort to reform U.S. sugar policy.

For the time being, mainland cane producers were pleased with the outcome of the 1971 Sugar Act. Proportionate shares for the 1972 crop were set at 240,306 and 356,916 acres for Florida and Louisiana, respectively; the total acreage was 13 percent higher than for the 1971 crop, with a 16.7 percent increase for Florida.[61] The proposed increase would bring the 1972 mainland cane acreage up to 606,222 acres, a significant increase over the previous high of 577,354 aces in 1964. Now, with sufficient proportionate shares allocated to Florida producers, Pawley was able to sell Talisman Sugar. It was purchased by St. Joseph's Paper Company, owned by Pawley's friend Ed Ball, who added another enterprise and more land to the more than a million Florida acres he already owned. Initially, Pawley retained 40 percent interest in Talisman but then sold his shares to Gulf and Western, "shares which St. Joe later bought to gain full possession of the farm and sugar mill" (Ziewitz and Wiaz 2004, 87). Five years later, reporting on Pawley's suicide, the *Miami Herald* called him a "Florida legend of industry, diplomacy, politics and international intrigue, . . . a swashbuckler in a gray flannel suit with a bit of a Midas touch" (quoted in Holland 2005, 39).

The intention of the CIEP to use the three years of the Sugar Act extension to study and then to overhaul U.S. sugar policy does not seem to have borne fruit. Secretary Hardin left the administration in 1971, to be succeeded by Earl Butz. In early 1974, sugar prices began a steep rise from 11.70 to 12.05 cents per pound, necessitating a sharp increase in the overall U.S. quota. By February, with the price of raw sugar more than 18 cents per pound, Secretary Butz outlined his position on the sugar program. He endorsed an extension of the Sugar Act, "but asked Congress for major changes, including abolition of domestic planting restrictions, subsidy payments to United States growers and a sugar processing tax" (*New York Times* 1974a, 46). He had previously suggested wiping out all marketing controls but now advocated keeping a floor price under sugar by regulating supply, including assigning quotas to foreign countries. However, he wanted to eliminate USDA authority for acreage restrictions "to allow a free shift of sugar planting between areas." (46). As the debate sharpened, sides were chosen; while "sugar men" advocated the extension of the Sugar Act, including domestic planting controls, "industrial users" urged a "freer market" (46). In May the House Agricultural Committee voted to extend the act by five years, but the bill died on the House floor, "freeing the market for the first time in 40 years from Government subsidies, import quotas, and complicated pricing formulas" (Rugaber 1974, 1). The existing act was not due to expire until December, however, so revival in the Senate was possible. Few in the Nixon administration were likely to be giving much thought to sugar legislation in the intervening months up to his resignation of the presidency on August 9, 1974.

Over the course of Gerald Ford's relatively brief tenure in office, the U.S. sugar market happened to be exceptionally volatile, with major structural changes dating to this period. First, prices continued upward, so that by late August a five-pound bag of sugar, which had cost $.89 the previous summer, retailed for $2.20 (Maidenberg 1974, 45). As a result, U.S. consumption patterns shifted, in part through organized consumer boycotts, in part because of household budget constraints, and perhaps also in response to Ford's request that "Americans reduce their use of sugar in cooking and cut in half the amount used with coffee and tea" (*New York Times* 1974c, 61). In November Ford announced that, in lieu of an extension of the Sugar Act, he would remove restrictions on domestic areas and replace individual country quotas with a global quota; the same week, Butz proposed that the United States resume trade with Cuba to reduce soaring sugar prices. In December a last-ditch effort was made to save the Sugar Act, which failed. Forty years old, its expiration was "regarded as a bitter development for the American sugar industry" (Barmash 1974, 39).

The demise of the sugar program is often attributed to its expiration during a period of high prices, but the domestic industry certainly did not want the end of the managed market. A more nuanced and complete explanation would include the ambitious plan of the Nixon administration to overhaul the entire program, which meant that the usual policy inertia did not hold. But no alternative plan emerged, for reasons unrelated to sugar and having more to do with Nixon's political demise. High prices did drive other changes, some temporary and others more permanent. During Ford's administration and through most of Carter's there was a palpable sense that the United States would draw closer to Cuba once again. Nixon, because of his conservative, anti-communist credentials, had been able to initiate relations with China but "could never quite bring himself to revise the Government's similarly anachronistic stance toward Fidel Castro's Cuba." Ford, "unburdened by emotional attachments," was ready to move in a new direction (*New York Times* 1975, 32). In addition to temporarily high prices, a more permanent, structural change in the U.S. sweetener market resulting from the introduction of high-fructose corn syrup (HFCS) also helped end the sugar program. In 1974 Coca Cola,

Royal Crown, and Dr. Pepper had approved use of a "high-fructose" product made from corn in their soft drinks. The two companies producing
HFCS, Clinton Corn and Staley, were "swamped with demand," and supplies were "generally unavailable to most bottlers" (Holsendolph 1974, 54).
Even so, Clinton and Staley promised the new sweetener would always be
priced below the market price of sugar. By 1976, "spurred by the record
sugar prices forced upon consumers by the reported shortfall in overseas
output in 1974," HFCS had already captured 25 percent of the U.S. sweetener market (Maidenberg 1976, 49).

While sugar prices remained high in early 1975, delegates from Latin
American and Caribbean countries that together controlled half of world
sugar exports met in the Dominican Republic to discuss a common pricing
policy for sugar. The group, which included Cuba, was characterized as em-
ulating OPEC. The weeklong meeting did not achieve the desired results
because of large differences among producers, especially Cuba and Brazil,
and for the simple reason that the commodity was ubiquitous. Several measures, however, were agreed upon, including the creation of a statistical bureau in Mexico. While the United States was not privy to these meetings,
developing international institutional structures for managing the sugar
market would, in the next presidential administration, become a central
plank of attempted U.S. policy reform.

By June 1975 sugar prices were headed downward. In September 1976,
when world market prices had collapsed from sixty to ten cents per pound,
Ford increased duties and began researching the possibility of implementing import restraints, while domestic growers of both beet and cane pressured him to reinstate import quotas. In the aftermath of the 1976 election,
a *New York Times* column entitled "What Ford Hopes Carter Will Keep"
listed, among other items, "a reassessment of United States sugar policy"
(Shabecoff 1976, F17). The sugar question that Carter inherited was particularly sticky, best described as chaos unbound. Foreign policy objectives and
domestic producers' expectations, always in conflict, were now on a colli-
sion course of historic proportions. Even before he took office, it was reported that because of the "perplexing problem" of HFCS, domestic cane
and beet growers were "pressing President-elect Carter to reimpose import quotas and high tariffs on foreign sugar." Because at the time sugar was
"the world's most depressed commodity," it was not only sugar producers
but also the HFCS industry that stood to benefit, since HFCS needed sugar
prices of at least fourteen cents per pound to remain profitable (Maidenberg 1977, 45). The strength inherent in the political geography of that coalition, as yet untested, was soon to be demonstrated.

"Carter, Coke, and Castro"

Carter's sugar dilemma was apparent from his first moments in office: "When President-elect Carter marched down Pennsylvania Avenue in his inaugural parade, a political and economic problem for his new Administration was already building up. It was the sugar problem" (Robbins 1979). The "usual conflict" between domestic and foreign concerns was intensified by the changing status of the United States in the world economy, by fears within and without the United States of inflation and recession, and, ultimately, by changes in the U.S. sweetener market, both in terms of sourcing and structure. As the United States lost market share in key commodities and began to run a trade deficit during the 1970s, Nixon officials introduced a trade bill that, when finally passed in 1974, renamed the Tariff Commission as the International Trade Commission (ITC) and gave circumscribed powers to the president to determine tariffs (Kunz 1997). In the fall of 1976 President Ford asked the ITC to review various imports, including sugar. Their findings, issued March 4, 1977, agreed with domestic sugar producers that they were jeopardized by rising imports. President Carter legally had sixty days to respond to the ITC recommendations, which included cutting by one-third the seven million tons of sugar imported to the United States. Similar recommendations on shoes and televisions raised the fear in the Carter administration and elsewhere that U.S. actions would trigger a wave of protectionism in a world economy verging on inflation and recession.

Since taking office, Secretary of Agriculture Bob Bergland had been floating ideas about price supports and import quotas, which Carter, committed to "free and fair trade," resisted. Now Bergland suggested subsidies and joined Carter and Katz in pinning his hopes on a new International Sugar Agreement (ISA) as the "cornerstone" of a national sugar policy. The ISAs of 1953, 1958, and 1968, and the 1973 protocol, set quotas for exporting members and obligated importing members to limit their imports from nonmembers. Neither the United States nor the European Economic Community joined the 1968 ISA (USDA 1978). What Carter and others envisioned was an economically stabilized *global* sugar system that would maintain sugar prices at levels between 13.5 and 23 cents per pound, which would obviate the need for protectionist measures. At the same time that Carter cast his lot with the ISA, Senator George McGovern of South Dakota was introducing an amendment to end the embargo with Cuba. National Security Advisor Zbigniew Brzezinski warned Carter to maintain a posture "not that of benevolent neutrality but rather that of skeptical neutrality" toward the amendment and to make sure it specified that trade would not result in undue hardship to sugar producers in the United States.[62] In May, meeting the ITC deadline, Carter rejected the recommendation to cut imports and instead proposed a subsidy of two cents per pound for domestic sugar growers. As the administration suggested subsidizing U.S. sugar producers and acceded to those who favored opening trade with Cuba, the president was attacked from the right of the political spectrum by commentator William Safire, who argued in a column entitled "Carter, Coke, Castro" that Carter's foreign policy and sugar program were structured to benefit the Coca-Cola Corporation, headquartered in Carter's home state of Georgia. Baseless as this accusation might have been, it stuck.

In July 1977 as the House prepared to amend the Farm Bill, the Carter administration was blindsided by what came to be known as the de la Garza Amendment, after the Democratic Congressman from Texas, Kika de la Garza, who introduced it. The amendment had actually been drafted by Horace Godfrey, and behind it stood Senator Robert Dole, Republican of Kansas, "whose weather eye [was] fixed on the Republican Presidential nomination in 1980" (King 1977, F1). Dole intended to leverage his way into the White House by exploiting the political geography of the sweetener coalition, which he had activated. And what a coalition! Alongside beet and cane farmers, sugar processors and Congressional representatives, stood corn farmers, processors, their representatives, the Corn Refiners Association, and the likes of agribusiness giants such as Cargill, Staley Manufacturing, Amstar Corporation, Anheuser-Busch, and Archer Daniels Midland. The amendment, which passed the House by a vote of eighty-one to three, sought to derail President Carter's proposed sugar program. It supported sugar at fourteen cents per pound (not coincidentally, just where HFCS producers needed it to be) and invoked a section of the Agricultural Adjustment Act of 1938 that required the USDA secretary to call for import quotas or duties if imports undermined the support program. In contrast, Carter wanted a lower price and greater flexibility to negotiate a "meaningful" international sugar trade agreement.

In negotiations with House and Senate conferees, the Carter administration requested and got a "self-destruct" measure added to the de la Garza amendment in the event that the United States were to ratify the ISA. Even so, Brzezinski was quite troubled by the situation Carter faced, as expressed in a memorandum classified top secret/sensitive:

> You are faced with a difficult perhaps a no-win decision on sugar by November 8. The Acting Secretary of Agriculture, John White, wrote a letter to Senator Dole committing the USG to implement the de la Garza amendment of the

1977 Agriculture Act by 1977. To raise the price of US sugar, however, it will be necessary to set limits on sugar imports. Such a decision will have an adverse impact on our relations with Latin America and the developing world, on your pledge to resist protectionism, and perhaps on the implementation of the International Sugar Agreement.[63]

With the addition of the "self-destruct" measure to the de la Garza amendment, the administration announced it would accept the bill, and in January Carter asked the Senate to ratify the ISA. Now however, Senators Frank Church of Idaho and de la Garza, both Democrats, announced they would not bring the ISA out of committee until a "satisfactory" domestic program was adopted. Their proposed program guaranteed a price of 17 cents per pound, 9 cents above the world price and substantially higher than the administration's target price of 14.4 cents. A similar situation occurred in the House, where the Agricultural Committee proposed a 16-cent price support and a country-by-country quota system (King 1978, D1). Commenting from Florida, Alfonso Fanjul wrote, "We are fighting it out in Washington to get proper sugar protection legislation to be able to survive."[64] Carter, wary about inflation, worried about protectionism, and concerned about the interests of consumers, continued to fight the fight. By August 1978 commodity analysts noted that "the struggle between the White House and Congress over sugar policies is considered the dominant factor in the market today" (Maidenberg 1978, D5). Meanwhile, after attending the inauguration of Dominican President Antonio Guzman, Secretary of State Cyrus Vance wrote to Carter, "Guzman is facing a severe economic crunch which poses serious problems to his government. Sugar dominates the Dominican economy and is, in turn, very vulnerable to our sugar policy. Our failure to ratify the Sugar Agreement has been devastating for the D.R." [65]

In December 1978 Carter signed a proclamation supporting the ISA, a symbolic gesture without Senate ratification. Long before that, evidence of the chaos in the U.S. sugar system had been apparent in Florida. When a warehouse in Belle Glade burst open at the seams with raw sugar "viscous as lava and as dark as motor oil," the Carter Administration was accused of ineptitude (Robbins 1979, A1). What had happened was that the legislation passed in the fall of 1977—"a document drafted hurriedly under pressure from Congress and without benefit of review by top sugar authorities"—was in effect for ten weeks, during which foreign producers were able to dump two million tons of sugar into the United States without payment of new import fees. In January 1979, two years after Carter's inauguration, the sugar program was still unresolved.

In February Senators Frank Church of Idaho and Russell B. Long of Louisiana, chairs of the Foreign Relations and Finance Committees, respectively,
introduced legislation that would guarantee domestic growers 17 cents per
pound. "The two senior senators are pursuing a strategy of threatening to
hold major pieces of the Carter Administration's legislative program hos-
tage to the domestic sugar growers" (King 1979a, D3). Faced with utter dis-
ruption of his entire agenda, including wage legislation, trade agreements,
and strategic agreements, "President Carter finally gave in," agreeing to
sponsor sugar price-support legislation, with a base price of 15.8 cents per
pound and a half-cent subsidy (King 1979b, D1). The Florida industry saw
the "Sugar Stabilization Act" of 1979 as their best hope, and turned to Governor Bob Graham for support. Graham consulted his advisors, who told
him, "The White House feels that, while it is not a perfect bill, the President
can generally support it and will not veto it."[66] Therefore, Graham was reassured that he "could support this bill without encountering any conflict
with the White House." Graham then wrote to the entire Florida delegation, urging their support: "During the last few years, foreign sugar producers, heavily subsidized by their governments, have dumped cheap 'home-
less' sugar into the United States depriving our growers of their traditional
markets. Florida's sugar industry has experienced one mill closing, and several of our growers have had 'walk-away' sales of their farms. Across the Nation, nearly twenty sugar mills have closed their doors."[67]

However, in October the House failed to pass the act and in doing so also
rejected approval of U.S. participation in the ISA. At this point, "the administration and key members of Congress took matters into their own hands"
(Mahler 1986, 170). After Secretary Bergland and Senator Church reached
agreement on key points, the Senate ratified the ISA, with a price objective of 15.8 cents per pound. Ironically, soon thereafter the market price
rose precipitously to 24 cents per pound due to a conjunction of factors,
including bad weather in the USSR and cane rust in Cuba. The rising sugar
prices had arrived too late for some, as *Florida Trend* reported, "Some of our
small sugar growers aren't going to make it" (Brown 1980, 78). Seven years
of "makeshift sugar support policies" had taken their toll.

Cutting Cane in Florida

The DARE report of 1964 had emphasized the importance of favorable legislation for the Florida industry, with regard to both quotas and maintaining legal access to an offshore labor force. With respect to the former, the
ensuing decade's production figures reveal a roller coaster, with the historic

high of the 1964–65 season of 219,800 acres declining to 153,600 acres in 1969–70, only to climb steadily to reach 258,400 acres in 1975, continuing upward to 320,700 acres in 1980–81. While it had required intensive political maneuverings to achieve, the output in 1975 matched the optimistic scenario forecast in the DARE report. With regard to the latter, the Florida industry was able to maintain access to Caribbean labor, though it took political effort to do so. Most of the time the geography of plantation production left hidden the living and working conditions of the sugarcane cutters, but when public attention did focus on the industry's labor relations, the picture was grim and opinion unfavorable.

By this time, the Florida industry had relied on the H-2 workers program for more than twenty years. As the industry expanded after the break with Cuba, harvest labor requirements grew accordingly. Meanwhile, in the context of President Johnson's "war on poverty," the use of foreign labor for agricultural work—especially in some of the poorest areas of the rural south—came under intense scrutiny in the administration and in Congress. In December 1964 Johnson's secretary of labor, W. Willard Wirtz, made public his opposition to the importation of farm workers from Mexico and the Caribbean, expressing his hope that the use of foreign workers "will be very greatly reduced and hopefully eliminated." As Wirtz persisted, "Senators from Florida and California warned they [might] try to get [him] fired," with Senator Holland cautioning that he "was prepared to get rugged" (*Miami Herald* 1965a).

In April Wirtz made a three-day tour of Florida's agricultural labor camps. His trip included a visit to one of USSC's plantations, where he "complimented the management" and compared conditions there favorably to the "appalling" accommodations of citrus workers. At that time, both the citrus and sugar industries were using offshore labor; following his visit, Wirtz continued to "put pressure on Florida growers to use more domestic labor, to cut down on the use of foreign labor" (*Miami Herald* 1965b). In the aftermath of Wirtz's visit, the *Miami Herald* described the accommodations for sugarcane workers as "expensive barracks," noting that the Sugar Cane Growers Cooperative had "over $700,000 tied up in camps" and that "Pawley's labor camp" at Talisman had been likened to a "country club." Between 1965 and 1966, the number of H-2 workers in Florida declined from 13,099 to 8,762, at which time they were almost exclusively employed cutting sugarcane.

In 1966 the Community Action Fund, a Florida nonprofit corporation formed to aid migrant workers, published a detailed study by Peter Kramer entitled *The Offshores: A Study of Foreign Farm Labor in Florida*. What Kramer found in his interviews with farmers, government officials, and workers did not support the industry's claims. Kramer documented poor living conditions and inadequate nutrition provided to workers who were performing some of the most physically demanding work in the country. In interviewing workers, Kramer found low wages to be their consistent complaint, followed by bad food, poor housing and dissatisfaction with liaison officers. The issue of liaison officers was critical, since they were the guarantor of workers' welfare and therefore of the program's legitimacy. However, in the words of a foreman for the Sugar Cane Growers Cooperative, "The liaison officer is the worker's man, but he has no choice except to back us in disputes. You see, we won't call a liaison officer into a hassle unless we're right. Since he's got a contract to uphold, the liaison officer ends up backing us. The liaison officer is a man caught in the middle" (Kramer 1966, 23).

Kramer's study demonstrated that the West Indian contract offered growers one irreplaceable advantage: "a formidable instrument of control over workers who are laboring in a foreign land, many miles from their homes." Company managers and foremen were well aware of this central fact of the employment relationship, explaining, "If he violates his contract we can send him home. So we've got leverage over that West Indian that we don't have over American workers" (Kramer 1966, 39). As one foreman elaborated, "They hear that the U.S. nigger has rights and they think they've got rights, too. They eat it up like slop. They don't know they ain't got rights in this country" (53). From his research, Kramer concluded that vegetable and citrus harvesting did not require offshore labor but that the "one crop in which there seems to remain a valid need for offshore farm labor at present is in Florida sugar cane. Due to a complex of reasons, Americans are unwilling—not unable—to harvest this crop, at least under present conditions" (91).

Labor Secretary Willard Wirtz reached the same conclusion, but California farm leader Cesar Chavez thought otherwise. Under his leadership, in 1972 the United Farm Workers Union (UFWU), A.F.L.-C.I.O., brought a class action suit against the administration officials who had certified the Florida sugar industry's requests for the importation of cutters from Jamaica. The central point of contention was whether the sugar companies were making "reasonable efforts" to recruit and employ domestic cane cutters; Dr. Marshall Berry, an economist, testified for the plaintiffs that there were eighty thousand unemployed farm workers in Florida and that efforts to recruit them had been "pro forma" (*New York Times* 1972a, 52). The union was also assisting a group of two hundred truck and tractor drivers, most of them Cuban, who were striking against Talisman Sugar Company. At the entrance to the Talisman plantation, a student of Marshall Berry's, Nan Freeman, was killed, by all accounts accidentally, as she and several other students helped to picket the mill. On the eve of selling Talisman, William Pawley stooped to publicly promulgating the pretense that Freeman had died elsewhere, claiming that picketers moved her body to place blame on the company.

The UFWU requested an emergency restraining order to prohibit the sugar companies from transporting ten thousand Jamaican workers to Florida. The Florida Sugar Cane League, including John Boy, Alfonso Fanjul Sr., William Pawley and Horace Godfrey among its officers and directors, marshaled industry resources to fight the union. The UFWU had already successfully organized orange pickers working for the Coca-Cola Company's food division, who fought for and won the first labor contract for migrant workers in Florida. In September, the executive vice president of Tropicana offered his assistance to sugar industry executives to arrange a meeting with Florida Governor Askew. By the time of their October meeting in Tallahassee, Federal Judge Peter Fay had already issued a preliminary finding denying the UFWU request to stop labor importation. The Florida Department of Commerce weighed in on the side of the industry, officially expressing "pessimism about finding labor anywhere in the United States willing to cut Florida sugar cane."[68] In preparation for the meeting, the league gave the governor "background information," including a "historical look at the use of off shore workers" and "sugar industry's domestic recruitment efforts." The historical brief explained that "the 'domestic' deserted the Florida sugar cane fields during World War II never to return" and that subsequent efforts "to utilize 'domestics' [had] never been successful."[69]

Judge Fay found in favor of the companies, saying the union had not only failed to substantiate charges that the companies were discouraging U.S. workers but also that there was "overwhelming evidence" to the contrary. Fay was seen as sympathetic to the plight of Florida's migrant workers, and he expressed bewilderment at the seemingly contradictory evidence of high unemployment and lack of domestic labor willing to cut cane, saying, "The court literally does not understand this situation or why it exists." He evinced one lingering concern, that while companies paid fifty dollars apiece to transport Jamaican workers, he knew "of no requirement that they pay the transportation, housing and food costs of American workers" (*Miami Herald* 1972, 1B).

Chavez did not give up; instead, he asked for a congressional investigation into the "exploitative, discriminatory and arbitrary" hiring practices of the Florida sugar industry (*New York Times* 1972b, 43). In March 1973, the renowned *New York Times* reporter, Philip Shabecoff, ventured into the EAA to interview workers who were at "the center of what is shaping up as an epic conflict between the sugar growers and the [UFWU]" (Shabecoff 1973, 24). One of his first findings was how difficult it was to talk to workers because their camps were "tucked away in the middle of cane fields, which stretch mile after mile across the flat landscape," where planters had placed physical discouragements, such as wire fences, no trespassing signs, and "watchful supervisors," and most of all, because of the workers' fear, "to a man," of being caught talking to strangers. Shabecoff described a conversation that took place one evening at the Saunders Camp, which belonged to the Glades County Sugar Growers Cooperative, with two workers who "flattened themselves against the dark wall, careful that the floodlights illuminating the labor camp did not touch their faces" (24). He witnessed workers packed into bare wooden structures without toilets or running water, and heard of a diet exclusively of rice served three times a day; meals for which deductions from pay were compulsory. While the UFWU charged that conditions were deliberately kept bad to discourage domestic labor, a spokesperson for the league claimed there was "a social taboo" among U.S. workers against cutting cane and that besides, it was "too hard" for them. Shabecoff saw evidence of more serious violations of workers rights in the pay stubs showing cash payments of less than a dollar per hour when the law mandated two. The practices by which workers were cheated out of half their pay included a seemingly complicated system of assigning a "row rate," the amount that a worker would be paid to cut a particular row and then manipulating actual hours to meet that "price." Thus, if an eight-dollar row required eight hours of labor, a worker's time card would be falsified to read four hours. In this way companies avoided paying what was known as "build-up," the difference between row rate and minimum wage. All in all, Shabecoff found the plantation "tangibly prison-like."

U.S. Department of Labor staff reports confirmed many of the UFWU's allegations. A team of researchers working under the direction of Saul Sugarman, the Labor Department wage and hour analyst, visited Florida to determine the sugar industry's compliance with wage regulations. Focusing on four of the largest companies, together responsible for 70 percent of the harvest, including USSC, Glades County Sugar Cane Growers Cooperative, the Atlantic Sugar Association, and Gulf & Western Industries, they found that all but the last were consistently violating minimum-wage regulations, undercounting workers' hours on average by 1.25 hours per day (McCally 1991). Sugarman was asked to retract the *Wage Survey for the 1973–74 South Florida Sugar Harvest*. After his refusal, the Department of Labor sent another investigator, who found no significant wage violations (Rothenberg 1998).

In 1981 a group of Haitian refugees sued, contending that the use of foreign workers kept wages too low for domestic workers. The issue had taken on "special urgency" because at the time Florida was "awash with refugees, an estimated 112,000" had arrived since January 1980 (Thomas 1981, 22). Even though some Haitian residents had previous experience cutting cane, they were unable to secure harvest jobs. Gregg Schell, an attorney for Florida Rural Legal Services in the Belle Glade office filed the case:

> The sugar companies had no interest in hiring the Haitians because the Haitians could essentially vote with their feet and leave the job. The problem was that if Haitians were available for the job, the H-2 workers would not be allowed to come in and the companies were very anxious to get rid of the Haitians because they for the first time represented a large group of domestic workers who would take these jobs. The irony was that Okeelanta was run by Gulf & Western and several of our Haitian clients showed us their I.D. cards from the Gulf & Western operations in the Dominican Republic where they had cut and they were perfectly suitable workers there and were perceived as not suitable here.[70]

Ultimately, a House subcommittee investigation found that the industry was not making a good-faith effort to recruit and retain this domestic labor force, which lacked the "deportability" of H-2 workers (U.S. House 1983). A former manager of the FSES office in Belle Glade noted that another critical difference between the two groups was that H-2 workers were provided housing and food, however rudimentarily: "I'd be visiting camps and see Jamaicans eating. . . . The Haitians had nothing. . . . If you dig ditches and work, you use up your body resources. The Haitians were dropping out like flies" (Thomas 1981, 22). The underlying problem, as earlier outlined by a UFWU organizer, was "a classic example of the poor people of one country being used against the poor of another" (Shabecoff 1973, 24). In similar fashion, the next round of U.S. sugar legislation found domestic growers once again pitted against foreign growers, with significant consequences for both Florida and the Caribbean.

Reagan "Rents" Sugar Votes for His Budget Bill

Initially, President Ronald Reagan opposed any form of price supports for sugar on the basis of consistency in "cost cutting" measures and "free" market principles. However, as sugar prices dropped through 1981, political pressure mounted for a system of comprehensive price supports. Needing votes for his budget, Reagan gave in on the question of price supports. When asked whether his vote for the budget could be bought with largesse toward sugar, John B. Breaux, Democratic Senator from Louisiana was widely reported to have replied, "No, but it can be rented." In December 1981 the House approved a Farm Bill of four years duration, the Agriculture and Food Act, which marked the return to a managed market. The program had two components: non-recourse loans and control of imports.

Shortly after the institution of the program, world prices fell again, and it became difficult to manage the market with the tools at hand. At this point, Reagan altered the program by adopting formal import quotas, which were potentially generous because they offered the U.S. price to foreign growers but were often quite restrictive in terms of the size of the quota. For example, in the case of the Dominican Republic, exports to the United States fell from 493,000 to 123,000 tons from 1981 to 1988. Though the import quota "solution" was supposed to be temporary, it became law in the Food Security Act of 1985. The 1985 law had three primary provisions: a minimum price, country-by-country quotas, and the requirement that it operate at no cost to the government. Generally speaking, it had four effects: maintaining the U.S. sugar price above world market price, providing the umbrella under which the HFCS industry gained market share, causing a decline in sugar imports, and conversely, an increase in domestic production. The law was devastating to the sugar industries of Caribbean countries and negated the purported gains of another Reagan program, the Caribbean Basin Initiative (Krueger 1993; MacDonald and Fauriol 1991). Sugar imports from the Caribbean to the United States declined between 1981 and 1988 from 1,552,000 tons to 357,000, with a corresponding decline in dollar values of $543,947,000 to $97,118,000, "an 82% direct loss in hard currency export earnings because of lower quotas" (McCoy 1990, 16). Hardest hit was the Dominican Republic, the leading sugar producer in the Caribbean with the exception of Cuba, and within the Dominican Republic it was "the large numbers of Haitian cane cutters who [bore] the brunt of adjustment" to U.S. sugar policy (17). Conversely, between 1981 and 1988, total acreage of sugarcane harvested in Florida jumped from 339,000 acres to 421,000 acres.

Florida producers lobbied hard for the 1985 Farm Bill, and it is easy to see why. Stability was the sweetest aspect of sugar production, and the years of market volatility had been rough. Dalton Yancey of the FSCL provided Governor Graham with written statements from key individuals supporting the bill. A member of the International Sugar Policy Coordinating

Commission of the Dominican Republic stated that HFCS, not U.S. producers, was damaging their market; the Jamaican minister of labor noted the benefits of the Florida industry to his country; and the principal of an elementary school in Clewiston highlighted infrastructure, such as playgrounds, that the industry provided to the community. Armed with these, Governor Graham was asked "to take a certain action on behalf of the industry, including contacts with a Florida congressman and certain governors whom they would designate."[71] Further instructions followed from "Horace Godfrey's people in Washington, who, based upon repeated visits, characterize the posture of selected Florida Congressmen," which were sorted into categories of "OK" or "free trader." Handwritten at the bottom of the memo was a quizzical note, "Who is Horace Godfrey?"[72]

And what had become of the various sugar enterprises that sprang onto the horizon two decades earlier? The Atlantic Sugar Association, which struggled to produce 27,080 tons of sugar in 1964–65, was producing more than 100,000 tons in 1982–83. The Sugar Cane Growers Cooperative followed a similar trajectory, producing almost 79,000 tons of sugar in 1962–63 and more than 240,000 by 1982–83. Yet by this time the two cooperatives had assumed quite different ownership structures. From fifty original members—farmers, doctors, investors—the Atlantic Sugar Association was down to nine large shareholders, including two companies owned by the chairman, Sam Knight, one owned by the Fanjuls, and one—Seminole Sugar Company—headquartered in New York. In contrast, the Sugar Cane Growers Cooperative, which began with fifty-one members, had fifty-two members in 1985. Although many of these were from the "traditional vegetable-growing families" of the Glades, this category included large, vertically integrated enterprises such as "sod-magnate A. Duda & Sons."[73]

Talisman's first crop of 325 tons in 1962–63 was quite unimpressive. However, by 1982–83 Talisman was producing more than 100,000 tons of sugar and had enlarged by acquiring several other plantations. One, as mentioned earlier, was the South Florida Sugar Company. The second was the Florida Sugar Corporation's land and mill, which Talisman purchased in 1971, soon thereafter shutting down the relatively small mill. The output of Osceola Farms Company—under the management of Alfonso Fanjul Sr. and Alfonso Fanjul Jr.—rose from a little more than 10,000 tons in 1961–62 to more than 130,000 tons in 1982–83.

However, those numbers do not capture the extent of the Fanjul family's growth in the Florida sugar industry. To do that, we must look at the fate of the Okeelanta Sugar Refinery, Inc., which, as noted earlier, in 1959 had purchased the Fellsmere facilities and mill, subsequently closing the mill in

1965. That year, the stockholders of Okeelanta sold all of the stock to the South Puerto Rico Sugar Company, and those shares were in turn acquired by Gulf & Western Industries, Inc. In October 1984 the Fanjul family, the majority stockholder of the Flo-Sun Corporation and owner of the Osceola mill, acquired "the sugar interests in Florida and the Dominican Republic, as well as all the other holdings in that island nation, of Gulf & Western Industries, Inc." (Salley n.d., 27). This expansion took place after the death of Alfonso Fanjul Sr. in 1980, following which Alfonso Fanjul Jr. became president of the corporation and his brother, José Pepe Fanjul, chairman and CEO. In 1982–83, Okeelanta produced more than 234,000 tons of sugar; thus with Osceola and Okeelanta, the Fanjul family was emerging as one of the largest producers in Florida, at that time second only to USSC, which had a record harvest of 467,000 tons in 1982–83. By 1985, the Fanjul family's holdings comprised the largest sugar company in Florida. Sugar was only part of their Flo-Sun Land Corporation, a diversified operation that included banking, real estate, and tourism development, including three resort hotels in the Dominican Republic. With 240,000 acres of sugar property there as well, they were "in a good position to benefit when the U.S. government uses its import quotas to help out Caribbean producers."[74]

CHAPTER SEVEN

Questioning Sugar in the Everglades

In the years following the Florida sugarcane region's explosive growth in the 1960s, the industry's labor problems simmered just below the surface of public attention. The severity and extent of labor abuses in the region occasionally rose to public visibility, as in the case of Kramer's 1966 investigation. Not until the 1980s, however, did labor conditions in the cane fields finally burst into popular consciousness and became a public relations nightmare for USSC and the industry in general. Politically, the industry continued to think of labor conditions in the EAA as its "Achilles heel." Technological advances finally made the mechanization of cane harvesting on muck soils economically feasible, leading to the end of the H-2 worker program in Florida in 1995. This did not end the industry's political and public relations problems, however, but merely allowed public scrutiny to focus on sugar's role in the degradation of Everglades ecology. Following the industry's second burst of expansion in sugarcane acreage in the 1980s, the extent of sugar's role in transforming the Everglades seemed even more evident and questionable in the public eye. The environment became big sugar's new Achilles heel.

This chapter completes the narrative's historical arc, with the sugar question and Everglades transformation still, after more than a century, at the center of public debates about the future of the south Florida region. The sugar question is now inextricably bound up with issues of globalization, regional trade pacts, and the neoliberal development policies that have dominated the global political economy and U.S. domestic politics for the past quarter century or more. Florida agriculture and the sugarcane agro-industry, especially, have played a central role in the domestic and international political-economic maneuverings around the formation of regional trade blocs, in the process eliciting arguments that echo those of earlier decades. Debates surrounding the transformation of the Everglades, on the other hand, have shifted significantly, and ideas about their landscape, ecology, and alteration have taken on new meaning. Once again, the Everglades have entered into the politics of presidential campaigns as an iconic landscape, with a key difference: transformation now means "restoration" rather than "reclamation." Sugar and Florida's sugarcane production region, as they had been during the era of reclamation, are the primary focus in the era of restoration.

Labor, the "Achilles Heel" of Florida Sugar

In the second half of the 1980s, mass media coverage catapulted the Florida sugar industry's labor practices into the arena of public debate. First, a 1986 pay dispute at Okeelanta erupted into what became known as the "Dog War." When workers refused to go to the fields until their grievance was settled, members of the Florida Highway Patrol and the Belle Glade Police Force entered the plantation, forced men out of barracks, and used dogs to disperse protesters. Several cutters were bitten (McCally 1991). That night hundreds of workers were loaded onto buses bound for Miami, many not even allowed to gather their personal possessions before leaving. From Miami, more than three hundred deportees were flown home (Florida Rural Legal Services 1994). The "deportability" of workers, of course, was the lynchpin in the industry's labor control machinery. The ease with which this company and the Florida industry in general could dispense with workers who were deemed uncooperative made patent why they chose not to hire skilled domestic workers, including the Haitian immigrants who lived nearby in Belle Glade. However, television and newspaper coverage, which made the circumstances of the workers' deportation more widely visible, began to weaken the political viability of this strategy.

Released in 1989, the film *H-2 Workers* documented the grim living and working conditions of cane cutters and their powerlessness to redress wage violations. The guerrilla-style tactics that documentary filmmaker Stephanie Black apparently had to employ to enter the barbed-wire-enclosed worker compounds—set in a landscape dominated by the single industry to which workers owed their livelihoods—underscored the cutters' vulnerability. Black confirmed and documented many of the findings of Shabecoff's (1973) investigation, detailed in chapter 6. The film depicted workers, isolated in plantation compounds, subject to time-card fraud, and forced to submit fewer hours than they actually worked. That same year, the *New Yorker* published an exposé of the industry's labor practices by Alec Wilkinson, a longer version of which was published as a book, *Big Sugar: Seasons in the Cane Fields of Florida* (Wilkinson 1989). Wilkinson's interviews with workers were reminiscent of Kramer's, describing, some twenty years later, inadequate food, poor housing, and lack of political representation. Together, the film, article, and book gave the interested public insights into the daily lives of Florida cane cutters and served as an indictment of the H-2 program. Wilkinson's book figured extensively in ABC's 1990 documentary, *Bittersweet Harvest*, which, from its opening words—"Tom Jarrel went where cameras aren't welcome" —stressed that plantations were off-limits to public scrutiny. The idea of nearly ten thousand men living half the year in single-sex barracks and deportable on company orders—the outcome of the social compromise at the heart of the post–World War II restructured labor market—was shocking to many viewers.

On February 15, 1991, the U.S. Commission on Agricultural Workers held hearings in West Palm Beach, Florida on the impact of the 1986 immigration law on agricultural workers and employers. The law amended the temporary foreign worker program to create separate agricultural and nonagricultural programs, with the former now called the H-2A program. A significant portion of the hearings concerned H-2A workers employed for sugarcane harvesting. Although the question was raised whether, theoretically, there was a wage at which U.S. workers would cut cane, one alternative was not discussed. That is, at what point would mechanical harvesting become "economic"? For example, when asked whether the industry attempted to recruit domestic workers, Leo Polopolus, an agricultural economist who had studied the Florida industry for decades, suggested that labor conditions were beyond the pale for North Americans at any wage. This, he explained, was due to the fact that cane grown on muck soil did not stand erect: "[B]ecause of that condition, you use a machete knife, a slash knife. . . . And so what it takes is a pretty macho guy with a machete knife that's going to go through muck soils and pound away at that cane. The experience is we don't have too many domestic workers who want to do that" (U.S. Commission on Agricultural Workers 1991, 289). When pressed further on the question of labor supply, Polopolus asserted that the domestic sugar industry had made "a bona fide effort to recruit domestic workers" but that because of "the nature of the work," the "domestic worker tries it and gives up" (289).

In contrast, Rob Williams, a Florida Rural Legal Services attorney, recommended that the government "get out of the business of sponsoring guest worker programs which guarantee growers a cheap, noncompetitive labor force" (U.S. Commission on Agricultural Workers 1991, 297). To stress his point, Williams described his recent visit to Osceola farms, part of Flo-Sun Corporation:

> There are over 700 workers in the camp, who come home from work each day with clothes filthy dirty from working in the fields; . . . the company has provided 20 wash basins and one dryer. . . . [I]f workers are caught trying to wash their clothes in the rest rooms, they risk being repatriated. . . . Unfortunately, under the H-2A program, there is no incentive for employers to make any improvements in working conditions, and that's the fundamental flaw in the H-2A program. (297)

The testimony of Dr. Marshall Berry—who, as noted in chapter 6, had testified twenty years earlier on behalf of the UFWU—echoed Williams but went even further, calling into question the whole enterprise of Florida sugar production. Berry forged explicit links between foreign policy, trade, labor, and environmental issues:

> Why should an industry like sugar get 10,000 Jamaicans to farm the Everglades with the ecological damage—when we pay more for sugar and there are no workers that are willing to do the work because it's so terrible? What's the advantage to taxpayers and consumers? We're trying to get these Latin American countries to pay us the foreign debt that they owe us, and we're taking one of their big cash crop markets from them, because they all grow sugarcane. (318)

In July 1991 a congressional committee reported that the U.S. Department of Labor had failed to enforce the rights of temporary workers in the Florida sugar industry. The committee found widespread under-reporting of hours worked and singled out Okeelanta Corporation for its 95 percent noncompliance rate (U.S. House 1991). Overall, the Fanjul's sugar operations were cited for violations more than any other in the report, leading the Labor Department to recommend civil penalties of more than $2.5 million. Coming to the Fanjul's defense was Al French Jr., a coordinator of agricultural labor affairs for the USDA, whose father—Allison French—had been instrumental in initiating the importation of workers from the Bahamas into Florida in 1943.[1] Al French Jr. claimed that most of the Fanjul's problems stemmed from "sloppy bookkeeping" or were mere "technicalities." That he had been previously employed by a labor management firm directed by Rafael Fanjul—Alfonso and Pepe Fanjul's uncle—did not, he claimed, influence his opinion (Mayer and de Cordoba 1991).

A final blow to the industry's labor recruitment and control was the June 1992 U.S. General Accounting Office (GAO) report, which focused on further violations of H-2A workers' rights and concluded that the Department of Labor had taken "minimal actions to enforce certain laws and regulations" (U.S. GAO 1992, 10). Specifically, deductions from workers' wages for transportation costs, health and life insurance, and enforced savings plans had either been improperly managed or taken without authorization. The vulnerability of workers was underscored by the fact that "[n]either H-2A workers nor U.S. workers employed by H-2A sugar cane producers [could] change employers during the course of the sugar cane harvest." The penalty for H-2A workers would be immediate deportation, as well as withholding of a portion of wages already earned. A U.S. worker who left "for whatever reason" would be "denied employment by all of the H-2A sugar cane employers in future years" (U.S. House 1991, 7).

"End of an Era for USSC": The Mechanization of Hand Harvesting

The unwelcome publicity on labor conditions compelled the industry to respond substantively and symbolically in an equally public manner. In December 1992 USSC's in-house publication, *The Company*, announced that "in a move unprecedented in the history of the 'H-2A' foreign worker program, U.S. Sugar and three leading farmworker advocacy groups announced 'labor peace' concerning the Company's sugarcane cutters at a press conference in Washington, D.C." USSC President Nelson Fairbanks stated his determination "that U.S. Sugar [would] be recognized as the nation's best agricultural employer" and that the company wanted "to save the H-2A program and jobs of hardworking cane cutters whose families depend on their earnings here at U.S. Sugar" (USSC 1992/1993, 1).

In conjunction with "labor peace," USSC initiated "Open Harvest," inviting news media to observe company operations in the hope that they would portray USSC as "the nation's best agricultural employer" beyond the confines of the plantation village. USSC sought to counter the sinister impression left by depictions in *Big Sugar* or *Bittersweet Harvest* of cutters as a captive workforce living in miserable quarters. As Fairbanks explained, "Our public relations advisor suggested we take a head-on approach to the allegations. We decided to open up the company from one end to the other" (quoted in Ruane 1991, 1). Access to workers in the company's rural villages was therefore a central focus of the event; freshly-painted and well-maintained village commissaries, recreation halls, houses, and barracks gave visitors an alternative view of plantation life, not unlike Bitting's idealized depictions more than fifty years earlier. USSC deemed the public relations campaign successful, resulting in "press coverage that extended across the nation" that offered "fair and balanced views of the company, the industry and the harvest" (USSC 1992/1993, 9). Public relations coup notwithstanding, the industry's labor problems continued. The October 1994 issue of the *Sugar Cane Workers News,* published by Florida Rural Legal Services (FRLS), was filled with articles concerning ongoing litigation and notices regarding payments due to workers, including a 1992 ruling mandating a $51 million back-wage settlement.

However, the end of manual harvesting was approaching, as machines began to replace men with machetes. Only fifteen hundred West Indian workers were hired for the 1994–95 season, all by USSC. The "labor peace" agreement protected cane-cutting jobs there, raised wages higher than those paid by other companies, and allowed USSC to avoid further legal problems. In contrast, Flo-Sun mechanized half of its 1992 harvest and the rest a year later, thereby eliminating two thousand cutting jobs (*Sugar y Azúcar* 1994). Farm worker advocates, including FRLS, had attempted to negotiate a guarantee of future employment for H-2A workers by offering these companies the prospect of smaller settlements for various ongoing lawsuits (Florida Rural Legal Services 1994). However, their effort failed because of a conjunction of factors; a Flo-Sun vice-president claimed it was "a normal technology-based transition in American industry" (23). Cane-cutting jobs would also soon decline at USSC, which had begun purchasing additional mechanical harvesters and assessing their profitability. Several independent growers identified reasons other than technological change for this transition: the fact that Belle Glade had the highest incidence of AIDS per capita in the United States "brought the sugar interests to a decision to see the advantages of mechanization in a new light"; thus, mechanization occurred "not because of external developments, but because of a perceived internal public health threat" (Heitmann 1998, 61).

On July 12, 1995, an article in the *Clewiston News,* headlined "End of an Era for USSC," reported that in the coming season the harvest would be fully mechanized. The demise of the H-2A program for Florida cane cutting seemed at once sudden and long in coming. After fifty years, the fact that offshore workers were still harvesting Florida cane with machetes appeared anachronistic, yet merely four years earlier close observers of the industry did not predict that complete independence from hand harvesting was so near. Many in Clewiston expressed nostalgia and regret regarding the end of this enduring, albeit paternalistic, employment relationship. Yet, even at the turn of the twenty-first century, Florida sugarcane was not entirely a "*machine made* crop": it still required hand labor to plant every acre that was not left to ratoon. Thus, a highly mechanized, industrialized, agro-production system was dependent on stoop labor to plant the crop row by row over tens of thousands of acres (fig. 7.1).

The counterfactual question—what would have happened had offshore workers not been available—is unanswerable. We can imagine a different agricultural labor market, in which domestic workers received appropriate wages for such demanding and arduous labor. This in turn might have bolstered the agricultural labor market more broadly, since south Florida provides counterseasonal employment at the national scale. Conversely, we can speculate that technological improvements would have occurred sooner had this agro-industry been limited to the domestic labor supply. History proves that the physical impediments to mechanization in Florida were not insurmountable. In the final analysis, the availability of domestic labor for cutting cane was not simply an economic question, but a social and political one as well. We can surmise from reading the historical record that no "white" person ever cut cane professionally in Florida, and that probably no one other than "black" workers ever did. Cane cutting was a racially inscribed job category and as such became associated with racist employment relations, rooted in the political culture of the Jim Crow South, that were repugnant to domestic workers. South Florida was not unique in its racism, but the racialized and gendered structure of the sugar plantation labor force was distinctive. Central to this structure were questions about who could and should cut cane and why, which was answered in the context of the intersection of the plantation system, corporate paternalism, U.S. agrarian and race relations, and Caribbean labor.

The Environment, Sugar's New Achilles Heel

With the mechanization of cane harvesting, the critical issue now facing the Florida sugar industry and EAA communities was the future role of agriculture in general, and big sugar in particular, in Everglades ecological restoration. Even as the industry's public relations machinery refashioned the image of labor conditions, company officials were acutely aware of a new challenge to sugar's place in the Everglades. Planning for "Open Harvest '92–'93," *The Company* predicted "that many of this year's Open Harvest guests will be more interested in the environment than in labor" (USSC 1992/1993, 9). This was not remarkably prescient, given that the sugar industry had been at the heart of a very public federal lawsuit filed in 1988 over Everglades water supply. USSC, which had finally brokered "labor peace," found that the environment had become their new Achilles heel. Though there had been ongoing public concern for the south Florida environment in general, and

Figure 7.1. Planting sugarcane in the twenty-first century still requires hand labor, much as it has for centuries. The planting practices in the top photo, taken on a Florida plantation in 1923, differ little from those in the bottom photo, taken in 1996. Top photo courtesy of the Historical Museum of Southern Florida. Bottom photo by the author.

the Everglades in particular, since at least as early as the 1920s, it was not until the mid-1970s that scientists and environmentalists placed the sugar industry at the center of this concern. To be sure, the question of the negative environmental impacts of sugar production in Florida had been raised in the 1930s when, John C. Gifford, a forester, expressed an interest in the "reclamation of the Everglades with trees." At a time when the nascent Florida industry was struggling to position itself as vital to national interests, he argued that by converting the region to sugarcane, "you will have large corporate interests dictating the politics and policies of this whole area," fixing "the level of the watertable to suit their own desires" (Gifford 1935, 25).

By 1947, when Marjory Stoneman Douglas famously described the industrial landscape emerging south of Lake Okeechobee, Gifford's "large corporate interests" had firmly established themselves in the Everglades.

> Everything was worked out with scientific exactitude, as directed by the Experiment Station or the laboratories of the sugar company, where soils are tested by light rays that cast a spectrum on a screen. The huge fields are set with dikes and irrigation ditches from which pumps bring up the water level to a required height every twenty-four hours. A tractor drags a cylinder six inches under the surface to make a covered drain like a long mole hole. Fertilizers and chemicals are added, plants dusted against insects. . . . The result is that more saw grass is burned and cleared for greater holdings, more ditches are dug, more water pumped from the lowered main canals. (Douglas 1988, 355)

As noted earlier, Douglas wrote *The Everglades: River of Grass* at a time of heightened awareness of the environmental problems caused by haphazard drainage.

The U.S. Army Corps of Engineers intended its C&SF Project to replace the haphazard drainage of the Everglades with a comprehensive, rationalized approach, but this did not solve the environmental problems. When the Corps completed the canal and levee system of the EAA in 1962, it effectively replaced the natural hydroperiod with an artificial water-regime, of which they were in charge. As it happened, the early 1960s were dry years, so that by 1965 an interim project was proposed to widen canals through the conservation areas in an effort to get more water into Everglades NP. When, after a severe drought in 1967, Governor Claude Kirk announced that continued release of water to the park was impossible, the *New York Times* editorialized "The choice is not between alligators and people. Rather, it is between farmers who will suffer a diminished crop and a park of national importance and unique quality" (*New York Times* 1967, E10). After a record low rainfall year, in September 1971 Governor Reubin Askew convened the

Governor's Conference on Water Management in South Florida. Askew declared a water crisis in terms of quantity, but also noted that "every major area in the South Florida basin . . . is steadily deteriorating in quality from a variety of polluting sources" (quoted in Blake 1980, 225). A task force was appointed to draft legislation that ultimately became the basis for the 1972 Florida Water Resources Act (FWRA), aimed at ensuring water quantity and quality.

In December 1972 Arthur Marshall, a marine biologist and founder of the applied ecology program at the University of Miami, led a team of scientists requesting that a "water quality master" for the K-O-E basin be appointed. The Corps' canalization of the Kissimmee River, a project that had reduced the forty thousand acres of wetlands north of the lake to less than nine thousand acres, had just been completed. The Marshall report also included a request that the legislature mandate restoration of the Kissimmee River's natural sinuosity. In response, the governor and legislature appropriated $1 million for the "Special Project to Prevent Eutrophication of Lake Okeechobee." The project report concluded that the Okeechobee Flood Control District operated mainly for the benefit of farming corporations, which consumed 58 percent of the water used in Palm Beach, Broward, and Dade counties but paid only 12 percent of the district's operating expenses. Recommendations regarding agriculture included weighing the ecological costs of flood control and drainage, imposing a user's tax, and converting sugar acreage to rice and other water-tolerant crops.

This report attributed the deterioration of Lake Okeechobee to three main causes: (1) the channelized Kissimmee River; (2) drainage projects north and northeast of the lake; and (3) the practice of backpumping used water from the EAA into the lake. Rejecting the restoration of the Kissimmee as too ambitious, the report recommended eliminating or minimizing backpumping by enlarging canals and sending surplus water to storage areas in the southern portion of the EAA. A thirty-thousand-acre tract known as the Holey Land was suggested for this purpose. Thus began the struggle over the geography of land use within the EAA that remains at the core of restoration politics. The special project report challenged powerful political interest groups: "The sugar growers . . . were threatened not only with a ban on backpumping but with exclusion from the Holey Land into which they had hoped to expand. Even more alarming was a possible change of state policy to discourage sugar culture and substitute the growing of rice and similar wetland crops" (Blake 1980, 265).

New water-management districts, which had been mandated by the 1972 FWRA and delineated where possible along watershed boundaries, came into being on January 1, 1977. The South Florida Water Management District (SFWMD), which replaced the Okeechobee Flood Control District, included the K-O-E system and extended to the coasts, including Fort Myers on the Gulf and Palm Beach on the Atlantic. That year, the Division of State Planning, Water Element published "The Florida State Comprehensive Plan," which included the following recommendation: "Encourage restoration of more natural hydrologic relationships in areas where development activities have significantly and detrimentally altered the natural hydrology beyond the extent necessary to support existing development and planned land use. Where practical, ecologically desirable, and where adequate documentation exists, the hydrologic conditions which existed prior to modification should be utilized as a guide for restoration efforts" (quoted in Blake 1980, 270). *Restoration*, a term used comfortably by USSC President Clarence Bitting thirty years earlier, was now becoming the goal of the Florida legislature and a growing threat to the Florida sugar agro-industry. It would also in the coming decade become part of the raison d'être of the U.S. Army Corps of Engineers.

"Save Our Everglades"

When environmental groups petitioned for a court order to enforce water-quality standards in Lake Okeechobee, the Florida Department of Environmental Regulation (DER) ordered that steps be initiated. In November 1978 the DER and the SFWMD agreed on a thirty-month permit to continue backpumping, but under carefully monitored conditions. In the meantime, the SFWMD agreed to devise a strategy for water-quality improvement, after which time EAA runoff would either have to be diverted or cleansed before being backpumped. Simply sending the water into canals and ultimately the ocean was deemed too wasteful; the alternative was to divert it into the conservation areas from where it would reach the park. The SFWMD biologists felt the nutrient-rich waters would not cause harm, but environmental groups were less sanguine and pushed to make water quality, not just flow and quantity, a central issue in planning. In the terms of its 1981 technical plan, the SFWMD agreed to make water quality a management goal. The plan included a program of best management practices (BMPs) for dairy farms north of the lake, a holding area in the Holey Land for discharge water from the EAA, and continued restrictions on backpumping into the lake.

While the SFWMD worked to improve its water-quality management, a new consortium of environmental groups, Friends of the Everglades, pushed the idea of restoration beyond questions of water quality and quantity. In their view, it was not just a matter of assuring that Everglades NP received so many acre-feet of cleaned up water, but that the whole geography of how that water reached the park needed to be considered. In a 1981 petition published by Marjory Stoneman Douglas, Friends of the Everglades urged that all federal and state agencies work toward "restoration of sheet flow to the greatest possible extent from the Kissimmee Lakes to Florida Bay" (Friends of the Everglades 1981, 1). They argued that water control had transformed the Everglades into a highly intensive, fossil-fueled system "which drastically displaces the solar-driven processes which produce wetland vegetation, peat and muck, potable water, fish, and wildlife" (2). Restored sheet flow is thus the key to reclaiming solar energy products from the Everglades.

A drought in 1981 followed by dry-season flooding in 1983 amplified the sense of crisis regarding the health of the Everglades ecosystem. In August 1983 Governor Bob Graham announced the Save Our Everglades (SOE) program, with the stated purpose that by the year 2000 the Everglades should look and function more as they had in 1900 rather than as they did in 1983. The program built on Arthur Marshall's plan (reiterated by Friends of the Everglades) and some of the initiatives already underway, such as the restoration of the Kissimmee River and of natural sheet flows to the park (Brumbeck 1990; Light, Gunderson, and Holling 1995). The SOE program was significantly different from the Marshall Plan, however, because, at least initially, it made no reference to nutrient problems that could degrade water quality (Light, Gunderson, and Holling 1995). In 1984 Governor Graham established a state resource and management committee and met with the environmental community to organize the Everglades Coalition. In addition, he instituted a Save Our Everglades report card to be made public annually.

A 1986 algae bloom on the lake "created a political fire storm" (Light, Gunderson, and Holling 1995, 144) in the aftermath of which the Florida legislature passed the Surface Water Improvement and Management (SWIM) Act of 1987. The act required Florida water-management districts to consider all of the water bodies within their boundaries and rank them in priority according to two categories. For those in pristine condition, districts were required to devise a strategy to maintain them, and for degraded water bodies, they were to develop a plan to restore and maintain them. SWIM set targets for how much phosphorous might enter Lake Okeechobee and specified a model to measure flows. It set up a technical advisory council to "study the effects of phosphorous on the WCAs [Water Conservation Areas] and other areas south of the lake" (John 1994, 136). The SWIM Act also included provisions regarding the park, specifying that water management districts "shall not divert waters to . . . the Everglades National Park in such a way that state water quality standards are violated [or] that nutrients in such waters adversely affect indigenous vegetation communities or wildlife" (quoted in John 1994, 136). The SFWMD conducted research showing that "that the diversion of nutrient-laden waters from the EAA had seriously damaged the biological integrity of the Everglades and, if left unchecked, could eventually damage the park. Native sawgrass and periphyton communities were being replaced by pollution-tolerant taxa" (Light, Gunderson, and Holling 1995, 145).

Meanwhile, the 1986 and 1990 U.S. Water Resources Development Acts (WRDA) authorized the use of federal funds for environmental restoration, the latter explicitly designating restoration projects as a mission of the Army Corps of Engineers. Whereas the Corps had earlier determined that no federal action should be taken to restore the Kissimmee, the 1990 WRDA directed the Corps to undertake a feasibility study of the project. The Corps, long targeted by environmentalists for the ecological havoc wreaked by its engineering projects, was now mandated to lead the way in integrating ecological restoration into federal resource management.

Taken together, the new laws, evolving institutional structures, increased knowledge base, and changing agency goals might have generated the social conditions necessary to "save our Everglades." As the SFWMD prepared the Everglades SWIM plan, however, the iconic wetlands once again became entangled in U.S. presidential campaign maneuverings. Dexter Lehtinen, acting U.S. attorney for the Southern District of Florida, filed suit on October 12, 1988, against the SFWMD and the State of Florida. Lehtinen, a Republican, did not give the customary thirty days' notice, nor did he inform his superiors of his intention to file the suit. His timing was in part intended to bring the Everglades situation to national attention during the 1988 presidential campaign, when the Democratic candidate, Michael Dukakis was seen as weak on environmental issues because of water quality in Boston Harbor (Hagy 1993; John 1994). In contrast, George H. W. Bush, then sitting vice-president, adopted as part of his campaign platform the newly drafted National Wetlands Policy Forum report in an effort to "green up [his] poor environmental record" (Vileisis 1997, 318).

The suit charged that the state had failed to enforce state law by allowing polluted water from sugarcane farms to flow into Everglades NP and

the Loxahatchee National Wildlife Refuge. The central question in litigation was the controversial scientific issue of how phosphorous affects the
park and the Loxahatchee refuge. On one side of the suit were Lehtinen,
the superintendents of the park and of the refuge, environmental groups,
and eventually the U.S. Department of Justice. On the other side were the
SFWMD, Governor Bob Martinez, the DER, and, though unnamed in the
suit, the sugar industry. As John explained, "When Lehtinen became acting U.S. attorney, in early 1988, he sought ways to expand his office's role in
environmental issues. He met with local environmentalists and asked them
what single most important thing he could do for the environment. Their
response was quick. The Everglades was the obvious issue, and the focus
should be the sugar industry and other agricultural uses" (1994, 139).

"Just Say No to Sugar"

The political relationships struck by the suit were unusual; in essence, the
federal government was suing the state and the district over the quality of
water being discharged by federally owned pumps. Lehtinen's suit avoided
the Clean Water Act, since it did not regulate agricultural runoff, and instead contended that the state was violating Florida water-quality laws,
thereby damaging the federally owned park and refuge. If the Corps, which
owned the pumps, had been named as a defendant, there would have been
no suit because the federal government cannot sue itself. Scientists and administrators working for the SFWMD, who were in the midst of preparing
a SWIM plan that many felt would have obviated the purpose of the suit,
expressed frustration at being its target. As the SFWMD's director of Everglades regulation, Paul Whalen, explained in an interview,

> This was ironic with the U.S. Department of Justice. They were saying that the
> state was not enforcing state law of federally subsidized, price-support assisted
> growers who discharge into a federally designed system, who we [SFWMD]
> operate under federal guidelines, and at the same standpoint, the Army Corps
> of Engineers, we tried to get them as a co-defendant, and the Department of
> Justice said "no," they wouldn't allow that because that would be quite awk-
> ward for one branch of the federal government suing another. They told the
> Soil Conservation Service, whose mission is to help growers develop conserva-
> tion plans, etc., to stay out of it for a total of five years because that would again
> be embarrassing for one branch of the federal government to do that. And they
> even tried to get the National Academy of Sciences, which is very unbiased,
> highly esteemed, to come in and define the problem for us. The U.S. Depart-
> ment of Justice told them not to.[2]

The FSCL requested status as interveners, but was refused. However, two towns, Clewiston and South Bay, were allowed to be parties because their municipal sewage contained phosphorous. Not surprisingly, the sugar industry paid the towns' attorney fees. To include Clewiston was tantamount to including USSC. Not only did USSC provide approximately twenty-six hundred full-time and another three thousand seasonal jobs, but it still owned virtually all the land in and around town, thereby controlling commercial and residential development. In what would become a familiar refrain in the "restoration" process, participants in the suit complained that bureaucratic and legal procedures consumed time and money without improving wetlands ecology. According to Whalen, "Fifty million dollars, by rough estimates, all parties; the district, the growers, the federal government, environmental groups . . . spent in legal fees. Not one ounce of phosphorous was reduced to the Everglades in that."[3] In the end, thirteen different parties joined the lawsuit, which, if nothing else, put the question of Everglades water quality on the table.

Lawton Chiles, then U.S. Senator, made the wastefulness of the lawsuit an issue during his successful 1990 Florida gubernatorial campaign. In his first speech after inauguration, made to the Everglades Coalition, he promised to settle the lawsuit within six months. Chiles asked each side in the lawsuit to designate its chief scientists, who were in turn asked to address the technical question of what level of phosphorous was harmful to Everglades NP and how large an artificial wetland would be required to bring EAA water to an acceptable level. Neither the sugar industry nor environmental groups were directly privy to these negotiations.

At this point legislation was being drafted that conceded the original point of Lehtinen's suit in requiring the SFWMD to apply to the DER for EAA pump permits. The bill, which required the district to prepare a SWIM plan to restore the Everglades hydroperiod, would grant the district power to condemn farmlands for use as wetlands and to raise funds to construct wetlands. The sugar industry successfully lobbied for a clause that required that the district base its assessment on a farm for cleanup costs on the amount of phosphorous the farm contributed to the total. The bill, which passed the Florida House and Senate in 1990, was named the Marjory Stoneman Douglas Everglades Protection Act; Douglas celebrated her hundredth birthday that year. Four years later, Douglas would request that her name be removed from the act because she felt that it favored the sugar industry. Reporting on her comments at a 1990 Earth Day rally, the *Miami Herald* quoted her views on sugar producers: "One of our great jobs is to get completely rid of them," she said (Morgan 1990, 4B).

In May 1991 Governor Chiles proclaimed to the U.S. District Judge hearing the federal lawsuit, "I have come to surrender. . . . I want to find out who I can give my sword to. . . . Let us use our troops to clean up the battlefield" (*Miami Herald* 1991, 12A). In July Chiles announced that Lehtinen, the federal agencies, and the SFWMD had agreed on a settlement that included a schedule for granting permits and wetland construction. However, the sugar industry appealed the settlement. Even as they did so, the FSCL was experimenting with methods of phosphorous removal and developing best-management practices. For the industry, the critical questions were whether they could meet the phosphorous standards, how much land would be taken out of production, and how much they would be required to pay toward restoration. For environmentalists, however, the issue was more fundamentally whether sugar had any place in the historic Everglades: "[Environmentalists] were convinced that they could defeat sugar again in the Florida legislature if they had to. Paul Parks of FOREVERGLADES expressed this feeling in telling an audience . . . at the 1992 Everglades Coalition conference that the settlement had created a 'new unity' against the 'common enemy': the sugar industry. His organization distributed bumper stickers that said 'Just Say No to Sugar'" (John 1994, 173). The stakes in the sugar versus the Everglades game were high and, to some observers, challenged more than the future of producers in the EAA. The December 1992 issue of the *Farm Journal* warned that the debate about phosphorous leaching from "the rich muck soil south of Lake Okeechobee holds implications for all U.S. agriculture. With environmental activists at battalion-strength for this showdown, it's clear they see the Everglades as a test case. If phosphorous standards are forced on farmers there, it could be done anywhere. The impact on farming could be huge" (Johnson 1992, B-4). Meanwhile, by late 1992 data showed that the various, newly implemented best-management practices had reduced phosphorous concentrations in EAA runoff by as much as 35 to 40 percent.

In February 1993 U.S. Secretary of the Interior Bruce Babbitt weighed in on the dispute and "got a standing ovation after speaking to the eighth annual Everglades Coalition conference." Babbitt proposed setting up a federal task force to coordinate the various agencies involved in the restoration as a step toward ecosystem-based management because, he explained, "we're all going to be coming back to the Everglades as the test case, for all federal agencies, for all park systems, for all states, for the entire country" (Parker 1993, 1). It was, thus far, a tremendously litigious "test case" that had already cost millions in legal fees and was expected to cost several million more because of lawsuits filed by the FSCL against the SFWMD (Hagy

1993). Litigation was now driving scientific research. According to one of SFWMD's employment representatives, it also drove the restructuring of that agency.

All of a sudden this governmental entity that's using tax dollars is now charged with defending itself, all of its scientific data, all of its environmental information, and you know, hey, you want to get your own expert witnesses up there, so all of a sudden you saw a very sharp increase in our numbers of scientists and very sophisticated upgradings in our computers, GIS stuff, data bases. We saw a huge increase in our number of attorneys, too.[4]

In 1992 Congress authorized the Corps to undertake a study to determine how the C&SF Project could be reengineered; now Secretary Babbitt brought Department of Interior and Corps officials together to explore how the Corps' "restudy" would fit in with the restoration process. In this way, Babbitt's proposed federal task force took shape, and in June 1993 the Department of the Interior convened the South Florida Ecosystem Restoration Task Force. The task force, comprised of representatives at the assistant secretary level from six departments and ten agencies, was formalized in September 1993 with the signing of the Federal South Florida Interagency Task Force Agreement. The task force was to undertake the original C&SF Project restudy, to design an ecosystem-based science program, and to provide support and coordination for endangered species recovery plans and various ongoing restoration projects. The task force then created a working group, which in turn established three subgroups: science; infrastructure; and management and coordination. The working group was asked to prepare a draft of an overall ecosystem restoration strategy by fall 1994 (FDCH 1994; U.S. GAO 1995).

At the February Everglades Coalition conference, USSC and Flo-Sun each had released a plan to resolve litigation. USSC said it would drop its lawsuits, publicly support the creation of thirty-five thousand acres of artificial wetlands and remove every pound of phosphorous that entered the water on its lands. Flo-Sun offered to spend $110 million for treating effluents chemically and for designating eighty thousand acres of public lands as a water-storage area. The critical question concerned how much money the sugar industry would and should pay. In negotiations between the state, district, and industry, the industry's share of the cost had been rising, until June 1993 when Lieutenant Governor Buddy MacKay, designated by Chiles to oversee the state's involvement, requested Babbitt's involvement.

Babbitt's appointed Everglades "Czar," George Frampton, and several other Interior Department officials met with MacKay and sugar industry executives to negotiate an agreement. At a press conference on July 13, 1993, Secretary Babbitt announced that they had agreed upon and drafted a "Statement of Principles." The sugar industry agreed to pay up to $322 million over twenty years to help build a system of wetlands, larger than had been called for by the previous settlement, and to withdraw its lawsuits after a ninety-day period. Forty thousand acres of farmland were to be converted into six artificial marshes. The agreement included a schedule for wetland construction but extended the deadline for water-quality standards by five years. "After Babbitt's announcement, Nelson Fairbanks of U.S. Sugar and Alfred Fanjul of Flo-Sun thanked Babbitt and MacKay. Americans had cast their votes in November 1992 for change, Fanjul said, and 'today, the Clinton administration delivers'" (John 1994, 182). Many environmentalists were displeased. Babbitt's compromise was called a "betrayal of the Everglades." Nineteen environmental organizations signed a letter expressing dissatisfaction with the statement of principles, asking that 70,000–120,000 acres of the EAA be set aside for wetlands. Lehtinen termed the agreement "disappointing" and "dangerous" because "[d]elay is the enemy of the Everglades" and because the agreement "essentially immunized big sugar from having to reach the final water restoration goal" of the original lawsuit (Rohter 1993, A1). The *Tampa Tribune* called it "too sweet" for growers (*Tampa Tribune* 1993, 8), while the *St. Petersburg Times* editorial questioned the administration's "benevolence to the sugar industry" (*St. Petersburg Times* 1993).

By December 1993 the "great optimism on the part of farmers." regarding the "Statement of Principles" was also fading because the agreement seemed to be falling apart. With the signing of the statement, USSC had expected "twenty years of environmental peace," but the tone of negotiations had changed so that "farmers believe that federal negotiators are now insisting on conditions that go far beyond the July Statement of Principles" (USSC 1993, 1, 3). Agriculture's contribution, which had been fixed for a period of twenty years, was now expected to increase in ten. Furthermore, according to USSC, farmers were being asked to bear the cost of financial overruns or technical failure in the construction of stormwater treatment areas (STAs). However, the devil was not only in the details. USSC's brief contended that much larger issues were being raised. Most troubling was the U.S. Environmental Protection Agency (EPA) announcement that STAs would be regulated as if they were urban water-treatment systems, thereby "imposing more severe standards than required of any other agricultural system in the nation and very time consuming" (3). More ominous yet was the draft report of the task force working group's science subgroup, submitted to the

Corps in November, which, in the company's reading, "calls for removal of the dike surrounding Lake Okeechobee and creation of a flow-away [*sic*] system through the heart of private farm lands to the south. The study also calls for the elimination of farming in South Florida and returning the land to a marsh stage that existed hundreds of years ago. Implementation of the report would flood three towns; Belle Glade, Clewiston, and South Bay" (4). Such publicized readings heightened a general sense of communities under siege in the EAA and the surrounding region, and fueled local opposition to restoration plans.

To this point, farmers and industry representatives had claimed that environmentalists wanted them out of the EAA; now, as they saw it, there were maps to prove it. With the release of the science subgroup's draft report, the struggle over sugar's role in restoration had become explicitly geographic. One regional newspaper reported that the draft recommended that the optimum recovery plan would involve "reestablishing sawgrass throughout the region where sugar cane grows" (McClure 1994, D1). Frampton's public comments concerning the future of sugar in the Everglades, including wildly speculative estimates of production cuts, heightened local anxiety. Farmers, he suggested, would have to remain flexible to change: "That does not mean that agriculture will go out of business. But the amount of cane sugar is going to be less. It may be 10 percent, or 20 percent, or 75 percent, or 28 percent less in 20 to 25 years" (quoted in Cushman 1993, 17). In mid-December the July agreement failed when negotiations reached an impasse, primarily over the sugar industry's demand for protection from further restoration programs.

In January 1994 the mayor of Clewiston petitioned Governor Chiles for a meeting "to discuss the economic and human consequences of the proposed plans to restore the Everglades." Published in the *Clewiston News* and cosigned by more than forty 'Glades area government officials, his letter explained the urgency of the request: "Recently, the Corps of Engineers visited our community to talk about a new plan to restore the Everglades. In addition, they distributed a report that calls for flooding the entire region south of the lake" (*Clewiston News* 1994, 4). The following Sunday the Clewiston Ministerial Association sponsored a prayer vigil at a local football field. According to the *Clewiston News,* the irrational actions of the federal government necessitated the vigil: "It isn't a logical sequence of events. That's why these ministers knew that a Higher Authority would need to intervene" (Chandler 1994). The ministers' summary statement said in part, "Our people are already weary from over three years of uncertainty concerning the negotiation of a solution to the environmental issues effecting Lake

Okeechobee and the Everglades. During these negotiations, the govern-ment secretly developed a plan that if imposed on the area would effectively end farming in South Florida. . . . Now our people feel they are fighting not only for their jobs, but for their lives and families" (3).

That week in his address to an Everglades Coalition meeting in Miami, George Frampton "backed away from a controversial federal proposal to eliminate South Florida's sugar-farming area and let the Everglades 'River of Grass' flow freely again there. But . . . [he] warned sugar farmers they bet-ter agree to pay for cleaning up polluted water flowing off their farms or they will face the wrath of the federal government." Saying it was unlikely that the government would carry out scientists' recommendation to abolish the sugar belt, Frampton suggested that turning a fifth of EAA agricultural land into filtration marshes was "by no means radical" (McClure 1994, D1). One day earlier, the U.S. government had reached an agreement with Flo-Sun regarding their part in the sugar industry's lawsuit: Flo-Sun agreed to pay between $80 and $120 million over a twenty-year period. Articulating a divide-and-conquer strategy, Frampton noted that the agreement would secure funds from Flo-Sun for restoration while "putting additional litiga-tion cost burdens on those who choose to delay the Everglades clean-up process in the courts" (*U.S. Newswire* 1994), most notably, USSC. Environ-mental groups attending the Everglades Coalition meeting agreed on a new proposal: a petition drive to place a referendum on the state ballot that would create a penny-per-pound tax on Florida sugar.

On March 10 Secretary Babbitt sent what came to be a widely reported memorandum to Secretary of Agriculture Mike Espy, which suggested the potential for governmental "wrath": "I would like to meet with you to dis-cuss the possibility that the sugar allotment program under the Agricultural Act of 1949 can be administered in a manner that takes into account the en-vironmental compliance of individual participating beneficiaries in cases involving important federal resources such as the Everglades, its surround-ing national wildlife refuge, and Florida Bay" (Bureau of National Affairs, Inc. 1994, 62). Babbitt's letter was interpreted as a threat "aimed most at U.S. Sugar and smaller growers," to which Bob Buker, a USSC vice president, publicly responded: "About the only rationalization he could have for this is to kill the Everglades bill. I think what it does is it convinces us that Bruce Babbitt is a scorched-earth, radical environmentalist" (Mitchell 1994, A1). Lieutenant Governor MacKay went on record in support of Babbitt's threat as a strategy to hasten negotiations and "as a catalyst to look at . . . archaic farm policies" (Bureau of National Affairs, Inc. 1994, 63).

The state legislature's 1994 deliberations on an "Everglades bill" produced the Everglades Forever Act, which Governor Chiles signed into law in May. The act, which abrogated further evidentiary hearings on the plan for implementing the settlement agreement, was based on the mediated technical plan, the draft of the SWIM plan, the July 1993 statement of principles, and the compliance schedule of the January 1994 agreement with Flo-Sun (U.S. GAO 1995). Legislators intended it to "authorize the district to proceed expeditiously with implementation of the Everglades Program" (State of Florida 1994, 3), which included STA construction, BMP development and implementation, water-quantity and quality monitoring and hydroperiod restoration, and monitoring and control of exotic species. The act mandated an Everglades Agricultural Privilege Tax on agriculturally classified land within the EAA; annual taxes began at $24.89 per acre for the period 1994–97 and increased to $35 per acre for the period 2006–2013. The tax was calculated to cover the cost of achieving, roughly, a 75 percent reduction in phosphorous run-off from the EAA through a combination of BMPs and constructed filter marshes.

The act was intended to achieve through legislation what had been impossible to negotiate: a timetable and funding for improving the quality of the water flowing into the Everglades. The agricultural privilege tax would provide funds for this. Though the intent of environmentalists had been to make "big sugar" pay, the tax also drove unintended land use change. The large sugar growers did pay the largest share of taxes, but smaller, diversified farms would be paying at the same rate per acre as would, for example, USSC and Flo-Sun. The tax was applied broadly, so that even pasturelands were subject to the agricultural privilege tax of twenty-five dollars per acre, which altered the economic parameters of farmers' decision making. Ironically, it encouraged the conversion of pasture—arguably the most ecologically benign land use in the EAA—to more intensively cultivated land. What a farmer or rancher needed under these new circumstances was a crop that would command an assured return per acre and one that was resilient in the face of market vagaries and cold snaps. Sugar fit the bill, providing one could obtain the milling rights. As one rancher noted, calculating that the privilege tax amounted to seventy-five dollars per head of cattle, "What they're going to do is force us into sugarcane."[5]

In the meantime, environmentalists' challenges to big sugar in the EAA found support in a 1993 GAO report reassessing the U.S. sugar program in light of recent changes in domestic and international conditions. The principal findings regarding the sugar program were that it had cost sweetener users approximately 1.4 billion dollars annually from 1989 through 1991 and that benefits to growers were concentrated among a relatively small percentage of farms. This was especially true in the cane-sugar industry "with 17 farms receiving over one-half of all cane grower benefits" (U.S. GAO 1993, 4). In the company newsletter, Fairbanks warned USSC employees, "We must pull together once more as the vote on the 1995 Farm Bill approaches, as this gives the environmentalists another chance to attack. I'm afraid they will use this as another means of attempting to drive farming out of this area completely" (USSC 1994, 2). The GAO's Sugar Program report provided environmentalists and sweetener users a trenchant document to use as a weapon in the upcoming battle.

The Everglades Forever Act had not foreclosed further legislation, regulation, and geographic restructuring, and it left unresolved two fundamental issues regarding the relationship between the EAA and Everglades restoration. One concerned the economic relationship between sugar production and restoration funding. EAA farmers were paying taxes for "cleanup" costs, but should sugar pay more towards other restoration goals? The other concerned the ecological relationship between sugar production and the goals of restoration. Were the negative externalities of sugar production such as to necessitate its removal from or extensive diminution in the EAA so as to achieve the respective goals of the Marjory Stoneman Douglas and Everglades Forever Act? The two issues, economic and ecological, would be linked in the debate surrounding the 1999 Farm Bill and the proposed penny-per-pound tax on sugar.

In the time since Lehtinen filed the 1988 federal lawsuit, the knowledge base concerning the Everglades ecosystem had increased considerably. A consortium of Everglades scientists collaborated in workshops and symposiums, one result of which was an edited volume, *Everglades: The Ecosystem and Its Restoration,* involving fifty-seven authors working "in the midst of a virtual war of politics and litigation over the future of the Everglades" (Davis and Ogden 1994, 6).[6] They observed that such a "large-scale ecosystem restoration requires an approach different from those more frequently practiced on a smaller scale" such as wetland creation or vegetation replanting (4). Noting that key attributes of the original Everglades system—spatial extent, heterogeneity, hydroperiod, and location within an undisturbed limestone basin—had been irrevocably altered, the authors collectively addressed the uncertainties of restoring a remnant ecosystem for which the "baseline of information still contains many important gaps" (789). Despite such complexity and uncertainty, restoration should proceed, since the Everglades would be lost before knowledge gaps could be filled (fig. 7.2).

This ecosystem approach to the Everglades "test case" was generating

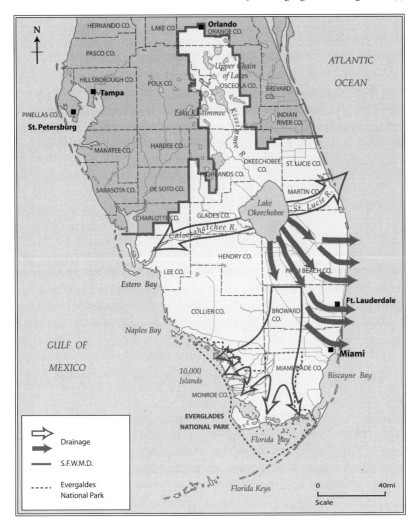

Figure 7.2. Planned "replumbing" of the water flow out of the Everglades Agricultural Area, a key aspect of Everglades restoration. Cartography by Mapcraft.com.

institutional complexities that seemed to mirror the ecological complexity of the K-O-E system. By 1995, twenty-six governmental agencies were involved in restoration planning. At the core of the restoration effort were the SFWMD, the Corps, the Governor's Commission for a Sustainable South Florida, the U.S. National Park Service, the U.S. Fish and Wildlife Service, the South Florida Ecosystem Restoration Task Force/Interagency Work Group, the Florida Keys National Marine Sanctuary, and the Florida Game and Fresh Water Fish Commission. Also involved were the USDA, the EPA, the U.S. Bureau of Indian Affairs, the U.S. Department of Justice, the U.S. Geological Survey, several other state and federal agencies, three regional planning councils, and the Departments of Natural Resources for Dade, Broward, and Palm Beach Counties. It was an enormous undertaking that stretched the limits of policymakers'—let alone the general public's—understanding of the complexities of the K-O-E system's political, ecological, and economic problems and the consequences of proposed solutions. Environmentalists, for their part, were able to distil the answer to ecological restoration down to two words: big sugar.

The Places of Big and Little Sugar in Restoration

By 1995, big sugar in Florida confronted industrial sweetener users, consumer groups, labor advocates, foreign governments, environmentalists, and government officials. As the politics of sugar sourcing and Everglades restoration merged, environmental organizations and major sweetener users formally united in the Coalition to End Welfare to Big Sugar. Hershey, Coca-Cola, the Sierra Club, the Audubon Society, and others joined forces, though their agendas differed. Domestic sweetener users wanted to buy sugar at the world market rather than U.S. price, whereas environmental organizations sought to downsize the Florida industry. In lobbying prior to the 1996 Farm Bill (the Federal Agriculture Improvement and Reform Act), which ultimately left the Sugar Program intact, the coalition published and distributed to lawmakers on Capitol Hill a mock tabloid, *The Bittersweet Times*, headlined "Aliens Earn Millions in Gov't Bonanza." Beneath the headline was a photograph of the Fanjul brothers, Alfonso and Pepe, captioned "These non-U.S. Citizens receive $65 million every year from the Gov't Sugar Program!" A highlighted quotation from *Forbes* magazine described the men as "greedy and ruthless" and "born in Cuba." Here and elsewhere, the Fanjuls in particular and Florida sugar industrialists in general were identified as "Sugar Barons." Thus the discourse of national identity was now deployed against the Florida industry as opponents portrayed it as greedy, foreign, and un-American, overlooking the history of protectionism that had fostered the transition of U.S. sugar sourcing from Cuba to Florida.

Such rhetoric, while scoring political points with anti-immigrant and xenophobic segments of the public, belies the social and political complexities of the Everglades sugar-production region. For one thing, the two largest companies that are most often identified as big sugar, USSC and Flo-Sun, are very different sorts of enterprises with very different relationships to place and community. USSC is now 49 percent employee-owned, with the remainder held by the Charles Stewart Mott Charitable Foundation. Flo-Sun Incorporated was established in 1970 when the Fanjul family restructured its Florida sugar operations. Whereas Flo-Sun is headquartered fifty miles away in urban Palm Beach, USSC is headquartered in Clewiston in the heart of sugar-producing country. USSC and the Sugar Cane Growers Cooperative are encompassed within the EAA and have a much stronger presence in the daily social life of their respective rural communities. Clewiston is a company town centered on USSC, socially, economically, politically, and spatially. The public park in the middle of town is flanked at each end by white-columned, plantation-style brick buildings, housing the corporate headquarters and the Clewiston Inn, also built and run by USSC. Walking across the park, one passes the library, youth center, playground, and swimming pool, evidence of the Mott Foundation's controlling hand in shaping the community. Production and place are fused in USSC, whereas the image of Flo-Sun is distant, corporate, and placeless.

As the sugar companies responded to environmental concerns, they chose widely disparate public relations and marketing strategies based on their distinct identities. USSC emphasized community, family, farming, and nation so as to counter the image of sugar barons.[7] In 2003, their Web site introduced an "integrated family" of agribusinesses and stressed two themes: the industrial aspects of sugar production and company ties to community. Wrapping sugar in the colors of the flag, USSC through Pillsbury ran a marketing campaign in 2000, heralding a "New Sugar Breakthrough!" The "new" product—granulated white sugar otherwise unchanged for more than a century—was adorned in red, white, and blue packages, signaling that the product was "American." Less able to claim such a homespun, community-oriented marketing campaign and well aware of consumer concerns for freshness, health, and environmentally sustainable production, Flo-Sun chose a very different strategy in the politically contested and economically competitive world of sweetener production.[8] In contrast to USSC's patriotic tints, Flo-Sun's 2003 earth-toned Web site depicted a stylized sun leading to topics such as "Natural Living," "Cooking Naturally" and "Environment" to emphasize consumer lifestyle and environmental sustainability rather than family, community, and agro-industry. Though Flo-Sun produced nearly 10 percent of the sugar consumed in the United States, its Web site did not emphasize refined white sugar but specialty sugars, such as Demerara, Light and Dark Muscovado, Milled Cane, and Certified Organic. Thus, the company strategically approached the high-end growth sector of the food market by developing organic and "heritage" sugars, a minimally processed granulated sugar, and organic cane syrup for use in industrial food processing. Through production practices as well as marketing—with packages depicting familiar symbols of a healthy Florida environment—they sought to reposition Florida sugar in the landscape, both symbolically and ecologically.

Adding to the social and political complexity of the EAA is the presence of a significant number of permanent and seasonal rural workers who depend on the sugar industry for their livelihoods. In 1996 both the Florida AFL-CIO and Jesse Jackson campaigned against the sugar tax on behalf of sugar industry workers (*Miami Herald* 1996; Zaneski 1996a). Many of the livelihood questions in the EAA pertain to agro-industrial and agricultural workers, including mechanics, engineers, heavy-equipment operators, truck drivers, mill workers, managers, and clerical workers. Although some jobs are seasonal, mills and plantations provide substantial year-round employment, because during the summer the mills are entirely rebuilt in preparation for the grinding season and there is extensive maintenance of equipment and fields. Since 1995, most of the fieldwork has been mechanized, with one important exception: cane planting. As a perennial, sugar is replanted every three years or so; thus, only a portion of the EAA sugarcane acreage is planted each year. Work crews employ both women and men, many of them Haitian or Mexican immigrants. Cane planting fills a critical gap in the seasonal labor schedule because it offers employment during the winter months, an otherwise slack time for agricultural workers.

Finally, big sugar is not the only producer in the EAA; it has coexisted with "little sugar" since the 1940s, when family ranchers and farmers in the area were encouraged to diversify by planting sugarcane. The relationship between the sugar corporations and the other farmers and growers in the region is necessary and symbiotic. Farmers emphasized the importance of sugar income to their diversified farming operations and the interdependence of big and little sugar. For example, farmers must have a contract from a mill to assure a market for their cane. As farmer Jim Kirk explained, "Before I planted my first stick of cane I had to go to Sugar [USSC], 'Will you grind?' You have to get a home for it."[9] His family's multigenerational farming operation dated to the 1940s, with 960 acres in cattle, 1,280 acres in cane, and 340 acres in citrus. They diversified from ranching in 1984, choosing cane over vegetables because the latter are vulnerable to frost and economically volatile. The decision to use sugar to stabilize income was typical. Another farmer explained why he had helped to found the Sugar Cane Growers Cooperative in 1960. "I was in the vegetable business . . . [but] sugar is more stable because of our sugar policy. I wanted to add some stability to my operations, so when this opportunity came along, we took it."[10] Many farmers felt that USSC had provided support for their operations above and beyond a milling contract. One farmer, a woman who is the primary operator of a five-hundred-acre farm, explained how, when the family dairy business failed in the 1980s, she learned to cultivate sugarcane: "The sugar company helped us a lot when we first got started. I was scared slap to death. I knew nothing whatsoever. Thank god for an old guy that worked for U.S. Sugar, he taught me how to raise cane."[11] Thus when multiple interests allied against sugar production in a restored future Everglades, big and little sugar joined forces to protect their mutual interests. They sought to demonstrate that sugarcane not only was improperly blamed for the entire water-quality problem, but also could be central to its solution.

The Place of the EAA in the Comprehensive Everglades Restoration Plan

On February 20, 1996, the *Miami Herald* was headlined "Hope for the Everglades: Gore Offers Plan for $1.5 Billion Rebirth." In the run-up to the 1996 presidential election, as the sitting vice president running for reelection with President Clinton, Al Gore appeared in Everglades NP to announce a "grand plan for doubling the pace of restoration of South Florida's battered natural environment" and to "make (restoration) a national priority for the first time." Dubbed "Gore's plan," it proposed to "slice away a huge section of the state's sugar growing region, flood it, and use it for marshes and water storage reservoirs" (Zaneski 1996b, 1A). In addition to announcing the retirement of as much as two hundred thousand acres of farmland, Gore also supported the proposed penny-per-pound tax on Florida sugar. Gore's plan was premised on the understanding that sugar was the central impediment to "restoring" the Everglades. Similar to the scenario represented in the map produced by the science subgroup of the federal task force, Gore's plan posed a direct threat to the viability of the sugar agro-industry and questioned its very existence in the Everglades. Reacting to the plan, the vice president of USSC stated, "The farmers think the Clinton Administration has stabbed them in the back" (quoted in Zaneski 1996b, 4A).

As did Gore's plan, the Everglades task force envisioned acquiring EAA land for purposes of water storage, and to that end, in early 1999 the federal government paid $133.5 million to the St. Joe Company for the Talisman plantation. In addition to the cash price, Talisman was granted rent-free use of the land for five years, farming rights that they in turn sold to USSC and Flo-Sun (Ziewitz and Wiaz 2004). The "complicated transaction" involved "numerous land swaps," with the federal government acquiring 45,114 acres for reservoirs and the state another 10,708 acres for pollution-cleansing marshes. The state also agreed to buy another 5,280-acre farm for storing water (Silva and Zaneski 1999, 1A). From the perspective of many environmentalists, it was a bitter irony that sugar would continue to be farmed on some of this property. Writing in support of the deal before it was finalized, the *New York Times* editors said of proposed leasing, "That would be a bad deal for the Government and for the Everglades. It would keep the poisonous phosphorous flowing southward for as long as the producers continued to grow cane" (*New York Times* 1997, A18). This view concerning the impact of sugarcane farming on the Florida landscape was widely shared.

Despite the overwhelming preoccupation with phosphorous runoff from agriculture, however, some were beginning to feel that sugarcane was not the sole, or primary, culprit in destroying the Everglades. A significant environmental problem posed by the EAA stems from the fact that drainage exposes organic soils that decompose and subside, thereby releasing nitrogen and phosphorous into the environment.[12] Subsidence is occurring at an average rate of one inch per year in the EAA and is directly proportional to the depth of the water table (Snyder 1994; see fig. 7.3). The problem of subsidence is thus related to key issues defining the relation between agriculture and the remaining Everglades: the quality, quantity, and timing of water leaving the EAA. Because agronomists recently have found that sugarcane can tolerate flooding, some have proposed develop-

Figure 7.3. Subsidence in the Everglades Agricultural Area is evidenced by this relic dirt clod on the fencepost, which marks the pre-drainage soil level. Courtesy of SFWMD.

ing sugarcane cultivation methods that would mimic natural hydrological systems by maintaining a very high water table during the historic flood season. "As the EAA evolves to a zero-subsidence agriculture, it would also be evolving to conditions more similar to its natural predrainage conditions" (Glaz 1995, 611). Because sugarcane requires relatively low levels of fertilizer, phosphorous concentrations in drainage water are slightly lower from sugarcane versus fallow drained plots (Izuno et al. 1995).

Thus some scientists involved in Everglades restoration see sugarcane as the lesser evil in the EAA landscape of today.[13] They argue that it offers ecological benefits that land retirement and permanent deep-water storage do not, such as the economic incentive and wherewithal to manage exotic vegetation, the possibility of mimicking seasonal water regimes associated with the historic Everglades, and the ability to maintain a landscape mosaic. According to a wildlife biologist who specializes in the study of alligators and crocodiles as indicator species, "you get a lot better combination of ecological benefits, or at least the potential for ecological benefits and economic productivity, with agriculture."[14]

In January 1999 the Corps unveiled the draft C&SF Comprehensive Review Study, which met with fierce criticism. Officials at Everglades NP "ripped the draft $7.8 billion plan," which fell "way short of its promised restoration of the Everglades and Florida Bay, and might worsen problems for a neighboring national park in Biscayne Bay" (Zaneski 1999a, 1A). An "all-star team of ecologists," including Edward O. Wilson, Paul Ehrlich, and Peter Raven, found the plan to be "riddled with deep systemic problems" and "based on a badly flawed computer model" (Zaneski 1999b, 1A). Four months later the Corps unveiled a revised plan that doubled the pace for critical projects and was therefore found more promising. By the fall of 2000, the U.S. House and Senate each had approved "one of the largest restoration projects in the nation's history" and were hurriedly moving toward compromise as the presidential campaign heated up: "In an unusual feat of bipartisanship, the project made allies of Vice President Al Gore; his Republican rival, Gov. George W. Bush of Texas, Mr. Bush's younger brother, Jeb, the Republican governor of Florida, and farmers and environmental advocates." Florida—considered a swing state—ranked fourth in electoral votes, and so "Mr. Bush and Mr. Gore are in a furious battle to sway voters there" (Schmitt 2000, A19).

On December 11, 2000, the morning on which the Supreme Court heard final arguments regarding the Florida vote count to determine the outcome of the U.S. presidential election, President Bill Clinton signed an updated WRDA, which committed four billion dollars to the Comprehensive Everglades Restoration Plan (CERP). Present at the signing was Florida Governor Bush, who had in May of that year signed the Everglades Investment Act, which committed the state of Florida to pay half the cost of restoring the Everglades, that is, an additional four billion dollars. Thus, with President Clinton's signature, the world's largest environmental restoration project—measured in terms of funding—was launched. Accounts of the event note that as he left the White House that day, when questioned about his brother's chance of occupying it, Governor Bush replied that "We're here to talk about something that is going to be long-lasting—way past counting votes. This is the restoration of a treasure for our country" (quoted in Zaneski 2001, 48).

The CERP outlined more than sixty projects, altogether estimated to take nearly forty years to complete, with the Corps and the SFWMD having primary responsibility for carrying out the plan. In addition, the EPA, the Florida DEP, and the Miccousukee and Seminole Indian tribes would participate in monitoring and enforcing water-quality standards. The CERP is best described as a replumbing rather than restoration of south Florida, with primary emphasis on holding water for various uses, including urban, agricultural, and environmental. Thus the plan calls for a subterranean water-storage system more than twenty times as large as any previously built, a method known as aquifer storage and recovery (ASR), involving a complex array of more than three hundred underground wells, built to store as much as 1.6 billion gallons per day with little evaporation loss (U.S. GAO 2000). Other types of projects in the CERP included more than 180,000 acres of surface storage reservoirs; more than 35,000 acres of stormwater treatment areas (in addition to 47,000 acres built by the state in the EAA); wastewater treatment plants; and the removal of barriers, such as canals and levees that were part of the original C&SF Project, in an effort to restore sheetflow. Even as the plan was hailed as a victory for bipartisanship, numerous concerns arose regarding its engineering emphasis. Some advocated starting with simpler alternatives, such as restoring a corridor of vegetation from the lake to Everglades NP; others noted that the methods were risky and untried, with the potential to lead to unintended consequences, such as the leaching of mercury and arsenic from deep rock into stored water (Revkin 2002).

Before long, the fragile coalition of political interests patched together to support the CERP began to unravel, and the initial celebratory rhetoric became muted. First, in 2003, as the deadline for water-quality standards set by the Everglades Forever Act loomed large, the Florida legislature passed a bill that seemingly maintained phosphorous limits but extended the deadline for achieving them by ten years—to 2016—and added the phrase "to the maximum extent practicable" in key places throughout the document. Press accounts of Governor Bush's signing of the bill into law never failed to note the strength of the sugar industry in supporting it. Second, at the 2004 annual Everglades Coalition meeting, the Sierra Club took the very public step of removing its support of the CERP and charged "the Bush administrations, in Tallahassee and Washington, with abandoning Everglades restoration as intended by Congress in 2000" (Sierra Club 2004). Soon after, the Natural Resources Defense Council joined the Sierra Club in a lawsuit seeking to overturn permits issued by the Army Corps of Engineers for rock mining in the Everglades, which would involve the destruction of at least 20,000 acres of wetlands over the course of thirty years. The Corps incorporated the mined quarries into the CERP in a project known as the "Lake Belt." A significant portion of the funds for Everglades restoration (at that time, one-eighth of the eight-billion-dollar budget) was allocated to turn the depleted quarries of the Lake Belt into water-storage facilities, as a part of the CERP. Evidentiary hearings revealed that the Corps was unable to answer accurately basic questions as to where mining was occurring and how much land was being excavated, and was relying on mining companies' reports for this information. In his ruling, Judge Hoeveler noted that over the course of the hearings, nothing "demonstrated that the Corps has since obtained a firm grasp of the number of acres being mined or impacted" (Morgan 2007a, 1B).

The "Lake Belt" plan illustrates the emphasis on engineering in the CERP. The plan builds into the CERP the existing and future waste pits created by mining limestone in southeast Florida. As written, the plan allows multinational cement and rock-mining companies to continue in the course of operations to mine out a series of eighty-foot-deep rock pits, destroying 5,000 acres of Everglades wetlands in the first ten years, with another 15,000 acres to be dug over the course of thirty years. Two former quarries west of Miami will become reservoirs covering 9,700 acres, using untested technology to prevent seepage, with unresolved water-quality issues and construction costs higher than conventional reservoirs (CROGEE 2005). Much of the area slated to become "lakes" abuts the boundary of Everglades NP, yet "no one knows if the plan and rock pits will hold water. The bottom of the pits will remain unlined and may leak. The walls of the reservoirs may collapse under fluctuating water levels and the reservoirs may actually promote seepage of water out of the Everglades and into the pits" (Clark and Dalrymple 2003, 556).

Critics claim that the emphasis of the CERP has shifted; initially, it gave priority to water for Everglades NP and more generally, ecosystem restoration. Part of the problem may be traced to the early planning process, which focused on phosphorous. As a key determinant of ecosystem health, phosphorus was a logical concern, but as a wildlife ecologist involved in restoration planning noted, "This allows space to be developed while people are fighting over phosphorous. I mean, what a great thing that's been for developers."[15] Thus, one of the constraints has been to find methods of restoring water-storage capacity that are not land-extensive. The CERP presented to Congress in 1999 stated that its "primary and overarching purpose" was to restore the south Florida ecosystem. Now, while the goal is still "getting the water right," many feel that water storage for urban and agricultural uses has become the overriding emphasis.

In 1999, the National Research Council established the Committee on Restoration of the Greater Everglades Ecosystem (CROGEE) with a mandate to advise the South Florida Ecosystem Task Force. The CROGEE raised numerous issues in its final two assessments of the CERP, among them the following. First, the CERP "relies very heavily on engineered solutions such as ASR and the Lake Belt storage system" although "experience suggests that natural restoration processes usually produce more satisfactory outcomes" (CROGEE 2005, 8). Second, ASR—which accounts for three-quarters of the water-storage capacity in the plan and does not require large amounts of land—is "an untested technology" when implemented "on as large a scale as envisioned in the Restoration Plan" (5). Third, the cost and quality of water retrieved from ASR wells and from the two proposed wastewater-reuse facilities are not specified. Fourth, uncertainties regarding CERP technologies and methods interact with systemic uncertainties such as climate change, timing of restored hydrologic and ecological conditions, human population dynamics, and the effects of invasive species (CROGEE 2005). Fifth, federal funding is less than what was planned, so that projects directed toward Everglades NP are being delayed. From 1999 through 2006, Florida contributed $4.6 billion, while the federal government's contribution of $2.3 billion fell short; meanwhile estimated costs since 2000 have increased 28 percent to nearly $20 billion. Cost estimates have increased, in particular for land acquisition, in part because federal delays in funding have occurred in the context of rising land costs. Moreover, estimates do not represent the likely costs because they do not include all the components of some projects and "because the full cost of most CERP projects is not yet known" (GAO 2007). Between the preliminary planning and the implementation stages, cost estimates for particular projects can easily double. Finally, the CROGEE identified two possible sites to be re-considered for water storage: Lake Okeechobee and the EAA.

In assessing the "variety of potential fates" facing the EAA, the CRO-GEE noted that the "worst from the point of view of Everglades restoration would be commercial, residential, and industrial development of the area." The assumption was that conditions for agriculture would deteriorate due to soil subsidence, so that production would become less than economic or would be maintained with treatments that would "make the area less amenable to restoration." Thus the CROGEE zeroed in on the EAA "to consider uses more aligned with restoration goals," such as "turning all or parts of it into a wetland" or "simply" flooding it for water storage and enhanced sheetflow (CROGEE 2005, 9).

"Death by a Thousand Cuts": Sugar in Regional Trade Pacts

While the restoration plan questioned sugar in the Everglades, the twenty-first century version of the sugar question was posing an equally daunting challenge to Florida's agro-industry. As the new century opened, globalization and free trade were the twin mantras for the world economy, now refereed by the World Trade Organization (WTO). Just as it had at the turn of the last century, the sugar question pitted free traders against protectionists, with added complications stemming from the existence of many more producers and the emergence of new regional trading blocks. These new regional trade pacts, including the North American Free Trade Agreement (NAFTA) and the European Union, have shifted the scale at which the sugar question is debated, often pitting single countries against economically powerful blocks of producers. For example, Brazil, now the world's largest national producer of sugar, took its case against the European Union's subsidization of its sugar farmers to the WTO in 2004 and won. U.S. sugar producers, in contrast, have opposed regional free trade pacts with mixed success. Robert Coker, USSC's senior vice president for public affairs, called such trade pacts "death by a thousand cuts" (quoted in Bussey 2005a, 25). Sugar ultimately became a key sticking point in efforts to create a Central American Free Trade Agreement (CAFTA) and to expand NAFTA into the Free Trade Area of the Americas (FTAA).

While the issue of global free trade versus national protectionism of agricultural commodities is an old one, in Florida it has taken an ironic new twist. Florida boosters of the early twentieth century pinned their hopes for the development of the state on wetlands drainage and the establishment of a sugar agro-industry. Boosters of the early twenty-first century, in contrast, push for the "restoration" of the wetlands and the opening up of trade relations with the Caribbean and the Americas. At least publicly, they appear willing to sacrifice Florida's sugar agro-industry to both. When the representatives from four of the proposed CAFTA countries met in Coral Gables, Florida, in April of 2005, the head of Florida's campaign to have the proposed FTAA headquarters located in Miami lashed out at the U.S. sugar industry for "economic terrorism" (quoted in Bussey 2005b, 1C). Such hyperbole indicates the sharp divisions within Florida's business community and their political allies over questions of trade.

When the United States and five Central American countries, plus the Dominican Republic, signed CAFTA in May 2004, the Florida sugar industry put its well-oiled lobbying and public relations machinery into high gear to make sure the U.S. Congress did not ratify it. Virtually identical guest opinions against CAFTA were published in the *Miami Herald* and *Palm Beach Post* in the run-up to the vote, one coauthored by Coker and Gaston Cantons, vice-president for corporate relations of Florida Crystals, and one by Don Carson, executive vice president of Florida Crystals (Coker and Cantens 2005; Carson 2005).[16] Both claimed that NAFTA had led to the loss of "35,511 jobs" in Florida, a "$200 million trade deficit" with Mexico, and an influx of "cheap, unregulated crops," which "devastated" Florida farmers. Arguing that CAFTA would lead to more of the same, the corporate VPs suggested that many Florida "sugar farmers" would not survive its implementation. Struggles over trade pacts lead to strange political alliances, and the sugar industry, long a popular symbol of corrupt agribusiness, found itself on the same side of the CAFTA debate with left-of-center Democrats and labor unions. In the end, though the vote in Congress was delayed for more than a year, too many other powerful interests were aligned in favor of CAFTA, and it was ratified by a narrow margin on July 27, 2005. Meanwhile, the FTAA initiative ran aground when proponents failed to navigate the shoals of national agricultural commodity support programs, and it splintered into a series of bilateral trade agreements.

In addition to environmental and trade pressures, several other difficulties plagued the Florida sugar industry and the sugar industry at large at the start of the twenty-first century. First, world sugar prices reached a twenty-year low for the 1999–2000 season. With a glut of sugar remaining on the market the following year, U.S. beet farmers destroyed 7 percent of their crop, plowing under 102,000 acres in exchange for title to surplus sugar held by the government (*Ft. Myers News Press* 2000). The Florida Sugar Cane Growers Cooperative and Flo-Sun's sugar companies—under the umbrella name of Florida Crystals and now including Okeelanta, Osceola, Atlantic, and Kloster Farms—forfeited seventeen million dollars worth of raw sugar to the government, in effect taking the crop loan amount rather than market price (McNair 2000). One factor that led to the U.S. sugar surplus was the 1996 Farm Bill, which reduced price supports for many commodities but not sugar, which in turn led to increased production of both beets and cane. Part of the explanation for sugar's political success in this regard was attributed to the diverse geography of its sweetener coalition and part to the fact that, although it accounts for only 1 percent of U.S. farm receipts, sugar is the single largest agricultural donor to political campaigns (Barrionuevo and Becker 2005). A more geographically specific explanation considers the political strength and acumen of the Florida companies, with Flo-Sun ranked forty-first on the national list of top corporate "soft money" and the Fanjuls' generosity to both parties well known (Davies 2001). For example, while Alfonso Fanjul Jr. has supported Democrats, Jose Fanjul contributed more than two hundred thousand dollars toward George W. Bush's reelection campaign (Barrionuevo and Becker 2005).

A second problem—from the point of view of sugar producers—was a renewed interest in reducing sugar consumption in the face of what was construed as a national "epidemic" of obesity. In the last quarter of the twentieth century, the U.S. diet, already rivaling U.K. standards, had become even sweeter, with a 26 percent increase in caloric sweetener consumption (Putnam and Allshouse 1998). Responding to increased rates of diabetes and health problems associated with obesity, in 2000 the USDA and the Department of Health and Human Services sought to issue stronger guidelines to "go easy" on added sugars. Heading off such dietary warnings was a job for the Sugar Association, the latest incarnation of the Sugar Research Foundation. In 2007, no longer sitting on the sidelines as it had in the 1940s, the Florida industry accounted for five of the fifteen member companies of the association and 25 percent of its board members (Sugar Association 2007). The association continued to use several strategies to maintain sugar consumption levels. One was "early surveillance and rapid response" (Sugar Association 2002). Thus, when the U.S. Surgeon General suggested a link between obesity levels and sweetener consumption, the association quickly issued press releases citing expert opinion that no scientific evidence could prove this link. A second strategy was more insidious: the Sugar Association is a member of the Dietary Guidelines Alliance, comprised of industry groups, the USDA, and the Department of Health and Human Services. From their seat at this table, the Sugar Association has been able to forestall any dietary advice that would go against its members' interest, while gaining the imprimatur of membership (Nestle 2002). Dietary guidelines, for example, were altered from "go easy on beverages and foods high in added sugars" to "choose beverages and foods to moderate your intake of sugars" (Marquis 2000, 24).

Third, the industry faced a similar problem in 2003 when the World Health Organization (WHO) posted a draft recommendation that sugar should account for no more than 10 percent of a healthy diet. Again, the Sugar Association took action, this time "threatening to bring the World Health Organization to its knees by demanding that Congress end its funding unless the WHO scraps guidelines on healthy eating" (Boseley 2003, 1). At the behest of the Sugar Association, the Bush administration requested an independent review. However, "despite threats from the administration and Congress to reduce American financial support" only modest changes were made in the report (Barrionuevo and Becker 2005, C1).[17]

A fourth problem facing the industry concerned the political geography of the sweetener coalition. As late as 1995, the sugar industry could be described as holding "an umbrella" over the heads of corn farmers and the HFCS industry. As one Florida grower emphasized, "They're very interested in us holding that umbrella because they don't want to get wet." However, by the turn of the century that umbrella was no longer indispensable, partly because of relative prices, but more because of a promising alternative destination for Midwestern corn—the burgeoning ethanol market that dispensed with surpluses and began to push up demand and price for corn, which was increasingly independent of the U.S. sweetener market. Now corn farmers most adamantly support U.S. tariffs on imported ethanol.

Resecuring Sugar in the Twenty-first Century

The Florida sugar companies faced a dynamic set of circumstances at the turn of the twenty-first century: Everglades restoration, the pressure to negotiate trade agreements, flattening prices, threats to consumption levels, and a shifting geography of commodity alliances. In response, USSC initiated Project Breakthrough, bringing twenty engineers and other sugar experts to Clewiston in 2004 for a two-week planning session to determine how to modernize the original 1927 mill. Using technology from Brazil, South Africa, Louisiana, Finland, and France, USSC embarked on an expansion project to build the largest mill in the United States and the third largest in the world. With the new mill, USSC was expected to be the lowest-cost sugar producer in the United States and competitive globally. The new mill also allowed USSC to close the Bryant mill in Canal Point, which was listed as the top releaser of cancer-causing air pollution in Florida in 2002 and fifth in the nation (Santaniello 2005). Once the mill opened, USSC continued its rationalization of the work force; laying off 30 percent of its administrative staff in 2004 and planning to cut 60 percent of its milling workforce, from 570 to 226 employees (Bussey 2005a; Salisbury 2007).

Florida Crystals and its parent company, Flo-Sun, have taken a somewhat different approach to globalization than USSC. This variation was foreshadowed by the diverging responses to a proposal made in "a closed-door meeting in Washington" of Florida Congressmen, members of Congress from beet-producing states, USDA Secretary of Agriculture Mike Johanns, and White House policy analysts (Salisbury 2005). The proposal—to subsidize a sugar-based ethanol program in a fashion similar to the corn-based ethanol program—was initiated to sweeten the deals that the Bush administration hoped to close with respect to CAFTA and ultimately, the FTAA. Spokeswomen for USSC and the Sugar Cane Growers Cooperative of Florida rejected the idea, whereas Florida Crystals' representative responded more positively. Shortly thereafter, "Governor Bush released the 2006 Florida Energy Act," which included a grant competition for renewable energy research. In 2007, eight awards were announced, among them a one-million-dollar grant awarded jointly to Florida Crystals and Florida International University (FIU) for ethanol research, to be matched by one million dollars from Florida Crystals. The purpose of the funded research was to develop a cost-effective pretreatment process to convert sugarcane bagasse to ethanol, as a step toward determining the feasibility of using Florida bagasse in a large-scale bio-energy plant. Pepe Fanjul was quoted on the FIU Web site as saying, "We hope that this effort with FIU will enable us to develop cellulosic ethanol from our sugar cane that will reduce our dependence on foreign oil" (FIU 2007).

Thus an old discourse was reconfigured with the potential to again elevate sugar production to the level of national security, and in this respect, among others, the Fanjuls found themselves very much in step with both Jeb and George W. Bush. As reported by the *Miami Herald*, "both the president and Gov. Bush praise ethanol as a way to wean Americans from their dependence on foreign oil" (Davis 2006, 8B). In April 2006 Governor Bush submitted to President Bush a position paper that had been drawn up by Florida FTAA, Inc., "15 by '15: A Hemispheric Wide Approach to Ethanol," which set a goal of U.S. consumption of fifteen billion gallons of ethanol annually by 2015. The paper also suggested that the United States cooperate with Brazil and reconsider ethanol tariffs. Accordingly, in his "last Miami appearance as governor," Jeb Bush joined with Roberto Rodrigues, a São

Paulo sugar grower "representing Brazilian agribusiness," and Luis Alberto Moreno, president of the Inter-American Development Bank, to launch the Inter-American Commission on Ethanol. The purpose of the commission was to "fund feasibility studies, promote the use of Brazilian ethanol technology and target production opportunities in Central America" (Bussey 2006, 1C). Hence Governor Jeb Bush's appearance at the Biltmore Hotel in Coral Gables in front of "Florida business and government representatives and a top agribusiness delegation from Brazil" prefigured the trip that President George W. Bush would take in March 2007.

The stated purpose of this tour—"the longest Latin American trip of his presidency"—was to promote wider use of ethanol throughout Latin America (Rutenberg and Rohter 2007, A1). A fair amount of press coverage focused on sparring between Venezuelan President Hugo Chavez and Bush, with ethanol as the focus—either freeing the region from Chavez's petroleum dictatorship or foisting upon it Bush's imperialistic plot to use food for fuel. Certainly the concern in Washington that Chavez's regional influence was bolstered by Venezuelan petroleum reserves pushed the United States into a more enthusiastic stance toward Brazilian ethanol. At the heart of the tour was the "ethanol agreement—signed by Secretary of State Condoleezza Rice and the Brazilian foreign minister," under which the United States and Brazil would "share technology to enhance ethanol production and push its development in other Latin American and Caribbean countries" (A7). Not up for serious discussion was the reduction of the U.S. tariff on Brazilian ethanol of fifty-four cents per gallon.

Together, Brazil and the United States account for more than 70 percent of global ethanol production, with the United States producing slightly more in 2005. U.S. ethanol production is based primarily on corn, whereas Brazil has been developing a sugar-based ethanol program since 1975, after the first energy crisis. By the 1980s, more than three-quarters of the cars made in Brazil ran on cane-based ethanol, which fell into disfavor there in 1989 when sugar prices spiked, leaving motorists without fuel. The recovery of the Brazilian industry was predicated on the development of the "flex fuel" motor, introduced in 2003, which allowed consumers to switch between fuels based on price and supply. Now, with flex fuel engines, rising petroleum prices, and increasing demand for ethanol fuel, Brazilian leaders have identified the U.S. tariff as an obstacle to significant foreign direct investment, from, for example, "the four international giants that control much of the world's agribusiness—Archer Daniels, Bunge and Born, Cargill, and Louis Dreyfuss—[who] have recently begun showing interest" (Rohter 2006, A1).

The goal of the ethanol agreement is therefore to enlarge ethanol fuel capacity—that is, the infrastructure of production, distribution, and consumption—at the hemispheric scale and beyond. While there is currently no public promise of tariff reduction from the United States, the agreement outlines a level of collaboration that has the potential to deliver the substantially enlarged market that would entice the big players in global agribusiness. Although it might seem counterintuitive, in this case scarcity does not improve price; rather, abundance and ubiquity would create the conditions for the technological transition necessary to develop sufficient markets for ethanol fuel, which would provide substantial demand and therefore buoy price. As the vice president of an ethanol equipment manufacturer explained, "We want ethanol to become a global commodity, and for that to happen, Brazil can't be the only producer" (Andrews and Rohter 2007, B9).

Domestic and multinational agribusiness firms, the U.S. and Brazilian governments, various corporate nongovernmental organizations, and local and state governments participated in this project to forge the ethanol agreement. For the Brazilian business community, which has significant expertise and investment in ethanol technology, an important aspect of the agreement is its emphasis on developing ethanol production in other countries through equipment sales. Of particular interest are Caribbean nations and signatories to the CAFTA, which are exempt from U.S. tariffs if they use their own crops to produce ethanol. Also, under the Caribbean Basin Initiative, Caribbean nations can import partially processed ethanol to finish it before exporting it to the United States in quantities limited to 7 percent of U.S. ethanol consumption. The director of the energy division of the Brazilian Foreign Ministry, Antonio Simoes, said of the ethanol agreement, "This is more than a document; it's a point of convergence in the relationship that is denser and more intense than anything we've seen in the last 20 or 30 years. Brazil will profit, the United States will profit, and so will third world countries" (Andrews and Rohter 2007, B9).

At the global scale, the ethanol agreement points to an emerging assemblage that does not, contrary to the quotation above, necessarily translate into advantages for the general population in particular countries. Rather, it exemplifies the sort of de-nationalization that Saskia Sassen (2006) has carefully delineated, which adds global capabilities to the nation-state, along with processes of deregulation and privatization. In this case, the U.S. and Brazilian governments are working hand in glove with multinational agribusiness to promote the expansion of ethanol production and consumption at the hemispheric scale and beyond, with Japan identified as a major market. As they do so, we see a shift from a discourse emphasizing the critical importance of a nationally organized food-commodity sector to a discourse promoting a "global assemblage" (Ong and Collier 2005; Sassen 2006) of states, domestic and transnational corporations, nongovernmental organizations, growers, consumers, and a set of interdependent technologies, including biotechnology, ethanol technology and automotive technology.

Precisely who will benefit is open to debate, but there is mounting concern regarding environmental and social costs at the hemispheric scale. For example, in response to Bush's "energy visit" of March 2007, the head of the United Nations Environmental Program voiced apprehension about impacts on the Brazilian Amazon, including accelerated deforestation. Indeed, Brazil's publicized plan to open one ethanol factory per month for the next six years has helped to create a resource frontier on lands suited to sugarcane farming. The position of the Forum of Resistance to Agribusiness, a consortium of South American NGOs, with regard to this emerging global assemblage—"the era of biofuels"—is that it "represents a grave threat to our region, our natural resources, and the sovereignty of our people" (quoted in Kenfield 2007). Of primary concern are trends toward increased concentration of ownership and control of the sugarcane industry and the expansion of monoculture, with an associated increase in landlessness, rural poverty, and food insecurity (Kenfield 2007; Wright and Wolford 2003).

Meanwhile, back in Florida, newspaper headlines championed "Ethanol a Boost for State? Florida May Be the Center of Brazil-U.S. Fuel Alliance" (Bauza 2007, 1A). Immediately upon leaving the governor's office, Jeb Bush became director of the Inter-American Ethanol Commission in Coral Gables, which he had only recently helped dedicate. The ethanol initiative that he had spearheaded seemed to lead the way out of the impasse brought on by the strength of a united commodity sector's political resistance to "free trade" agreements. Ethanol was the perfect weapon to divide and conquer. USSC, the Florida Sugar Cane Growers Cooperative, and smaller growers remained skeptical—sensing a Trojan horse, perhaps—while Flo-Sun—with extensive holdings not only in Florida but also in the Dominican Republic—saw a moment of political and economic opportunity.

Thus, at the turn of this century, Florida sugar producers such as the Fanjuls stood to gain—as their predecessors did—from the federal government's interest in providing infrastructural support for corporate expansion. Previously it was water and labor control. Now, with the CERP, water control is maintained and water storage expanded. Furthermore, the ethanol agreement enhances the potential to develop a transnational commodity network based on using sugarcane for fuel. The Fanjuls were in a particularly advantageous position to profit because the U.S. market is open to Caribbean ethanol, for which they have economies of scale because they can export from their plantations in the Dominican Republic. Though USSC is not producing sugar for ethanol, a subsidiary of USSC, Southern Gardens Citrus, is designing a citrus-waste biomass ethanol plant in Hendry County with the help of a $2.5 million Florida state grant awarded to a Clewiston-based company, Citrus Energy LLC.

Throughout the twentieth century and into the twenty-first, the Florida Everglades and sugar bowl have been caught up in politics in myriad ways, through discursive practices and material interests operating at multiple scales and through linkages—cooperative and competitive—among places of production. At the national scale, the politics of agriculture, trade, and environment have been critical, resounding at the highest level of office. The most recent examples are characterized by an unusual convergence of interests, with brothers occupying the White House and the Florida governor's mansion for an overlapping period of six years. Two generations of the Bush family have greened their presidential or gubernatorial campaigns by discursively favoring Everglades restoration, while Florida in turn has supported their political ambitions with its large electorate. For the Bush brothers, the ethanol initiative seemed to provide several political-economic advantages—further greening through the promotion of alternative fuels, supporting the interests of multinational corporations, and transforming the politics of sugar in order to break down place-based resistance to free trade by dividing the interests of the Florida sugar industry. According to the Interamerican Ethanol Commission—chaired by Jeb Bush—President Bush "has even been dubbed as ethanol's 'Promoter-in-Chief'" (Interamerican Ethanol Commission 2007).

After a century, the sugar question still flourishes. Sugar geopolitics, once at the heart of colonial endeavors, are now thoroughly implicated in postcolonial global restructuring. U.S. sugar interests continue to frame their claims to protection with reference to European producers and EU sugar policies, and have more recently cited Japanese sugar policies as another example in support of their cause. However, now Brazil rather than Cuba is the largest and lowest-cost producer in the world; and now, as Brazil seeks access to the U.S. market, it does so not only through hearings in the U.S. Congress concerning domestic quotas, but also at the global scale in negotiations surrounding the FTAA and through petitions to the WTO. Thus the moral discourse has shifted from a question of bilateral bonds—What does the United States owe to Cuba?—to the issue of fairness in globalization, which the global ethanol assemblage is well poised to articulate. After decades of near silence on the sugar question, these recent geopolitical maneuverings have prompted a response from Fidel Castro, who "chided the Bush administration for its support of ethanol production for automobiles, a move that he said would leave the world's poor hungry" (*New York Times* 2007, A6).

As it has in the past, the latest version of the sugar question employs the discourse of security, which over time has grown more encompassing. First used during wartime with reference to sugar supplies and sugar's role in increasing soldiers' endurance, the securitizing of the sugar question was enhanced during World War II, when sugar became an ingredient of "every normal bomb and bullet" (Gervasi 1945, 20). The sugar industry's propaganda of that time—deployed to counter nutritionists' advice, to demonstrate sugar's wartime role, and to build support for postwar demand—seems prescient now. An advertisement produced by the Sugar Research Foundation depicted a futuristic family's personal helicopter with the accompanying instruction: "Give it a lump of SUGAR, Mary, we're flying out to the country. Far fetched? Maybe. But sugar experts don't think so. Making gasoline from sugar is quite possible they'll tell you. Here is the Research Foundation's confident prediction: *Sugar will become as common a source of raw materials for industry as coal or petroleum*" (emphasis in original).[18]

Certainly during the Cold War the sugar question, situated at the intersection of U.S. foreign and domestic policy, was framed in the context of national security. In its latest incarnation, security remains central to the sugar question, but now it is enhanced by modifiers such as "energy" and "environment." In outlining the "strategic partnership with Brazil" based on ethanol, a State Department spokesman explained "Our goal is to advance global energy security by helping countries diversify their supply" (quoted in Bachelet 2007, 9A). While in Brazil, President Bush promoted the agreement by invoking security: "If you're dependent on oil from overseas, you have a national security issue. In other words, the dependence on energy from somewhere else means you're dependent on the decisions of somewhere else" (quoted in Chang 2007, 9A). By no means is this limited to the U.S. administration; for example, the British government has proposed subsidies for "energy crops" for purposes of "environmental security" (Maynard 2007). But for the United States in particular, the resecuritized sugar question revitalizes the Cold War discourse, now placing Venezuela together with Cuba outside the realm of the emerging global ethanol assemblage.

How is the sugar question playing out in the EAA? More than a decade ago, it was possible to envision an agro-ecological resolution to the dilemma of the EAA in the form of "zero-subsidence" agriculture. Not only is sugarcane capable of growing as a wetland crop, but—as agronomists explained and the archives attest—insufficient funding for water control led to frequent flooding at the Canal Point research station, which meant that they were inadvertently selecting for flood-tolerant varieties. And because the dry seasons of the historic Everglades happen to coincide with the planting and harvesting seasons of sugarcane, it would have been possible to maintain cultivation schedules. However, subsidence has been a touchy topic for decades, because it brings into question the whole enterprise of sugar farming as mining the muck. Speaking from experience, USDA agronomist Barry Glaz noted, "Agricultural scientists since the 1920s have brought up this issue and EAA growers have told them to please not bring it up for years."[19]

Perhaps if the sugar question had been less fraught in Florida, a transition to a more sustainable— at least with respect to muck soils—cane farming would have happened in the EAA. Instead, having for decades flatly denied and disputed that they were mining the muck, today the sugar companies and farmers are prepared to concede the point. Lobbyists for the Sugar Cane Growers Cooperative have worked with road contractors, development interests, and mining companies in an effort to rewrite state mining laws to do away with local regulation of land use by "blocking counties or municipalities from enacting or enforcing any rule that 'prohibits or prevents' operation or construction of quarries on land zoned or designated for mining—an area that covers the 700,000 acre Everglades Agricultural Area." They claim that in the face of "foreign competition" and thinning soils, Florida sugar farmers need the "profitable option" to mine limestone without interference from local communities (Morgan 2007b, 1B). In addition, the Fanjul family has plans to develop thousands of acres of their Florida Crystals landholdings in Palm Beach County into a residential subdivision, and it is all but certain that they are not alone among the major growers in contemplating substantial housing development in the EAA (Sorrentrue 2005). Thus the distinction between developers and agricultural interests in the EAA—always a fragile proposition—is rendered moot. It is a wonder that the agrarian myth served so long and so well to undergird the interests of these corporations, whose concentrated ownership and control of massive amounts of Florida real estate is now—as they diversify from sugar farming to rock mining and residential construction—rendering the imagined geography of a "restored" Everglades even more fantastical.

Biographical Sketches of Key Figures in the Transformation of the Florida Everglades

Hamilton Disston (1844–96)	Early Florida real estate developer who initiated drainage in central Florida. From a Philadelphia family, owners of Disston & Sons Saw Works. Contracted in 1881 with Florida's Internal Improvement Fund Trustees to pay $1 million toward the state's Civil War and Reconstruction debt. The agreement required Disston to drain the state's wetlands in exchange for half of the drained lands.
Samuel Lupfer	Came from Pennsylvania to St. Cloud, Florida, where Disston had established a sugarcane plantation; he was superintendent there at the time of Disston's death. A landowner, farmer, and businessman, he became an outspoken advocate of Everglades drainage during Governor Broward's administration.
Rufus E. Rose	Engineer for the Disston Drainage Co. and later state chemist of Florida. In 1885 Rose began a sugarcane plantation in St. Cloud, which eventually became part of Disston's holdings. As state chemist, he was a vocal advocate of wetlands drainage to unlock what he viewed as Florida's unlimited potential for sugarcane production.
Harvey W. Wiley (1844–1930)	Appointed USDA Chief Chemist in 1883 and later became widely known for his work with the Good Housekeeping Institute (funded by the magazine of the same name) and for his public opposition to food additives, which he publicized through contributions to *Good Housekeeping*. His advocacy was instrumental in the passage of the 1906 Pure Food and Drug Act. As the USDA's chemist in the 1880s, he analyzed Everglades muck soils, touting their fertility and advocating systematic drainage for sugarcane production.

Herbert Myrick *(1860–1927)*	Editor of several farm press publications at the end of the nineteenth century, including the *American Agriculturalist, Orange Judd Farmer, New England Homestead,* and *Farm and Home.* He was an officer of the American Sugar Growers' Society and a tireless booster of the U.S. beet- and cane-sugar agro-industry. He was director of the Good Housekeeping company from 1900 to 1911.
Napoleon Bonaparte *Broward (1857–1910)*	An advocate of Florida sugarcane farming and Everglades drainage, active in both the Interstate Sugar Cane Growers Association and the National Drainage Congress, an organization devoted to the exchange of drainage information and to promoting federal drainage. Elected Governor of Florida in 1904 on a platform of rational drainage of the Everglades. During his term of office (1905–9) dredging was begun, and miles of new canals were dug. He was the Democratic nominee for the U.S. Senate race in 1910, but died before the election.
Bror G. Dahlberg *(1882–1954)*	Early investor in Everglades sugarcane production and President of Celotex Corporation. Dahlberg purchased the Southern Sugar Company in 1925 with the notion of using sugarcane residue to manufacture Celotex, a building material. Under Dahlberg, sugarcane expansion and wetlands drainage were part of a geographically diverse, vertically integrated plan for interregional agro-industrial development.
Manuel Rionda *(1854–1943)*	Immigrated to Cuba from Spain as a teenager and became a major player in the international sugar trade. His family owned sugarcane plantations and mills in Cuba. Founded the Czarnikow-Rionda Company, a major sugar brokerage firm, in New York and the Cuban Trading Company, the firm's trade name in Havana. He was succeeded as company president by his nephew, Bernardo Braga Rionda.
Charles Stewart Mott *(1875–1973)*	An industrialist and philanthropist whose Weston-Mott Company, which originally manufactured wire wheels for bicycles, began producing wheels for automobiles in Flint in 1906. Weston-Mott became part of General Motors Corporation in 1913, and Mott became executive vice-president of GM in 1920. In 1929, along with associates in the auto industry, Mott bought controlling interest of the Southern Sugar Company in Clewiston, Florida, which was renamed the U.S. Sugar Corporation.
Clarence Bitting *(b. 1891)*	President (1930–46) and major shareholder in the U.S. Sugar Corporation, he came to Florida in 1930 from New York, where he ran a management firm, Bitting, Inc., with his brother William. He was a business associate of Fred Fisher of the Fisher Body Corporation, which became part of General Motors

Corporation's holdings. As USSC president, Bitting served as an effective company spokesperson through congressional testimony, press releases, and well-researched corporate publications.

John E. Dalton   Author of the 1937 monograph *Sugar: A Case Study of Government Control* and one-time faculty member of the Harvard School of Business Administration. He became chief of the Sugar Section of the U.S. Agricultural Adjustment Administration, 1934–35 and, in 1937, president of the National Sugar Refining Company of New Jersey. He also served as executive secretary of the United States Cane Sugar Refiners Association.

Harold D. Cooley   U.S. Representative from North Carolina (1934–66) and chair-
(1897–1974)   man of the House Committee on Agriculture (1949–66). Known as the "Sugar Czar" for the power he wielded as committee chairmen in allocating production quotas. He played key roles in the development of the Farmers Home Administration, the Soil Conservation Service, the Crop Insurance Program, and the Tobacco Program.

William Pawley   Businessman, U.S. ambassador, special presidential envoy, and
(1896–1977)   occasional covert operative for the CIA. His investments ranged widely and included real estate in 1920s Florida, Cuba's first commercial airline in the 1930s, municipal transport in Havana in the 1950s and Miami in the 1960s, and sugarcane in the Everglades in the 1960s. He was a staunch anticommunist whose close political associates included John Foster and Allen Dulles, Dwight Eisenhower, George Smathers, and Richard Nixon. He used these and other connections to try to protect his investments in Talisman Sugar Corporation and to lobby for an increased Florida quota and expanded Florida acreage.

Ernest Graham   Member of the Florida Senate (1937–44), father of Washington
(1886–1957)   Post Publisher Philip Graham and Florida Governor and U.S. Senator Bob Graham. He operated a short-lived sugarcane enterprise, the Pennsylvania Sugar Company, in Dade County.

Claude Pepper   Longtime Congressman from Florida, first as U.S. Senator (1936–
(1900–1989)   51) then as U.S. Representative (1962–89). Elected to the Senate as a stalwart New Dealer, Pepper used his access to the Roosevelt administration to promote higher quotas for Florida sugarcane in the 1940s. Later, as U.S. Representative, he lobbied the USDA for policies favorable to the expansion of Florida sugarcane acreage in the years following the Castro-led revolution in Cuba.

Spessard Holland   Florida Governor (1941–45) and U.S. Senator (1946–71). As
(1892–1971)   governor, he negotiated for the purchase of wetlands in 1944

that led to the creation of Everglades National Park in 1947. In the 1960s he lobbied the USDA on behalf of William Pawley's efforts to protect Florida sugar quotas and expand the sugarcane acreage.

George Smathers
(1913–2007)

Florida U.S. Representative (1947–51) and later U.S. Senator (1951–69). He was a vocal anticommunist and cold warrior and close friend of both John F. Kennedy and Richard Nixon. Known for his internationalist politics, he initially defended Cuba's access to U.S. sugar markets, but later advocated for increasing Florida's quota at Caribbean producers' expense. He supported the efforts of William Pawley (a close political associate) to expand Florida sugar in the 1960s.

Daniel Robert "Bob"
Graham (b. 1936)

Florida Governor (1979–87) and U.S. Senator (1987–2005), known within the state as a pro-environment politician. Graham was a tireless booster of Florida agriculture, including the sugar industry. He also initiated the Save Our Everglades program in 1983 that promoted the restoration of the K-O-E watershed.

Horace Godfrey
(1916–1998)

A sugar lobbyist known as "Mr. Sugar" for his efforts to protect American sugar quotas and price supports for more than twenty years on Capitol Hill. The Florida Sugar Cane League regularly retained the services of his company, Godfrey Associates, Inc. Before his lobbying career, he spent twenty-seven years with the USDA, beginning in 1934 with the Agriculture Adjustment Agency and later as national director of the Agriculture Department's Stabilization and Conservation Service.

Dexter Lehtinen
(1946–)

As U.S. District Attorney, sued the state of Florida in 1988 for failing to control pollution from sugarcane plantations. He later became attorney for the Miccosukee Tribe in their water-quality suit against the state of Florida. He is married to Florida's U.S. Representative Ileana Ros-Lehtinen.

Alfonso "Alfy"
Fanjul Jr. and
José "Pepe" Fanjul

The sons of Alfonso Fanjul Sr. and Lillian Gomez-Mena, whose families founded the Czarnikow-Rionda Company (New York, Havana, and London) and the New Gomez-Mena Sugar Company (Cuba), respectively. After the Castro-led revolution, the Fanjul family came to Florida, where Alfonso Sr. bought four thousand acres near Lake Okeechobee. After Alfonso Sr.'s death in 1980, Alfy and Pepe took over and expanded the company, Flo-Sun, Inc., which now includes subsidiaries Atlantic, Osceola, and Okeelanta as well as Florida Crystals, its direct marketing brand of sugars. Their sugar holdings have grown to more than four hundred thousand acres in Florida and the Dominican Republic.

Lawton Chiles (1930–1998)	Florida Governor (1991–98) and U.S. Senator (1971–89). As governor, he signed the Everglades Forever Act of 1994 and negotiated a settlement of the federal lawsuit over Everglades water quality in 1991.
Marjory Stoneman Douglas (1890–1998)	Miami-based writer and environmental activist, author of *The Everglades: River of Grass* (1947), member of the original committee that lobbied for the establishment of Everglades National Park, and founder of the watchdog group, Friends of the Everglades. Her public campaigning for wetlands protection in Florida spanned nearly six decades.

Key Legislation, Trade Agreements, and Policies in the Transformation of the Florida Everglades

McKinley Tariff, *1890*	This bill admitted sugar free of duty, but it also provided a two-cent-per-pound bounty to American sugar growers. The bounty supported the expansion of domestic cane and beet production, and the absence of import duties encouraged the Cuban industry to expand. It also set the average ad valorem tariff rate for imports to the United States at 48.4 percent in order to protect domestic agriculture. However, domestic farmers were adversely affected because of retaliatory tariffs by foreign countries, which made U.S. agricultural exports less competitive.
Wilson-Gorman *Tariff, 1894*	This law slightly reduced the U.S. tariff rates from those set by the McKinley Tariff in 1890 and marked a return to the pre-1890 system. It had particularly harmful consequences for Cuba's industry, because sugar was now subject to import duties. Cuba retaliated with duties on U.S. imports, putting its citizens in a double squeeze of declining export revenue (and associated reductions in wages and employment) and rising import prices. The "Sugar Trust" of refiners was the biggest beneficiary of the legislation.
Dingley Tariff, 1897	The Dingley Act of 1897 raised tariffs to counteract the Wilson-Gorman Tariff, which had lowered rates. It placed duties on imported sugar equivalent to the duties paid to sugar producers in foreign countries. The tariff remained in place for the next seventeen years, during which the beet industry flourished. It further disadvantaged Cuban sugarcane producers, however, by providing incentives for U.S. beet-sugar producers to supply the domestic market.

Food and Fuel Control Act, 1917	Also known as the Lever Act, it granted President Woodrow Wilson certain powers during World War I, which he exercised to create the U. S. Food Administration. The Food Administration was established to assure the supply and distribution of food; facilitate transportation of food and prevent monopolies and hoarding; and maintain control of food using voluntary agreements and licensing. Future U.S. President Herbert Hoover was appointed head of the U. S. Food Administration, wielding power delegated to him by Wilson.
U.S. Sugar Equalization Board, 1918	Using the authority provided under the 1917 Food and Fuel Control Act, President Wilson created the U. S. Sugar Equalization Board, with Herbert Hoover as chairman. Its aim was to secure foreign-produced sugar in cooperation with the allied countries. The board, which was dissolved in 1920, was authorized to purchase the Cuban sugar crops of 1918 and 1919.
Tariff Acts of 1921 and 1922	Known as the Emergency Tariff Act and Fordney-McCumber Act, respectively, these tariffs raised the duty on raw sugar from Cuba from 1.0048 cents a pound to 1.6 and then to 1.7648 cents a pound. This encouraged increased production in the continental United States and its island possessions, especially the Philippines. Despite the decline in exports to the United States, Cuba was able to maintain its production by filling shortages in other regions.
Smoot-Hawley Act, 1930	This legislation raised tariffs on more than twenty thousand imported goods, including sugar. It raised the duty on Cuban sugar to two cents per pound. The subsequent drop in U.S. demand devastated Cuba's production, which declined by 50 percent. The tariff offered only a modicum of protection to U.S. sugar producers; by 1932 the pre-duty price of sugar was the lowest in history, the duty was the highest since 1890, the duty-paid price was the lowest on record, and consumption had declined.
Agricultural Adjustment Act, 1933	This act was intended to regulate the declining terms of trade experienced by American farmers during the Great Depression. The goal was what eventually became known as "parity"; that is to achieve an equitable exchange relationship between agriculture and industry and between on-farm and off-farm citizens. Indices of prices paid for goods and services in relation to prices of agricultural commodities during the base period (initially 1910–14) were developed to estimate parity. The instruments to accomplish parity included direct payments for the voluntary reduction of acreage and the use of tax revenue to expand markets and reduce agricultural commodity surpluses. The act created the Agricultural Adjustment Administration to oversee

the program. Congress ammended it in 1935 to insure that imports did not interfere with the domestic farm program. The U.S. Supreme Court declared the law unconstitutional in 1936.

Jones-Costigan Act,
1934

Also known as the Sugar Act, it added sugarcane and sugar beets to the AAA list of basic agricultural commodities. It applied a new method for regulating the domestic sugar industry and controlling sugar imports that lasted for forty years. The act required the secretary of agriculture to determine sugar consumption for the continental United States. Once consumption was determined, the quantity of sugar required was divided among domestic and foreign producers through a quota system.

Sugar Act, 1937

Retained much of Jones-Costigan, but placed certain conditions on the distribution of benefit payments, such as the elimination of child labor and not producing above the alotted quota. In addition, it raised the quota of mainland cane by more than 50 percent above that of the 1934 act, largely to account for increased production potential. The effect was to separate domestic sugar prices from those in the rest of the world.

International Sugar
Agreement, 1937

Twenty-one countries accounting for 85 to 90 percent of world sugar production signed an agreement that limited expansion of importing countries' domestic sugar industries. In exchange, exporting countries pledged to observe their quotas. The essential aim of this and subsequent agreements was to stablize the price of sugar in the global "free market" by limiting the quantity exported. World War II made the agreement inoperative.

Agricultural
Adjustment Act,
1938

This act rectified the unconstitutionality of the 1933 act and combined some of the successful features of the 1936 Soil Conservation and Domestic Allotment Act. It was designed to deal with price and income crises resulting from surplus production. Instruments included crop insurance, parity payments, nonrecourse loans to farmers to keep prices stable, and marketing quotas for certain commodities. The 1938 act remained the foundation for farm support programs for the next three decades.

Wartime Measures,
1939–1947

See appendix C.

Sugar Act, 1948

Replaced the 1937 Sugar Act, which was due to terminate on December 31, 1947, unless amended. It retained the basic features of the previous acts, but allocated domestic quotas as tonnages rather than percentages of total projected consumption. The bill gave special consideration to Cuba's role in supplying allies during the war. Thus Cuban producers benefited from a provision granting that any consumption increases would be filled by foreign producers and another that gave Cuba 95 percent of any deficit in the Philippine's quota. Cuba's share of the latter amounted to nearly two million tons of sugar over the first five years.

International Sugar Agreement, 1953   Similar to the 1937 agreement in that it assigned basic quotas to sugar producers for sugar exported to the "free" market. Trade exempted by the agreement included imports into the United States and USSR (from certain East European countries). The largest change in the agreement was an increase in Cuba's quota of 140 percent.

Sugar Act, 1956 amendments   Introduced extensive changes in the quota system, notably that domestic producers' share of consumption increases went from zero in the 1948 act to 55 percent. It additionally benefited mainland cane growers through the government's purchase of a hundred thousand tons of sugar to be distributed to developing countries. Cuba's quota share was reduced to less than the 1948 allotment.

International Sugar Agreement, 1958   This agreement increased the number of member nations, including Brazil for the first time. Declining sugar prices forced a reduction of 80 percent in basic quotas in 1959 and of 85 percent in 1960. The quota provisions were suspended in 1961 after Cuba, under the Castro government, exported a quantity of sugar that exceeded its permitted quota.

Sugar Act, 1960 and 1961 amendments   Following the 1959 Castro Revolution in Cuba, the law was amended to give the president the authority to determine the size of Cuba's quota. By proclamation, the president reduced Cuba's share to zero the same day that he signed the amended law. Mexico, the Dominican Republic, and the Philippines were among the biggest foreign beneficiaries. The 1961 amendment included a provision that gave special quota consideration to Western Hemisphere countries and countries purchasing U.S. agricultural commodities.

International Sugar Agreement, 1968   Neither the United States nor the European Economic Community were members of this agreement. This meant that EEC countries could export as much as they wished. Cuba's quota was nearly double that of any other exporting member, and its exports to Communist countries were exempt from the quota.

Sugar Act, 1971 amendments   The act was amended to cover the period through December 31, 1974. The major change concerned the way the secretary of agriculture estimated consumption requirements. Specifically, the secretary was required to adjust the consumption estimates when sugar prices rose or fell 4 percent or more above or below the annual price objective. The quota system that had been in place for forty-one years ended when the law was allowed to expire at the end of 1974.

Florida Water Resources Act, 1972	This legislation created five water-management districts in Florida. In 1976, a voter-approved state constitutional amendment gave these districts the authority to levy property taxes to help fund management activities. The South Florida Water Management District oversees the K-O-E watershed and operates and maintains approximately eighteen hundred miles of canals and levees and twenty-five major pumping stations. It is a key agency in the Everglades restoration plan.
Agriculture and Food Act, 1981	The 1981 Farm Bill was intensely fought as the newly elected Reagan Administration attempted to curtail agricultural spending. Though the bill was less expensive than the 1977 version, it retained the system of target prices and loans and included a new sugar price-support program.
Save Our Everglades Program, 1983	Launched by Florida Governor Bob Graham to make "the Everglades look and function by the year 2000 more as it did at the turn of the century." Central to this goal is restoring freshwater flow to the lower Everglades, including the national park. In 1984 Governor Graham established a state resource and management committee and met with the environmental community to organize the Everglades Coalition.
Food Security Act of 1985	The 1985 Farm Bill was sometimes referred to as the "Swampbuster," because of provisions that denied program benefits to producers who converted wetlands after December 23, 1985. The bill provides price supports for sugar producers, primarily through import restrictions and nonrecourse loans. Like previous farm bills, it requires the secretary of agriculture to estimate domestic sugar consumption in determining the level of imports allowed.
Florida Surface Water Improvement and Management (SWIM) Act of 1987	This act required Florida water-management districts to consider all of their water bodies within their boundaries and rank them in priority, according to two categories. For those in pristine condition, districts had to devise a strategy to maintain them, and for degraded water bodies, they had to develop a plan to restore and maintain them. SWIM set targets for how much phosphorous might enter Lake Okeechobee and specified a model to measure flows. It also contained provisions to monitor and maintain the quality of water flowing into Everglades National Park.
Federal water quality lawsuit against Florida, 1988	The federal government, through the action of acting U.S. Attorney Dexter Lehtinen, sued Florida's Department of Environmental Regulation and the South Florida Water Management District for failing to enforce water-quality laws in the Everglades. The suit was eventually dropped after water-quality

concerns were addressed in Florida's Marjory Stoneman Douglas Everglades Protection Act, 1991.

Food Agricultural Conservation and Trade Act of 1990

The 1990 Farm Bill continued price supports for sugar producers through import restrictions and nonrecourse loans. Import quotas keep prices above loan rates. If the secretary of agriculture determines that the amount of sugar imported is less than 1.25 million short tons for any fiscal year, the secretary must establish marketing allotments on domestically produced sugarcane and sugar beets at a level that will raise sugar imports to the specified level. If marketing allotments are imposed upon domestic sugar, they also must be established for crystalline fructose made from corn.

Florida Marjory Stoneman Douglas Everglades Protection Act, 1991

This bill mandated that the South Florida Water Management District implement a surface-water improvement and management plan for improving water quality in the Everglades. Douglas asked to have her name removed from the legislation four years after its passage because she felt it was too favorable to the sugar industry.

Florida Everglades Forever Act, 1994

The Everglades Forever Act, passed in 1994, builds upon the plan of the preceding act by establishing a restoration program based on construction, research, and regulation. Specifically, the 1994 act calls for a concerted effort to control the growth of exotic species and for the creation of stormwater-treatment areas to filter phosphorous from agricultural runoff before it reaches the Everglades. In addition, farmers must use best-management practices to minimize the amount of nutrients used on or discharged from their fields.

Water Resources Development Acts, 1986, 1990, 1996, and 2000

This series of acts directs the U.S. Corps of Engineers on its projects. The WRDA of 1986 is considered the omnibus act; most of the provisions in subsequent acts either amend or add to its sections. Of the many WRDAs, three have particular importance for Everglades restoration. The 1990 WRDA directed the Corps to undertake a feasibility study regarding the restoration of the natural watercourse of the Kissimmee River. A section of the 1996 act, entitled "Everglades and South Florida Ecosystem Restoration," directed the Corps to plan and implement projects for the restoration of the K-O-E system. It gives the Department of the Interior authority to participate in restoration projects and designates the secretary of the interior as the chair of the newly created South Florida Ecosystem Restoration Task Force. The 2000 WRDA approved the CERP as the organizing framework for modifying the C&SF Project so as to restore the K-O-E system. In general, it directed that the plan be implemented to protect water quality and quantity in the "South Florida eco-system." It provided federal funds for a number of pilot and ini-tial projects in the restoration process.

Federal Agriculture Improvement and Reform Act, 1996

The 1996 Farm Bill offered sugar producers nonrecourse loans and eliminated domestic sugar marketing assessments. In gen-eral, the sugar industry was exempted from the agricultural sub-sidy reforms implemented by this bill and continued to receive federal price supports. Marketing allotments that were part of the 1990 Farm Bill were not continued.

Farm Security and Rural Investment Act of 2002

This act provides for nonrecourse price-support loans as un-der the 1996 Farm Bill. It uses marketing allotments, used in the 1990 but not the 1996 act, to keep domestic production at spec-ified levels, with import shortfalls divided between domestic beet and cane. Like prior farm bills, the 2002 Farm Bill uses im-port quotas to restrict the supply of sugar that enters the United States.

Chronology of Principal U.S. Government Wartime Sugar Controls, 1939–47

DATE	ACTION
September 11, 1939	The president suspended sugar quotas under the Sugar Act of 1937.
December 26, 1939	The president restored sugar quotas.
January 5, 1942	Ceiling price of sugar was raised to 3.74 cents per pound in New York City, with small differentials for other refining ports.
January 28, 1942	The U.S. government contracted for the purchase of the entire 1942 Cuban sugar crop, with the exception of that needed for Cuban consumption.
April 14, 1942	The president again suspended sugar quotas under the Sugar Act of 1937.
May 1, 1942	Sugar rationing was established for industrial and institutional users.
May 5, 1942	Sugar rationing was established for household users.
June 9, 1942	The U.S. president and the U.K. prime minister jointly authorized the creation of the Combined Food Board to recommend international allocations of sugar and other foods.
April 3, 1943	The U.S. government contracted for the purchase of 2.7 million tons of Cuban raw sugar. The contract provided that Cuba would limit its total 1943 production of sugar to not more than 3,225,000 tons of raw sugar.
September 22, 1943	The U.S. government contracted for the purchase of the 1944 Cuban sugar crop, with the exception of 200,000 tons for consumption in Cuba.
April 26, 1945	The U.S. government contracted for the purchase of the entire 1945 crop of Cuban sugar, less 454,320 tons for consumption in Cuba and free export, chiefly to Latin America.

July 1, 1946   The International Emergency Food Council took over the activities of the Combined Food Board.

July 16, 1946   The U.S. government contracted for the purchase of the 1946 and 1947 crops of Cuban sugar, less 704,196 tons in 1946 and 738,270 tons in 1947 for consumption in Cuba and free export, chiefly to Latin America.

June 11, 1947   Household sugar rationing ended.

July 28, 1947   Industrial and institutional sugar rationing ended. This was the last commodity removed from ration control during World War II.

September 23, 1947 The International Emergency Food Council announced that sugar-importing countries would be permitted to exceed their previously recommended allocations of sugar. This effectively ended international sugar allocations.

October 31, 1947   All price ceilings on sugar were removed, ending all World War II price controls except rent.

Source: Ballinger, 1975, 97

Notes

CHAPTER ONE: *From Everglades to Sugar Bowl and Back Again?*

1. The Convention on Wetlands, signed in 1971 in Ramsar, Iran, is an intergovernmental treaty that provides the framework for international cooperation and national initiatives for wetland conservation. The Ramsar List of Wetlands of International Importance includes 1,675 wetland sites totaling approximately 150 million hectares (about 370 million acres). See 2007 information from the Ramsar Convention Secretariat at the Web site of the Ramsar Convention on Wetlands, http://www.ramsar.ogr.

2. These include specifically materials obtained from the U.S. National Archives in College Park, Maryland, the Clewiston (Florida) Public Library, the Miami-Dade Public Library, the Florida State Archives in Tallahassee, and the Manuscript and Special Collections section of the University of Florida P. K. Yonge Libraries, in particular, the Braga Brothers Collection. In interviews, persons responding in their official capacities are identified by name, while private individuals have been given pseudonyms.

CHAPTER TWO: *The Sugar Question in Frontier Florida*

1. The term *muck* refers to one of the original classifications of soils found in the Everglades. Muck is defined as composed of well-decomposed organic matter in contrast to peat, composed of raw, un-decomposed material. See Snyder (1994) for further discussion of Everglades soils.

2. Deerr notes of the U.S. beet-sugar industry that "on the technical side it was watched over by Harvey Wiley and on the economic side it was protected by James Wilson, Secretary of Agriculture for sixteen years in the administrations of McKinley, Roosevelt and Taft" (Deerr 1950, 485).

CHAPTER THREE: *Securing Sugar, Draining the 'Glades*

1. William S. Jennings was a cousin to William Jennings Bryan and "shared Bryan's hostility to corporate power. Jennings and Bryan had somewhat similar careers. Both were

born in small Illinois towns, attended Illinois colleges, and studied law in the Union College of Law in Chicago. Both . . . moved to frontier communities" (Blake 1980, 92).

2. Broward's midterm recommendations to the state legislature included a resolution directed to the U.S. Congress "to purchase territory, either domestic or foreign, and provide means to purchase the property of the Negroes, at reasonable prices, and to transport them to the territory purchased by the United States. The United States to organize a government for them . . . and to prevent Negroes from migrating back to the United States" (Proctor 1993, 252).

3. R. Rose to Governor Broward, April 30, 1906, box 9, Broward Manuscript Collection, Special Collections, University of Florida Library.

4. Subsequent investigation revealed that of the two dollars per acre that Bolles "paid" for this land, only fifty cents went to the IIF; the remainder went toward the improvement of Bolles's land (Manuel 1942a, 12871).

5. N. B. Broward, "Draining the Everglades," 3–4, *Florida Everglades Land Company,* Colorado Springs, CO, c. 1908, box 9, Broward Manuscript Collection, Special Collections, University of Florida Library.

6. Spencer's booklet gained a wider audience through W. F. Blackman, who had been a professor at Yale and was a former president of Rollins College in Winter Park, Florida. In his published address, "Sugar and Cane Syrup in Florida" (1921), which he presented in Chicago to the American Association for the Advancement of Science and reprised for a corporate audience in Columbus, Ohio, Blackman identified *The Sugar Situation* as the "exhaustive study" on the topic.

7. S. Rionda to M. Rionda, January 4, 1922, and H. Bowerman to M. Rionda, January 11, 1921, Misc. Manuscripts, box 83, Special Collections, University of Florida Library.

8. Dye notes the difficulty of assigning nationality to such an anonymous medium as money, explaining that among those listed as "North American" were Cubans invested in New York capital markets, whereas some American investments might be buried in listings classified as domestic due to the nature of business partnerships.

9. Sugar refining in the United States was an industry that had matured in the late nineteenth century from a competitive to a centrally controlled form that was the subject of extensive investigations aimed at "trust-busting" (Eichner 1969).

10. The potential of the Cuban industry inspired its own U.S.-based sugar boosterism, such as the publication *Cuban Cane Sugar—a sketch of the industry from soil to sack, together with a survey of the circumstances which combine to make Cuba the Sugar Bowl of the World* (Wiles 1916).

11. Among those involved in the negotiations between the USSEB and President Wilson, Taussig's intervention was seen as critical. "It is not publicly known, but generally believed, that it was this dissenting view that was responsible for the delay in acting by Washington" (unpublished manuscript, History of the Sugar Equalization Board, Braga Brothers, RG 11, series 10c, box 103, Sugar Equalization Board.

12. Signatories included Cuba, Java, Germany, Poland, Hungary, Belgium, Czechoslovakia, Yugoslavia, and Peru.

13. Telegram from Graham to F. C. Elliott, January 16, 1922, Ernest Graham Papers,

box 32, Everglades Drainage and Water Issues, Special Collections, University of Florida Library.

14. Carbon copy of letter from George Earle to Governor John W. Martin et al., Ernest Graham Papers, box 32, Everglades Drainage and Water Issues, Special Collections, University of Florida Library.

15. E. W. Brandes, principal pathologist in charge, Division of Sugar Plant Investigations, USDA, to Prof. H. H. Hume, assistant director, Agricultural Experiment Station, University of Florida, December 9, 1932, University of Florida Archives, Public Records, Experiment Station.

16. E. W. Brandes, senior pathologist in charge, Sugar Plants, USDA, to F. E. Bryant, August 11, 1927, University of Florida Archives, Public Record Collection, series 30b, box 10, E. W. Brandes, Sugar.

17. There were several other failed attempts to develop commercial cane plantations. Notable was that of James F. Jaudon, the Dade County tax assessor instrumental in securing funding to build the Tamiami Trail across the Everglades, completed in 1928. Soon thereafter he established the Royal Palm Sugar Cane and Planting Company in Ochopee, located on the Trail and far to the south of the present-day industry, where he hoped to produce sugar and rum (Dovell 1952; Jaudon 1934).

18. The company faced other problems as well: it was too far from energy supplies and attempted to burn muck as fuel, and moonshiners started tapping molasses tanks at night (Manuel 1942b, 12957).

19. Nolen helped establish the planning programs at Harvard University and Massachusetts Institute of Technology, and was one of the founders of the professional organization that would become the American Planning Association.

20. Clewiston's founding mother, Marian Horwitz O'Brien, hired Nolen. She came to the area in 1917, when "nature took a hand in the proceedings and . . . sent five successive 'dry' years—from 1917 to 1922—and farms sprang up everywhere." With the backing of A. C. Clewis, O'Brien planned the railroad and purchased the land for the town. "Determined that the new town should be orderly and beautiful, she employed John Nolen, of Boston, the foremost city planner in the United States, to design the city plat." Nolen received ten thousand dollars for the plan (*Clewiston News* 1953, A2).

21. Fred L. Williamson, Statement Relative to Everglades Flood Control Matters Made Nov. 5, 1929, at West Palm Beach, on behalf of the Florida Flood Control Association, Fred L. Williamson Manuscript Collection, Special Collections, University of Florida Library.

CHAPTER FOUR: *Wish Fulfillment for Florida Growers*

1. Barro y Segura presents a different view of the Jones-Costigan Act: "After two years of Hawley-Smoot treatment, Cuba was engulfed in a triple crisis: economic, political and social, from which we would have found it utterly impossible to emerge . . . had not the Roosevelt Administration, partly for inter-American political reasons, but, fundamentally, to protect continental United States beet and cane production from

American-owned offshore areas, decided to change radically the tariff policy it had inherited from its predecessor, the Hoover Administration. And so, whilst saving its own beet and sugar industries, the Roosevelt Administration prevented Cuba's ruin" (Barro y Segura 1943, 13).

2. Ickes argued for higher quotas for offshore areas such as Puerto Rico, where people "are starving, because of the strict quota under which sugar is held" (Ickes 1954, 3).

3. Memorandum re: The United States Sugar Corporation Digest of Report and Comparison with Francisco-Elia Year Ended June 30, 1937. October 25, 1937. Braga Brothers, Record Group 2, Series 10C, box 51, Special Collections, University of Florida.

4. Memorandum re: salient points of the U.S. Sugar Corporation Annual Report for fiscal year ended June 30/37, September 27, 1937. Braga Brothers, Record Group 2, Series 10c, box 136, Special Collections, University of Florida Libraries.

5. Regional studies distinguish between the "Old South," referring generally to the antebellum slave economy of the plantation system, and the "New South," a term that has been used variably to refer to the nineteenth-century postbellum period, the 1930s "New Deal" period, and the post–World War II Sunbelt boom. Historians concur that the origins of the New South lie with the end of Reconstruction in 1877, when white conservative Democrats completed their takeover of the region from a biracial coalition of northern-supported Republicans (see Ayers 1992; Woodward 1971).

6. Jim Crow was first associated with southern state's laws, including Florida's, which mandated racial segregation on trains in the 1880s. It became over the decades a complex set of laws, policies, and practices that institutionalized racial difference and the political and economic subjugation of blacks in the New South. Jim Crow rule played a key role in controlling black labor and keeping the region's wages low during the first half of the twentieth century, particularly in rural areas. First, violence and the threat of violence against southern blacks were important in controlling the South's black agricultural labor force in the 1930s and 1940s. Second, the cultural practices of the period served to define and reinforce racial identities and interracial power relations. (see Woodward 1971; Hoelscher 2003).

7. C. Bitting to Senator Claude Pepper, June 4, 1937. Franklin Delano Roosevelt Library.

8. M. E. Von Mach, Personnel Director, United States Sugar Corporation, Clewiston, Florida, form letter to the Managing Editor, April 18, 1940. In author's possession.

9. Baldwin speculated that "[p]erhaps no other aspect of chronic rural poverty attracted greater public attention than the problems of migratory farm laborers in the 1930's" (Baldwin 1968, 221).

10. For example, agricultural wage rates in 1941 in Iowa were $3.10 per day, $1.55 in Florida, $1.25 in Alabama, and $1.15 in Georgia. USDA, Bureau of Employment Security, Farm Labor Report, October, 1941, *Records of WMC, Bureau of Placement, Rural Industries, 1940–1943*, RG 211, Entry 197, box 1, National Archives.

11. L. Levine, "Florida—Report of visit on January 14 through January 26, 1942 to United States Employment Service Office," Labor Market Survey Reports, Florida, Reports of Field Visits—Florida Folder, Bureau of Employment Security, RG 183, box 68, National Archives.

12. William V. Allen, Report, Palm Beach County, Florida State Employment Service, March 14, 1941. Records of the War Manpower Commission, RG 211, Bureau of Placement, Rural Industries, State of Fla., 1941. National Archives.

13. J. Tiedtke (USSC), *Transcripts of the Hearings of the State WFA and USDA Wage Boards, 1942–1946*, Nov. 16, 1942, RG 224, Entry 10, box 10, folder: Florida, Belle Glade, Labor, National Archives.

14. USES, Research and Statistics Unit, Agricultural Labor Report by Counties and Crops as of May 13, 1942, RG 211, Entry 199, box 21, folder: Florida 1942, *Farm Labor Market Reports, 1941–1943*, National Archives; and USES, Research and Statistics Unit, Agricultural Labor Report by Counties and Crops as of April 21, 1943, RG 211, Entry 199, box 21, folder: Florida 1943, *Farm Labor Market Reports, 1941–1943*, National Archives.

15. A. French (Field Supervisor, USES, West Palm Beach), "Weekly agricultural report, March 5," RG 211, Entry 199, box 21, folder: Florida 1943, *Farm Labor Market Reports, 1941–1943*, National Archives.

16. L. Levine, "Florida—Report of visit on January 14 through January 26, 1942, to United States Employment Service Office," RG 183, box 68, Bureau of Employment Security, Labor Market Survey Reports, Florida, Reports of Field Visits—Florida Folder, National Archives (emphasis added).

17. A. French, "Summary, farm labor report, Southeastern Florida Area, week ending 16 April" (1943), Records of the War Manpower Commission, RG 211, Records of F. W. Hunter, National Archives.

18. A. French (Field Supervisor, USES, West Palm Beach), "Summary of daily reports of farm placement activities, week ending January 29" (1943), RG 211, Entry 199, box 21, folder: Florida 1943, *Farm Labor Market Reports, 1941–1943*, National Archives.

19. A. French, "Summary of daily reports of farm placement activities, week ending February 13," RG 211, Entry 199, box 21, folder: Florida 1943, *Farm Labor Market Reports, 1941–1943*, National Archives.

20. A. French, "Agricultural Report Summary," Florida reports (postmarked March 5); RG 211, Entry 199, box 21, folder: Florida 1942, *Farm Labor Market Reports, 1941–1943*, National Archives.

21. Ibid.

22. The USES agricultural report of January 20, 1943, reported that in Palm Beach, Hendry, and Glades Counties, bean pickers were earning $12 per day versus $4 per day for cane cutters. WMC, USES, Weekly Agricultural Labor Report for Florida, 1/20/43, National Archives.

23. A. French, "Summary of daily reports of farm placement activities, week ending December 4," RG 211, Entry 199, box 21, folder: Florida 1943, *Farm Labor Market Reports, 1941–1943*, National Archives.

24. A. French, "Summary of daily reports of farm placement activities, week ending December 4," RG 211, Entry 199, box 21, folder: Florida 1943, *Farm Labor Market Reports, 1941–1943*, National Archives.

25. A. French, "Summary of daily reports of farm placement activities, week ending December 4," RG 211, Entry 199, box 21, folder: Florida 1943, *Farm Labor Market Reports, 1941–1943*, National Archives.

26. A. French, "Agricultural Report Summary," Florida reports (postmarked March 5); RG 211, Entry 199, box 21, folder: Florida 1942, *Farm Labor Market Reports, 1941–1943*, National Archives.

27. Ibid.

28. A. French, "Summary of daily reports of farm placement activities, week ending December 4," RG 211, Entry 199, box 21, folder: Florida 1943, *Farm Labor Market Reports, 1941–1943*, National Archives.

29. A. French, "Summary, Daily Report, Farm Placement Activities, February 13, 1943," Entry 199, box 21, *Farm Labor Market Reports 1941–1943*, folder: Florida 1943, National Archives.

30. A. French, "Summary, Farm Labor Agricultural Report, Southeastern Florida Area, week ending 16 April" (1943), Records of the War Manpower Commission, RG 211, Records of F. W. Hunter, National Archives.

31. Work Agreement (Jamaican Worker), OFA Form OL-601-W, revised April 21, 1943, RG 224, Entry 8, box 11, folder: Jamaica 194, Laborers, 18, *Records of the Office of Labor, Gen. Corr., 1945–1947*, National Archives.

32. A. French, "Summary, Farm Labor Agricultural Report, Southeastern Florida Area, week ending 2 April" (1943), Records of the War Manpower Commission, RG 211, Records of F. W. Hunter, National Archives.

33. Carter explains that the district was acting on information that would later be published, in 1948, as Bulletin 442 of the Florida Agricultural Experiment Station, "a major document in the history of Everglades water management" (Carter 1974, 90). One important finding was that soils deep enough for agricultural development extended only approximately twenty-five miles south of Lake Okeechobee.

34. Edwin Messinger to U.S. Senator Spessard Holland, March 16, 1948. Holland Papers, box 287.61 (Flood Control), Special Collections, University of Florida Library.

35. It is unclear whether Patterson was an interim president; McGovern (1981) says that Charles Wetherald was appointed president in Dec. 1947 and remained in office until 1958. Mott noted in his diary that Wetherald had previously been production manager of Chevrolet and was "an expert in agriculture, cattle raising, and factory management. He has already greatly improved operations and reduced costs, and this year our sugar production will be greater than we ever produced, probably 100,000 tons" (quoted in Young and Quinn 1963, 169).

36. Fred L. Williamson, Statement Relative to Everglades Flood Control Matters Made Nov. 5, 1929, at West Palm Beach, on behalf of the Florida Flood Control Association, Fred L. Williamson Manuscript Collection, Special Collections, University of Florida Library.

CHAPTER FIVE: *The Cold War Heats up the Nation's Sugar Bowl*

1. Margaret Mead, meeting transcripts, Liaison Session of the Committee on Food Habits and Food Nutrition Board, National Research Council, March 15, 1942, National Academies Archives, Washington D.C.

2. The Sugar Research Foundation was a veritable who's who of the industry, with members such as Ody Lamborn, president of Lamborn and Company and president of the Coffee and Sugar Exchange of New York, and Joseph Abbott, chief executive of the American Sugar Refining Company, who served as president of the foundation. Louis V. Placé Jr. had been the director of the Sugar Institute, an industry trade association organized in 1927 that was found in violation of antitrust laws by the Supreme Court in 1937 (Genovese and Mullin 1999; Braga Brothers Collection, RG 4, box 11, McCahan Sugar Refining and Molasses, Co., Records of Louis Placé, Special Collections, University of Florida Library).

3. Sugar Research Foundation, Inc., June 28, 1943, Certificate of Incorporation, New York, Braga Brothers Collection, RG 4, Series 74, box 8, McCahan Sugar Refining & Molasses Co., Records of Vice President Louis Placé.

4. W. D. Outman, Director of the Florida Economic Advancement Council, Letter to W. A. Johnson, Project Division, War Production Board, April 25, 1945. Series 532, box 10, Florida State Archives.

5. Nathan Mayo, Commissioner of Agriculture, Telegram to W. D. Outman, April 28, 1947, Series 532, box 10, Florida State Archives.

6. Secretary to W. D. Outman to Nathan Mayo, Commissioner of Agriculture, April 28, 1947, ibid.

7. Antonio Barro to Bernardo Braga, December 13, 1948, Braga Brothers Collection, RG 3, Series 50, box 19, Special Collections, University of Florida Library.

8. Anonymous, *A Report to the Sugar Industry Task Group*, 1950 (5 pp.), Braga Brothers Collection, RG 3, Series 50, box 12, Special Collections, University of Florida Library.

9. Excerpt from *U.S. News and World Report* of April 8, 1955, quoted in telegram from Czarnikow-Rionda Company, New York, to company offices in Havana, April 5, 1955, Braga Brothers Collection, RG 3, Series 47, box 5, Special Collections, University of Florida Library.

10. Telegrams: Havana office of Czarnikow-Rionda Company to Czarnikow-Rionda Company, New York, August 20, 1953; August 24, 1953; August 28, 1953; September 10, 1953. All documents from Braga Brothers Collection, RG 3, Series 50, box 12, Special Collections, University of Florida Library.

11. From advertisements contained in the Braga Brothers Collection, ibid.

12. Telegram, Havana offce of Czarnikow-Rionda Company to Czarnikow-Rionda Company, New York, September 10, 1953, Braga Brothers Collection, RG 3, Series 50, box 12, folder: Henry George Atkinson, Advertising Campaign, Sugar, Special Collections, University of Florida Library.

13. Telegram (Havana) to Miklos Szako-Pelsocizi, VP, Czarnikow-Rionda Company, May 14, 1956, Braga Brothers Collection, RG 3, Series 47, box 6, Special Collections, University of Florida Library.

14. Laurence A. Crosby, Chairman, United States Cuban Sugar Council, May 25, 1956, Supplement to the Digest of the American Chamber of Commerce, "United States Sugar Legislation, 1954–1956," Braga Brothers Collection, RG 3, Series 47, box 6, Special Collections, University of Florida Library.

15. The World Sugar Agreement, negotiated in London among seventy-eight partici-

pating countries, had allocated slightly under half of the whole free market to Cuba for a period of five years (Thomas 1971, 846).

16. Lawrence Myers, Director, USDA Commodity Stabilization Service to True D. Morse, August 19, 1960, General Correspondence, Sugar 1960, boxes 3502–4, RG 16, National Archives.

17. James C. Hagerty, Press Secretary to the President, The White House, Statement by the President, December 16, 1960, ibid.

18. USDA, *Special Study on Sugar,* Notice of Opportunity to Submit Written Views and Information by Interested Parties, October 24, 1960, ibid.

19. Anonymous, "An Analysis of the Sugar Problem," Memorandum forwarded by Secretary of Agriculture Freeman to Bob Lewis, February 19, 1961, Sugar 1961, boxes 3670–72, RG 16, National Archives.

20. Robert G. Lewis, Deputy Administrator, Price and Production, Agricultural Stabilization and Conservation Service, USDA, to Secretary Freeman, June/July 1961, ibid.

21. Harold W. Cooley, Chairman, Committee on Agriculture, U.S. House to Orville Freeman, Secretary of Agriculture, September 7, 1961, ibid.

22. Willard W. Cochrane, Director, Agricultural Economics, Cover letter to the Secretary and the Undersecretary of Agriculture, September 22, 1961, accompanying confidential memorandum regarding U.S. sugar policy, ibid.

23. G. C. Chappell and Orlin J. Scoville, SEG, confidential memorandum regarding U.S. sugar policy addressed to Willard W. Cochrane, Director, Agricultural Economics, September 20, 1961, ibid.

24. Ibid.

25. Hon. Charles E. Bennet, House of Representatives, to Marvin L. McLain, Asst. Secretary of Agriculture, October 15, 1960, General Correspondence, Sugar 1960, boxes 3502–4, RG 16, National Archives.

26. "Cuba Sugar Setup for Belle Glade and 21.8 Million Land Deal Called Biggest Ever," *Miami Herald,* September 1, 1961.

27. Ibid.

28. The Okeelanta mill, which included a refinery, was purchased in 1952 by investors headed by Salustiano Garcia, Manuel de Quintana, and Stewart Macfarlane, associated with sugar interests in Cuba of Garcia-Diaz and Company. At that time refining was discontinued; the refinery was rebuilt in 1982 (Salley n.d.; news clippings from the Miami Dade Public Library, Florida Collection, "Industry—Sugar" file).

29. Hon. Joseph M. Montoya, House of Representatives, to James T. Ralph, Assistant Secretary of Agriculture, November 15, 1961, Sugar 1961, boxes 3670–72, RG 16, National Archives.

30. James T. Ralph, Assistant Secretary of Agriculture to Hon. Joseph M. Montoya, House of Representatives, November 29, 1961, ibid.

31. H. Olson, President, Olsen and Dickey Advertising, to Governor Farris Bryant, January 9, 1961, Series 756, box 139, Florida State Archives.

32. H. T. Vaughn, Chairman, Florida Sugar Committee, to Governor Ferris Bryant, April 14, 1961, ibid.

33. Florida State Legislature, House Memorial No. 2963, June 15, 1961, ibid.

34. Orville Freeman, Secretary of Agriculture, to Doyle Connor, Commissioner, Flor-

ida Department of Agriculture, November 16, 1961, Sugar 1961, boxes 3670–72, RG 16, National Archives.

35. Orville Freeman, Secretary of Agriculture, Memorandum to Charley Murphy, John Duncan, James Ralph, Horace Godfrey, and Robert Lewis, December 26, 1961, ibid.

36. H. T. Vaughn, Chairman, Florida Sugar Committee, Memorandum to all Florida sugar companies, December 21, 1961, Series 756, Box 139, Florida State Archives.

37. R. Richardson, President, Richardson Tractor Company, to Governor Farris Bryant, February 16, 1962, ibid.

38. H. T. Vaughn, Chairman, Florida Sugar Committee, Telegram to Florida's U.S. Congressional delegation, March 2, 1962, ibid.

39. Fred Dickinson, Chairman, Florida Council of 100, Telegram to Orville Freeman, Secretary of Agriculture, March 8, 1962, ibid.

40. Governor Ferris Bryant to President John F. Kennedy, March 8, 1962, ibid.

41. H. T. Vaughn, Chairman, Florida Sugar Committee, to Doyle Connor, Florida Commissioner of Agriculture, March 2, 1962, ibid.

42. Senator Allen J. Ellender, Chairman, Committee on Agriculture and Forestry, U.S. Senate, to Hon. Charles S. Murphy, Undersecretary of Agriculture, April 23, 1962, General Correspondence 1962, boxes 3837–39, RG 16, National Archives.

43. Ibid.

44. Orville Freeman, Secretary of Agriculture, to Harold B. Cooley, Chairman, Committee on Agriculture, U.S. House, May 29, 1962, ibid.

45. Testimony of W. S. Chadwick, Mainland Cane Sugar Area Hearing, November 15, 1962, New Orleans, Louisiana, Braga Brothers Collection, George Atkinson Braga, Series 19, box 2, Special Collections, University of Florida Library.

46. Testimony of George Wedgeworth, 1964 Proportionate Share Hearing, April 10, 1963, Braga Brothers Collection, ibid.

47. Testimony of W.S. Chadwick, Mainland Cane Sugar Area Hearing, November 15, 1962, New Orleans, Louisiana.

48. Memorandum to the Secretary from the Asst. Secretary for Stabilization, Subject: "Increase in sugarcane proportionate shares—1963 Crop," March 12, 1963, General Correspondence 1963, boxes 4023–25, RG 16, National Archives.

49. Lawrence Myers, Director, Sugar Policy Staff, Agricultural Stabilization and Conservation Service, to Undersecretary Murphy, May 1, 1963, ibid.

50. Hubert Kelly, *Washington Star*, n.p., n.d., enclosure accompanying memorandum from Myer Feldman to Undersecretary Murphy, May 8, 1963. Secretary of Agriculture, General Correspondence 1963, RG 16, box 4025, National Archives.

51. Orville Freeman, Secretary of Agriculture, to Myer Feldman, Deputy Special Counsel to the President, May 16, 1963, ibid.

52. Howard Hjort to William W. Cochrane, November 6, 1963, ibid.

53. Horace D. Godfrey, Administrator, Memorandum to the Secretary and the Undersecretary, U.S. Department of Agriculture, January 18, 1964, General Correspondence 1964, RG 16, box 4184, National Archives.

54. Congressman Thomas "Tip" O'Neill to Orville Freeman, Secretary of Agriculture, February 7, General Correspondence 1964, boxes 4183–85, RG 16, National Archives.

55. N. E. Coward, Secretary, Harris County AFL-CIO Council, to Congressman Jack Brooks, June 8, 1964, ibid.

56. Congressman Claude Pepper to Orville Freeman, Secretary of Agriculture, May 27, 1964, ibid.

57 Orville Freeman, Secretary of Agriculture, to Congressman Claude Pepper, May 7, 1964, ibid.

58. Orville Freeman, Secretary of Agriculture, Memorandum for file, June 17, 1964, ibid.

59. Memorandum, "Sugar Legislation," from Secretaries of State and Agriculture, draft, August 12, 1964, ibid.

60. Harold W. Cooley, Chairman, Committee on Agriculture, U.S. House, to Orville Freeman, Secretary of Agriculture, August 21, 1964, ibid.

61. Charles Murphy, Undersecretary of Agriculture, to Myer Feldman, Counsel to the President, August 18, 1964, ibid.

62. Ibid.

63. Finding Aid, Braga Brothers Collection, Special Collections, University of Florida.

64. The New Tuinucú Sugar Company, Inc., n.d., Braga Brothers, Michael J. P. Malone Papers, Special Collections, University of Florida. The New Tuinucú Sugar Company comprised two mills, more than ten thousand acres of land, a dock, a distillery, and a yeast plant; in August 1960 the Cuban government expropriated these, and in July 1961 the company presented a formal claim of more than twenty-one million dollars to the U.S. State Department. After adjusting for the loss of its Cuban assets, the company was valued at just over half a million dollars. The ranch was a joint venture between the company and Malone.

65. Letters from James Monahan, Reader's Digest, to Michael J. P. Malone, November 2, 1962, and August 7, 1963, Braga Brothers Collection, Malone Papers, box 4, Special Collections, University of Florida.

66. Memorandum, December 20, 1961, Conversation held on December 14th, 1961, Molasses, Mr. George Wedgeworth and Michael J. P. Malone, ibid., box 23.

67. Memorandum Re: Molasses 1962 Crop, May 11, 1962, from Michael J. P. Malone to George Wedgeworth, and Letter, Michael J. P. Malone to J. Y. Edwards, Cargill, Inc., August 21, 1962, ibid., box 23.

68. Memorandum, Conversations held on December 14–16, 1961—Molasses, between Mr. George Wedgeworth and Michael J. P. Malone; Letter from Mario Leao to Alfred L. Webre Jr., October 28, 1961; Memorandum Re: Molasses—1962 crop, May 11, 1962, from Michael J. P. Malone to George Wedgeworth, ibid.

69. Letter from Alfonso Fanjul to Michael J. P. Malone, Vice President, Czarnikow Rionda, May 22, 1962, Braga Brothers Collection, Malone Papers, box 10, Special Collections, University of Florida.

70. Senator Spessard Holland to Orville Freeman, Secretary of Agriculture, December 30, 1963, General Correspondence 1964, boxes 4183–85, RG 16, National Archives.

71. Orville Freeman, Secretary of Agriculture, to Senator Spessard Holland, January 16, 1964, ibid.

72. Ken Birkhead to Orville Freeman, Secretary of Agriculture, April 13, 1964, ibid.

73. Orville Freeman, Secretary of Agriculture to C. Murphy, H. Godfrey, and T. Murphy, April 15, 1964, ibid.

74. Memorandum from Big B Sugar Corporation, n.d., ibid.

75. Sam Knight, President Atlantic Sugar Corporation, to Orville Freeman, Secretary of Agriculture, September 25, 1964, ibid.

76. Handwritten margin comments of Orville Freeman, Secretary of Agriculture, on the correspondence of Sam Knight, President Atlantic Sugar Corporation, to Orville Freeman, Secretary of Agriculture, September 25, 1964, ibid.

77. Orville Freeman, Secretary of Agriculture, to Sam Knight, President Atlantic Sugar Corporation, October 19, 1964, ibid.

78. Alfonso Fanjul to George A. Braga, President, Czarnikow-Rionda, March 25, 1965, Braga Brothers Collection, Series 19, George Atkinson Braga, box 1, Special Collections, University of Florida.

79. Tom Murphy, Director, Sugar Policy Staff, to H. D. Godfrey, USDA Administrator, November 17, 1964, General Correspondence 1964, boxes 4183–85, RG 16, National Archives.

80. Ibid.

81. Letter from R. C. Lee, Marketing Relations Manager, Sugar Cane Growers Cooperative of Florida, to Michael J. P. Malone, Vice-President, Czarnikow-Rionda Company, December 16, 1964, Braga Brothers Collection, Malone Papers, box 23, Special Collections, University of Florida.

82. Horace Godfrey, USDA Administrator, Memorandum, "Impact of the Sugar Program on Federal Budget," prepared for Charles Murphy, Undersecretary of Agriculture, July 28, 1965, General Correspondence 1965, boxes 4389–90, RG 16, National Archives.

83. John Schnittker, Act. Secretary of Agriculture to Undersecretary, January 19, 1965, ibid.

84. Horace Godfrey, USDA Administrator, Memorandum, "Impact of the Sugar Program on Federal Budget."

85. Rev. George Speidel, Atlantic Sugar Association, to Orville Freeman, Secretary of Agriculture, April 22, 1965, General Correspondence 1965, boxes 4389–90, RG 16, National Archives.

86. Sam Knight, President Atlantic Sugar Corporation, to Orville Freeman, Secretary of Agriculture, April 20, 1965, ibid.

87. Tom O. Murphy, Director, Sugar Policy Staff to John A. Schnittker, Director, Agricultural Economics, May 4, 1965, U.S. Department of Agriculture, General Correspondence 1965, RG 16, box 4389, National Archives.

88. H. D. Godfrey, USDA Administrator, to Orville Freeman, Secretary of Agriculture, December 13, 1965, General Correspondence 1965, boxes 4389–90, RG 16, National Archives.

89. John Schnittker, Act. Secretary of Agriculture, Memorandum, "Allotment of Sugar Quota Mainland Cane Sugar Area," December 15, 1965, ibid.

90. Alfonso Fanjul to Richard Liddiard, December 28, 1961, Braga Brothers Collection, RG 3, Series 63, box 10, Special Collections, University of Florida Library.

CHAPTER SIX: *A Restructured Industry*

1. Possible Additional Measures Against Cuba. Department of State, Doc. CK 3100131884, n.d., from the Web site of the Gale Group, Declassified Documents Reference System: http://galenet.galegroup.com/servlet/DDRS.

2. McGeorge Bundy to the Secretary of Agriculture, May 15, 1963, White House, National Security Action Memorandum No. 244, confidential, The Future of the World Sugar Market, Doc. no. CK3100431911, ibid.

3. Ibid.

4. U.S. Department of Agriculture, The World Price of Sugar, Secret Memorandum, May 24, 1963, declassified June 3, 1980, doc. no. CK3100429599, ibid.

5. Summary Record of NSC Standing Group Meeting No. 10/63. July 16, 1963, ibid.

6. INR Thomas L. Hughes to the Secretary of State, April 9, 1964.

7. Ibid.

8. Gordon Chase, The White House, to McGeorge Bundy, April 25, 1964, from the Web site of the Gale Group, Declassified Documents Reference System: http://galenet .galegroup.com/servlet/DDRS.

9. Gordon Chase, The White House, to McGeorge Bundy, April 28, 1964, ibid.

10. Douglas Dillon, Department of State, Memorandum of Conversation, U.S.-Dominican Relations, Acting Secretary Dillon, Senator George Smathers, Mr. William Pawley, L. D. Mallory, Acting Assistant Secretary for Inter-American Affairs, May 16, 1960. Douglas Dillon, Department of State, Memorandum for the President, United States-Dominican Relations and Their Impact at Home and in Latin America, May 12, 1960, ibid.

11. Walter J. Stoessel Jr., Director, Executive Secretariat, Department of State, to Brig. Gen. A. J. Goodpaster, The White House, Memorandum to the President from the Dominican Dissidents Concerning Dominican Sugar Imports, September 9, 1960, ibid.

12. Christian A. Herter, Memorandum for the President, December 8, 1960, Application of Additional Economic Measures against the Dominican Republic, ibid.

13. Secretary of State Dean Rusk to President Kennedy, Situation in the Dominican Republic Created by Sugar Legislation, July 2, 1962. Confidential. Declassified February 22, 1977, ibid.

14. W. W. Rostow to President Johnson, Allocations to the Dominican Republic of Sugar Shortfalls During the Remainder of 1966, Confidential Memorandum for Action, August 1, 1966, ibid.

15. W. W. Rostow to President Johnson, Special Sugar Quota for the Dominican Republic, Confidential Memorandum, May 11, 1967, ibid.

16. Letter of March 17, 1968, to President from President Balaguer of the Dominican Republic (official translation), ibid.

17. Senators Spessard Holland and George Smathers to Orville Freeman, Secretary of Agriculture, March 20, 1968, General Correspondence 1968, box 4871, RG 16, National Archives.

18. Jess Ferrill, Vice President of Exchange National Bank of Tampa, to Orville Free-

man, Secretary of Agriculture, May 28, 1968, General Correspondence, box 4870, RG 16, National Archives.

19. Embassy, Santo Domingo, to Washington, D.C., Department of State Telegram. Priority: Special Sugar Legislation. May 1968.

20. Chairman, Board of County Commissioners, Palm Beach County, to The President, May 27, 1968, General Correspondence, box 4870, RG 16, National Archives.

21. Telegram to the President from Cuban Revolutionary Nationalist Front, Napoleon Vilabon, General Delegate; II Front Alpha 66, Andes Nazario Sargon, Secretary General; Students Revolutionary Directory, Juan Manuel Salvet, General Secretary; Christian Democratic Movement of Cuba; Cuban Liberation Army, Higinio Diaz, General Delegate; May 30, 1968, ibid.

22. Horace Godfrey, National Administrator, Agricultural Stabilization Service, to Orville Freeman, Secretary of Agriculture, June 19, 1968, ibid.

23. Ibid.

24. John Schnittker, Undersecretary, to Orville Freeman, Secretary of Agriculture, June 27, 1968, box 4870, RG 16, National Archives.

25. Horace Godfrey, National Administrator, Agricultural Stabilization Service, to Orville Freeman, Secretary of Agriculture, August 9, 1968, ibid.

26. Irvin A. Hoff, President, United States Cane Sugar Refining, et al., to the President, September 11, 1968, ibid.

27. Acting Administrator, Agricultural Stabilization Service, to Orville Freeman, Secretary of Agriculture, September 19, 1968, ibid.

28. Doyle Connor, Commissioner of Agriculture, Florida, to Orville Freeman, Secretary of Agriculture, September 17, 1968, ibid.

29. Orville Freeman, Secretary of Agriculture, to Doyle Connor, Commissioner of Agriculture, Florida, October, 2, 1968, ibid.

30. Clarence D. Palmby, Assistant Secretary for International Affairs and Commodity Programs, to Clifford Hardin, Secretary of Agriculture, June 27, 1969, General Correspondence, box 5104, RG 16, National Archives.

31. S. N. Knight, President, Atlantic Sugar Association, to Clifford Hardin, Secretary of Agriculture, April 3, 1969, ibid.

32. Reverend Speidel to President Richard Nixon, March 28, 1969, ibid.

33. Under Secretary Phillip Campbell to Reverend Speidel, June 16, 1969, ibid.

34. Robert N. Giaimo, Rep., to President Richard Nixon. April 4, 1969, ibid.

35. Kenneth Frick, Administrator of the Agricultural Stabilization and Conservation Service to Under Secretary Cambell. May 19, 1969, ibid.

36. William D. Pawley to Richard Nixon. January 15, 1965, ibid.

37. William D. Pawley to Richard Nixon. September 6, 1968, ibid.

38. William D. Pawley to President-Elect Richard Nixon. November 19, 1968, ibid.

39. William D. Pawley to President Richard Nixon. March 18, 1969, ibid.

40. John D. Ehrlichman, Counsel to the President, to Honorable William D. Pawley, Miami, Florida. March 27, 1969, ibid.

41. John D. Ehrlichman, Counsel to the President, to Clifford Hardin, Secretary of Agriculture, Confidential Memorandum. March 27, 1969, ibid.

42. Clifford Hardin, Secretary of Agriculture to John D. Ehrlichman, Counsel to the President. May 6, 1969, ibid.

43. William D. Pawley to Mr. I. Lee Potter, Executive Director, Republican Congressional Boosters Club, Washington, D.C., May 19, 1970, General Correspondence, box 5294, RG 16, National Archives.

44. Jake Esterline, CIA Project Director for the Bay of Pigs Invasion, found William Pawley's views on the Cuban situation "highly personal and rigid" and "inimical to the best interest of the United States" (Pfeiffer 1970, 256). The CIA chose not to cut communication with Pawley because it wanted to maintain "a window" into his groups' activities and because Pawley had duplicate channels, anyway.

45. Bryce N. Harlow, Counselor to the President, to Clifford Hardin, Secretary of Agriculture, June 2, 1970, General Correspondence, box 5294, RG 16, National Archives.

46. Clarence Palmby to Bryce N. Harlow, Counselor to the President, June 11, 1970, ibid.

47. Agricultural Stabilization and Conservation Service Memorandum to Clifford Hardin, Secretary of Agriculture, July 18, 1970. ibid.

48. From the online directory at the Web site of the National Archives, http://www.archives.gov.

49. Deane Hinton to Peter G. Peterson, Executive Director, Council on International Economic Policy, March 23, 1971, Records of the Council on International Economic Policy, Security Classified Records Relating to U.S. Sugar Policy, box 268, RG 429, National Archives. (This group of records was declassified in August 2005 at the author's request.)

50. Deane Hinton to members of the Council on International Relations, Memorandum, Subject: Sugar, April 15, 1971, ibid.

51. Deane Hinton to Peter G. Peterson, Executive Director, Council on International Economic Policy, April 26, 1971, ibid.

52. Ibid.

53. Ibid.

54. Deane Hinton to Peter G. Peterson, Executive Director, Council on International Economic Policy, April 27, 1971, ibid.

55. Clifford Hardin, Secretary of Agriculture, to President Richard Nixon, April 29, 1971, box 268, RG 429, National Archives.

56. Tom Koroglos through Clark MacGregor at the White House to Peter G. Peterson, Executive Director, Council on International Economic Policy, April 30, 1971, ibid.

57. Peter G. Peterson, Executive Director, Council on International Economic Policy, to President Richard Nixon, May 3, 1971, ibid.

58. Ibid.

59. A. Butterfield, White House, to Peter G. Peterson, Executive Director, Council on International Economic Policy, May 6, 1971, box 268, RG 429, National Archives.

60. Memorandum for the President, Council on International Economic Policy, October 11, 1971, ibid.

61. Administrator, Agricultural Stabilization and Conservation Service, to Clifford Hardin, Secretary of Agriculture, October 22, 1971, box 5481, RG 16, National Archives.

62. Zbigniew Brzezinski, National Security Advisor, to President Jimmy Carter, April 27 and May 2, 1977, White House, Memoranda on Senator McGovern's Memorandum on Cuba, from the Web site of the Gale Group, Declassified Documents Reference System: http://galenet.galegroup.com/servlet/DDRS.

63. Zbigniew Brzezinski, National Security Advisor, to President Jimmy Carter, National Security Weekly Report #35, Top Secret/Sensitive, November 4, 1977, ibid.

64. Fanjul went on to say, "I am confident that help will come as beet and sugar cane within the USA and Hawaii only produce 55% of the USA consumption, and if we drop that production, the USA will be in the hands of other countries, creating a situation similar to the oil and the OPEC" (Alfonso Fanjul, Palm Beach, Florida, to George Braga, Alpine, New Jersey, May 30, 1978, Braga Brothers Collection, Series 19, George Atkinson Braga, box 3, Special Collections, University of Florida Library.

65. Cyrus Vance, Secretary of State to President Carter, Memorandum for the President, August 17, 1978, from the Web site of the Gale Group, Declassified Documents Reference System: http://galenet.galegroup.com/servlet/DDRS.

66. Joe Thomas to Governor Bob Graham, Sugar Stabilization Act, Series 656, carton 7, Florida State Archives.

67. Governor Bob Graham to Senator Richard B. Stone, Sugar Stabilization Act, ibid.

68. The quotation is from Claude Smith, "State Commerce Official Pessimistic on Attracting Florida Cane Cutters," *Tribune,* n.d., included in the file material from a meeting between Florida Governor Askew and representatives of the Florida Sugar Cane League, October 19, 1973, Series 942, Carton 14, Florida State Archives.

69. These memoranda were included in the file material from a meeting between Florida Governor Askew and representatives of the Florida Sugar Cane League, October 19, 1973, ibid.

70. Interview with Greg Schell, July 21, 1995, Florida Rural Legal Services, Belle Glade, Florida.

71. Dalton Yancey, Florida Sugar Cane League, to Governor Bob Graham, May 29, 1985, Sugar Policy Series, carton 25, Florida State Archives.

72. Governor Bob Graham to Ash Williams, June 18, 1985, ibid.

73. R. Resnick, Big Firms Now Run Co-ops Built by the Small Farmers, *Miami Herald,* October 28, 1985, n.p., news clipping from the Miami Dade Public Library, Florida Collection, "Industry—Sugar" file.

74. R. Resnick, Powerful Fanjuls Use Sugar as Springboard, ibid.

CHAPTER SEVEN: *Questioning Sugar in the Everglades*

1. The relationship between Allison French and Al French Jr. was confirmed in a phone call made by Professor Marc Linder, College of Law, University of Iowa, to Al French Jr. at the USDA on April 5, 1999, at the request of the author.

2. Interview with Paul Whalen, Director, Everglades Regulation, SFWMD, August 23, 1995, South Florida Water Management District Headquarters, West Palm Beach, Florida.

3. Ibid.

4. Interview with Karen Ansell, Employment Representative, SFWMD, December 15, 1995, South Florida Water Management District Headquarters, West Palm Beach, Florida.

5. Interview with Bob Paige, August 22, 1995, Paige Ranch.

6. For example, the superintendent of Everglades National Park attempted to prevent the publication of the volume.

7. Interview with Judy Sanchez, USSC Public Relations, 14 July, 1995, Clewiston, Florida.

8. Interview with Pete Rosendahl, Environmental Public Relations, Flo-Sun, 3 May, 1996, West Palm Beach, Florida.

9. Interview with Jim Kirk, 20 December 1995, Clewiston, Florida.

10. Interview with Hans Schmidt, 22 August 1995, Belle Glade, Florida.

11. Interview with Donna Bentley, 15 December 1995, Clewiston, Florida.

12. Phosphorous-laden run-off from the EAA is blamed for the shift in species composition in adjacent water conservation areas, where cattails are replacing sawgrass. Cattails are not an exotic, but their replacement of sawgrass has adverse implications for habitat and signals eutrophication (McCormick et al. 2002; Sklar et al. 2002).

13. Because sugarcane requires low levels of applied fertilizer (no nitrogen and small amounts of phosphorous), more phosphorous left the fields in biomass than was applied. Thus, one of their recommendations is that "vegetable field drainage water should be stored and filtered through sugarcane fields" (Izuno et al. 1995, 743). For example, a high-input crop such as escarole requires thirty-six times more phosphorous fertilizer than sugarcane. Interview with Barry Glaz, agronomist, USDA, May 3, 1996, Sugarcane Field Station, Canal Point, Florida.

14. Interview with Frank Mazzotti, Wildlife Scientist, University of Florida, Broward County Extension Office, 6 May 1996, Davie, Florida.

15. Ibid.

16. Gaston Cantens served two terms as a Republican member of the Florida House of Representatives and was the coalition chair of Hispanics for Bush during the 2004 Bush-Cheney presidential campaign.

17. The position of the Sugar Association at the national level was similar, at the global level, to that of the International Life Sciences Institute—comprising Coca-Cola, Pepsi-Cola, General Foods, Kraft, and Proctor and Gamble—which gained accreditation with the WHO and the UN Food and Agriculture Organization.

18. The ad is in the Braga Brothers Collection, RG 4, Series 74, box 11, McCahan Sugar Refining & Molasses Co., Records of Vice President Louis Placé.

19. Interview with Barry Glaz, agronomist, USDA, May 3, 1996, Sugarcane Field Station, Canal Point, Florida.

References

Albert, Bill and Adrian Graves. 1984. *Crisis and Change in the International Sugar Economy, 1860–1914*. Norwich, England: ISC Press.

Allen, J. 1999. Spatial Assemblages of Power: From Domination to Empowerment. In *Human Geography Today,* ed. D. Massey, J. Allen, and P. Sarre, 194–218. Cambridge: Polity Press.

Allison, R. V. 1956. The Influence of Drainage and Cultivation on Subsidence of Organic Soils Under Conditions of Everglades Reclamation. *Soil and Crop Science Society of Florida Proceedings,* 16:21–31.

Andrews, E., and L. Rohter. 2007. U.S. and Brazil Seek to Promote Ethanol in West. *New York Times,* March 3, A1–B9.

Avance. 1961. Cuban Investors Guide Sugarbowl Expansion in Florida. October 6. (Author's translation from Spanish original).

Ayala, César. 1999. *American Sugar Kingdom: The Plantation Economy of the Spanish Caribbean, 1898–1934.* Chapel Hill: University of North Carolina Press.

———. 2001. From Sugar Plantations to Military Bases: The U.S. Navy's Expropriations in Vieques, Puerto Rico, 1940–45. *Centro: Journal of the Center for Puerto Rican Studies* 13, no. 1:22–44.

Ayers, E. L. 1992. *The Promise of the New South: Life After Reconstruction.* Oxford: Oxford University Press.

Bachelet, P. 2007. Brazil to Forge Ethanol Alliance. *Miami Herald,* February 5, 9A.

Baldwin, Sidney. 1968. *Poverty and Politics: The Rise and Decline of the Farm Security Administration.* Chapel Hill: University of North Carolina Press.

Baldwin, W. 1941. Pressure Politics and Consumer Interests: The Sugar Issue. *Public Opinion Quarterly* 5, no. 1:102–10.

Ballinger, R. 1975. *A History of Sugar Marketing through 1974.* USDA, Economics and Statistics Service, Agricultural Economic Report No. 382. Washington, DC: USDA.

Barmash, I. 1974. Expiration Due Dec. 31. *New York Times,* December 7, 39, 41.

Barrionuevo, A., and E. Becker. 2005. Fewer Friends in High Places for This Lobby. *New York Times,* June 2, C1.

Barro y Segura, Antonio. 1943. *The Truth About Sugar in Cuba.* Havana: Ucar, García & Cía.

Batterbury, S. 2001. Landscapes of Diversity: A Local Political Ecology of Livelihood Diversification in South-Western Niger. *Ecumene* 8, no. 4:437–64.

Bauza, R. 2007. Ethanol a Boost for State? Florida May Be the Center of Brazil-U.S. Fuel Alliance. *South Florida Sun-Sentinel,* March 9, 1A.

Benjamin, J. 1990. *The United States and the Origins of the Cuban Revolution: An Empire of Liberty in an Age of National Liberation.* Princeton, NJ: Princeton University Press.

Bernhardt, Joshua. 1920. *Government Control of the Sugar Industry in the United States.* New York: Macmillan.

Birger, Larry. 1961. Florida Sugar Mills Ready for Long Grind. *Miami News,* October 1, 8A.

Birt, G. 1961. 100 Million $$ Being Invested in 'Sugar Area.' *Palm Beach Post,* November 5, 14.

Bitting, C. 1936. *Florida Sugar.* New York: B. H. Tyrrel.

———. 1937a. *Sugar in the Everglades.* New York: B. H. Tyrrel.

———. 1937b. *Some Notes on Cuba.* New York: B. H. Tyrrel.

———. 1938a. *Some Notes on Offshore Conditions.* New York: B. H. Tyrrel.

———. 1938b. *Controlled Output.* New York: B. H. Tyrrel.

———. 1940. *The Fruit of the Cane.* Clewiston, FL: United States Sugar Corporation.

———. 1943. Report on the Everglades and Contiguous Areas. February 26. Includes The Problem of the Everglades and How it Affects the Entire Southern Peninsula of Florida and cover letter to His Excellency, the Governor. Clewiston, FL: C. R. Bitting.

Blackman, W. F. 1921. *Sugar and Cane Syrup in Florida.* Jacksonville, FL.

Blair, W. 1971. Target Date Set for a Sugar Bill. *New York Times,* June 1, 53.

Blake, Nelson M. 1980. *Land into Water—Water into Land: A History of Water Management in Florida.* Tallahassee: University Presses of Florida.

Bonsal, Philip. 1971. *Cuba, Castro, and the United States.* Pittsburgh, PA: University of Pittsburgh Press.

Boseley, S. 2003. Sugar Industry Threatens to Scupper WHO. *Guardian,* April 21, 1.

Bourg, C. J. 1942. Sugar and War Consciousness. *Sugar Bulletin,* May 15, 1.

Broward, Napoleon Bonaparte. 1906. The Drainage of the Everglades. In *Proceedings of the Fourth Annual Convention of the Inter-State Sugar Cane Growers Association,* Inter-State Sugar Cane Growers, 89–99. Mobile, AL, February 7–9.

Brown, J. 1980. High Farm Costs and the Inability of Congress to Pass a Sugar Price-Support Bill Spell Pain for Sugar Growers. *Florida Trend,* March, 78–83.

Browne, W. 2003. Benign Public Policies, Malignant Consequences, and the Demise of African American Agriculture. In *African American Life in the Rural South, 1900–1950,* ed. R. Hurt, 129–51. Columbia: University of Missouri Press.

Brumbeck, Barbara C. 1990. Restoring Florida's Everglades: A Strategic Planning Approach. In *Environmental Restoration: Science and Strategies for Restoring the Earth,* ed. John J. Berger, 352–61. Washington, DC: Island Press.

Bryant, F. E. 1919. Letter to the editor. *The Florida Planter* 1, no. 2:5.

Bureau of National Affairs, Inc. 1994. Babbitt Explores Tying Sugar Subsidies to Industry Role in Everglades Cleanup. *BNA Washington Insider,* March 15, 62–64.

Burrows, Geoff, and Ralph Shlomowitz. 1992. The Lag in the Mechanization of the Sugarcane Harvest: Some Comparative Perspectives. *Agricultural History* 66, no. 3:61–75.

Business Week. 1961. Florida Bets on a Boom in Sugar. October 14, 58–62.

Bussey, J. 2005a. U.S. Sugar Corp. Gets a Wake-up call. *Miami Herald,* Business Monday, March 14, 24–25, 27.

———. 2005b. CAFTA Backers Lash Out At Sugar Interests in the U.S. *Miami Herald,* April 16, 1C, 3C.

———. 2006. Gov. Bush Throws Support Behind Ethanol Initiative. *Miami Herald,* December 19, 1C.

Buttel, Frederick H., and David Goodman. 1989. Class, State, Technology and International Food Regimes. *Sociologia Ruralis* 29, no. 2:86–92.

Cammack, L. H. 1916. *What About Florida?* Chicago: Laird and Lee.

Carlebach, Michael, and Eugene F. Provenzo Jr. 1993. *Farm Security Administration Photographs of Florida.* Gainesville: University Press of Florida.

Carlton, D. 1928. Letter from Florida Governor Carlton to B. Dahlberg, President, Southern Sugar Company, Dec. 19, 1928. Reprinted in Reese 1929.

Carter, Luther J. 1974. *The Florida Experience: Land and Water Policy in a Growth State.* Baltimore, MD: Johns Hopkins University Press.

Carson, D. 2005. No to CAFTA: As Happened with NAFTA, U.S. Farmers Will Suffer. *Palm Beach Post,* July 10.

Chalmin, P. 1984. The Important Trends in Sugar Diplomacy Before 1914. In Albert and Graves 1984, 9–19.

Chance, G. W. 1919. Sugar Centrals and Plantations for South Florida. *Florida Grower,* June 21, 6–7.

Chandler, I. 1994. Glades Residents Seek Comfort. *Clewiston News,* January 12, 1, 3.

Chang, J. 2007. Bush Kicks Off Tour, Talks Biofuels with Brazil. *Miami Herald,* March 10, 9A.

Clark, A., and G. Dalrymple. 2003. $7.8 billion for Everglades Restoration: Why Do Environmentalists Look So Worried? *Population and Environment* 24, no. 6:541–69.

Clewiston News. 1953. City's Founding by a Woman in 1920 Is Unique. February 12, A1–A2.

———. 1994. Mayor Determined to Save Glades. January 5, 4.

Coker, R., and G. Cantens. 2005. CAFTA Is a Sour Deal for Sugar Industry. *Miami Herald,* May 12, 25A.

Colburn, David R., and Jane L. Landers, eds. 1995. *The African American Heritage of Florida.* Gainesville: University Press of Florida.

Coppin, Clayton. 1990. James Wilson and Harvey Wiley: The Dilemma of Bureaucratic Entrepreneurship. *Agricultural History* 64, no. 2:167–81.

Crampton, Charles A. 1899. The Opportunity of the Sugar Cane Industry. *North American Review* 168 (March): 276–84.

———. 1901. Sugar and the New Colonies. *Forum* 32:283–291.

Crane, Verner W. 1956. *The Southern Frontier 1670–1732*. Ann Arbor: University of Michigan Press.

Crawford, J. S. 1902. Can We Raise Our Own Sugar? *Gunton's Magazine* 22 (January): 37–58.

Crispell, B. L. 1999. *Testing the Limits: George Armistead Smathers and Cold War America*. Athens: University of Georgia Press.

CROGEE. 2005. *Re-Engineering Water Storage in the Everglades: Risks and Opportunities* (Executive Summary). National Research Council, Committee on the Restoration of the Greater Everglades Ecosystem. Washington, DC: National Academies Press. Online at the Web site of the National Academies Press: http://www.nap.edu/catalog/11215.html.

Cronon, William. 1991. *Nature's Metropolis: Chicago and the Great West*. New York: W. W. Norton.

Cushman, John. 1993. Everglades Cleanup Agreement Fails. *New York Times*, December 19, 17.

Dacy, G. H. 1929. Florida Will Fill America's Future Sugar Bowl. *Florida Grower* 37, no. 2:1, 8, 28.

Dahlberg Corporation of America. 1929. *The Dahlberg Sugar Cane Industries*. Chicago: Dahlberg Corporation.

Dalton, John E. 1937. *Sugar: A Case Study of Government Control*. New York: Macmillan.

———. 1938. Sugar and Public Opinion. *Public Opinion Quarterly* 2, no. 2:287–94.

DARE. 1964. *Report of the D-A-R-E Meetings Sugarcane Session, May 19–20*. Gainesville: University of Florida Press.

Davies, F. 2001. 'Big Sugar' Gears Up to Defend Subsidy. *Miami Herald*, August 22, 16A.

Davis, P. 2006. State Bets Big on Ethanol. *Miami Herald*, July 6, 8B.

Davis, Steven N., and John C. Ogden. 1994. *Everglades: The Ecosystem and Its Restoration*. Delray Beach, FL: St. Lucie Press.

DeAngelis, D. 1994. Synthesis: Spatial and Temporal Characteristics of the Environment. In *Everglades: The Ecosystem and Its Restoration,* ed. S. Davis and J. Ogden, 307–20. Delray Beach, FL: St. Lucie Press.

Deerr, Noel. 1950. *The History of Sugar*. Vol. 2. London: Chapman and Hall.

Dodson, P. 1971. Hamilton Disston's St. Cloud Sugar Plantation. *Florida Historical Quarterly* 49, no. 4:356–69.

Douglas, Marjory Stoneman. 1988. *The Everglades, River of Grass*. Rev. ed. with afterword by Randy Lee Loftis with Marjory Stoneman Douglas. Sarasota, FL: Pineapple Press. (Orig. pub. 1947.)

Dovell, J. E. 1947a. *The Everglades—Florida's Frontier*. Part 1. University of Florida, Economic Leaflets, vol. 6, no. 5, April. Gainesville, FL: Bureau of Economics and Business Research, College of Business Administration, University of Florida. Reprinted as part of The Everglades, a Florida Frontier. *Agricultural History* 22, no. 3 (1948): 187–97.

———. 1947b. *The Everglades—Florida's Frontier*. Part 2. University of Florida, Economic Leaflets, vol. 6, no. 6, May. Gainesville, FL: Bureau of Economics and Business Research, College of Business Administration, University of Florida. Reprinted as part of The Everglades, a Florida Frontier. *Agricultural History* 22, no. 3 (1948): 187–97.

———. 1952. *Florida: Historic, Dramatic, Contemporary.* Vols. 1 and 2. New York: Lewis Historical Publishing.

Dulles, J. F. 1954. Memorandum to Secretary of Agriculture Benson, June 4. In *Foreign Relations of the United States, 1952–1954,* ed. William Z. Slany. Vol. 4. *The American Republic,* ed. N. Stephen Kane, 900–902. Dept. of State Pub. 9354. Washington, DC: U.S. Government Printing Office, 1983.

Dye, Alan. 1998. *Cuban Sugar in the Age of Mass Production: Technology and the Economics of the Sugar Central, 1899–1929.* Stanford, CA: Stanford University Press.

Dykers, A. W. 1962. Up Front with the League. *Sugar Bulletin* 40, no. 10:106, 111.

Eichner, Alfred S. 1969. *The Emergence of Oligopoly: Sugar Refining as a Case Study.* Baltimore, MD: Johns Hopkins University Press.

Eisenhower, D. 1987a. Memorandum from the President to the Secretary of Agriculture (Benson), January 27, 1955. In Glennon 1987, 6:781–82.

———. 1987b. Memorandum of a Conversation, Ambassador's Residence, Panama City, July 23, 1956. In Glennon 1987, 6:832–33.

Ellis, R. A. 1905. Sugar Cane as a Staple Money Crop. In *Proceedings of the Third Annual Convention of the Inter-State Sugar Cane Growers Association,* Inter-State Sugar Cane Growers, 85–88. Montgomery, AL, January 25–26.

Emerson, Charles Stafford. 1919. C. Lyman Spencer: Author, Scientist and Man of Research. *Florida Planter* 1, no. 2:5.

Fabel, Robin. 1988. *The Economy of British West Florida, 1763–1783.* Tuscaloosa: University of Alabama Press.

FDCH. 1994. Capitol Hill Hearing Testimony, September 20. Federal Document Clearing House, Inc.

Federal Writers Project. 1984. *WPA Guide to Florida.* New York: Pantheon. (Orig. pub. 1939.)

FIU (Florida International University). 2007. FIU and Florida Crystals to Develop Ethanol Technology. Florida International University, Applied Research Center, Newsletter Archive. Online at the Web site of Florida International University, FIU Media Relations: http://www.arc.fiu.edu/news_20070222.asp. (Accessed February 28, 2007.)

Florida Grower. 1932. Sugar Properties in the Everglades Reorganized. 40 (January): 6.

Florida Rural Legal Services. 1994. *Sugar Cane Workers News.* October.

Fogel, R. 1989. *Without Consent of Contract: The Rise and Fall of American Slavery.* New York: W. W. Norton.

Fogg, N. H. 1905. Sugar Cane in Florida. In *Proceedings of the Third Annual Convention of the Inter-State Sugar Cane Growers Association,* Inter-State Sugar Cane Growers, 101–3 Montgomery, AL, January 25–26.

Ford, Robert N. 1956. *A Resource Use Analysis and Evaluation of the Everglades Agricultural Area.* PhD diss., Department of Geography Research Paper No. 42, University of Chicago.

Fox, William Lloyd. 1980. Harvey W. Wiley's Search for American Sugar Self-Sufficiency. *Agricultural History* 54:516–26.

Friedmann, Harriet. 1982. The Political Economy of Food: The Rise and Fall of the Postwar International Food Order. In *Marxist Inquiries: Studies of Labor, Class, and States,* ed. Michael Burawoy and Theda Skocpol. (Supplement to *American Journal of Sociology* 88:S248–86).

Friends of the Everglades. 1981. Repair the Everglades. *Newsletter of the Friends of the Everglades* (Spring). Coconut Grove, FL.

Ft. Myers News Press. 1926. Okeechobee Men Favor Waterways. August 6. Clipping file, Fred L. Williamson Manuscript Collection, Special Collections, University of Florida.

———. 2000. Farmers Destroy 7 Percent of Crop to Trim Sugar Glut. Nation & World, December 10, 1E.

Ft. Pierce News-Tribune. 1923. Vital Matters to be Discussed. November 14. Clipping file, Fred L. Williamson Manuscript Collection, Special Collections, University of Florida.

Galloway, J. H. 1989. *The Sugar Cane Industry: An Historical Geography from Its Origins to 1914.* Cambridge: Cambridge University Press.

Gates, P. W. 1960. *The Farmer's Age: Agriculture, 1815–1860.* Vol. 3. *The Economic History of the United States.* New York: Harper & Row.

Genovese, David, and Wallace P. Mullin. 1999. The Sugar Institute Learns to Organize Information Exchange. In *Learning by Doing in Markets, Firms and Countries,* ed. Naomi R. Lamoreaux, Daniel M. G. Raff, and Peter Temin, 103–43. Chicago: University of Chicago Press.

Genovese, Eugene D. 1979. *From Rebellion to Revolution.* Baton Rouge: Louisiana State University Press.

Gerber, D. 1976. The United States Sugar Quota Program: A Study in the Direct Congressional Control of Imports. *Journal of Law and Economics* 19, no. 1:103–47.

Gervasi, F. 1945. Where's All the Sugar? *Collier's,* September 8, 20, 67–68.

Gifford, J. 1935. *The Reclamation of the Everglades with Trees.* New York: Books, Inc.

Gilmore, R. 2002. Fatal Couplings of Power and Difference: Notes on Racism and Geography. *Professional Geographer* 54, no. 1:15–24.

Glaz, B. 1995. Research Seeking Agricultural and Ecological Benefits in the Everglades. *Journal of Soil and Water Conservation* 50, no. 6:609–12.

Glennon, J. P., ed. 1987. *Foreign Relations of the United States, 1955–1957.* Vol. 6. Washington, DC: U.S. Government Printing Office.

Goluboff, R. 1999. "Won't you please help me get my son home?" Peonage, Patronage, and Protest in the World War II Urban South. *Law and Social Inquiry* 24, no. 4:777–806.

Goodman, D., and M. Watts. 1997. *Globalizing Food: Agrarian Questions and Global Restructuring.* London and New York: Routledge.

Graham, K. 1998. *Personal History.* New York: Vintage Books.

Grantham, Dewey W. 1995. *The South in Modern America.* New York: HarperPerennial.

Grosvenor, M. B. 1947. Cuba—American Sugar Bowl. *National Geographic Magazine,* January, 1–56.

Grubbs, Donald H. 1961. The Story of Florida's Migrant Farm Workers. *Florida Historical Quarterly* 40, no. 2:103–22.

Gunton, George. 1902. Some Free Sugar Fallacies. *Gunton's Magazine* 22 (February): 137–46.

Hagy, J. 1993. Watergate. *Florida Trend*, March, 32–37.

Hahamovitch, Cindy. 1997. *The Fruits of Their Labor: Atlantic Coast Farmworkers and the Making of Migrant Poverty, 1870–1945.* Chapel Hill: University of North Carolina Press.

Hanna, Alfred Jackson, and Kathryn Abbey Hanna. 1948. *Lake Okeechobee: Wellspring of the Everglades.* Indianapolis, IN: Bobbs Merrill.

Harner, Charles. 1973. *Florida's Promoters: The Men Who Made It Big.* Tampa, FL: Trend House.

Harvey, David. 1996. *Justice, Nature and the Geography of Difference.* Cambridge, MA: Blackwell.

Heitmann, John. 1987. *The Modernization of the Louisiana Sugar Industry, 1830–1910.* Baton Rouge: Louisiana State University Press.

———. 1998. The Beginnings of Big Sugar in Florida, 1920–1945. *Florida Historical Quarterly* 78, no. 1:39–61.

Heston, Thomas Janey. 1975. Sweet Subsidy: The Economic and Diplomatic Effects of the U.S. Sugar Acts, 1934–1974. PhD diss.. Department of History, Case Western Reserve University.

Hoelscher, D. 2003. Making Place, Making Race: Performances of Whiteness in the Jim Crow South. *Annals of the Association of American Geographers* 93, no. 3:657–86.

Holland, Max. 1999. A Luce Connection: Senator Keating, William Pawley, and the Cuban Missile Crisis. *Journal of Cold War Studies* 1, no. 3:139–67.

———. 2005. Private Sources of U.S. Foreign Policy: William Pawley and the 1954 Coup d'Etat in Guatemala. *Journal of Cold War Studies* 7, no. 4:36–73.

Hollander, G. 2003. Re-Naturalizing Sugar: Narratives of Place, Production, and Consumption. *Social and Cultural Geography* 4, no. 1:59–74.

———. 2005. Securing Sugar: National Security Discourse and the Establishment of Florida's Sugar-Producing Region. *Economic Geography* 81, no. 4:339–58.

Holling, C., L. Gunderson, and C. Walters. 1994. The Structure and Dynamics of the Everglades System: Guidelines for Ecosystem Restoration. In Davis and Ogden 1994, 741–56.

Holsendolph, E. 1974. Big Sugar Users Study Substitutes. *New York Times*, August 13, 45, 54.

Hudson, John C. 1994. *Making the Corn Belt: A Geographical History of Middle-Western Agriculture.* Bloomington: Indiana University Press.

Ickes, Harold L. 1954. *The Secret Diary of Harold L. Ickes.* Vol. 3. *The Lowering Clouds, 1939–1941.* New York: Simon and Schuster.

Interamerican Ethanol Commission. 2007. Online at the Interamerican Ethanol Commission Web site, http://helpfuelthefuture.org/ourhistory.htm.

ISCGA (Inter-State Sugar Cane Growers Association). 1903. Call for Convention. In *Proceedings of the Interstate Sugar Cane Growers First Annual Convention*, Interstate Sugar Cane Growers, 9–11. Held in Macon, GA, May 6–9. Macon, GA: Smith and Watson.

Izuno, F. T., A. B. Bottcher, F. J., Coale, C. A. Sanchez, and D. B. Jones. 1995. Agricultural BMPs for Phosphorous Reduction in South Florida. *Transactions of the American Society of Agricultural Engineers* 38, no. 3:735–44.

Jaudon, James F. 1934. Letter from Jaudon to Bureau of Weights and Standards, February 15. Available online at the Web site of the Publication of Archival Library Museum Materials, in the Florida Heritage Collection: http://fulltext10.fcla.edu. (Accessed on November 19, 2006.)

Jenks, Leland Hamilton. 1976. *Our Cuban Colony: A Study in Sugar.* St. Clair Shores, MI: Scholarly Press. (Orig. pub. 1928.)

John, DeWitt. 1994. *Civic Environmentalism: Alternatives to Regulation in States and Communities.* Washington, DC: Congressional Quarterly Press.

Johnson, C. 1992. Phosphorus Follies: An Outbreak of Cattails in the Everglades Could Cost 125 Farmers up to $600 Million. *Farm Journal* 116, no. 13:B4–B5.

Johnston, William. 1928. New Sugar Lands in the Florida Everglades. *Facts about Sugar: A Weekly Journal of the Sugar Industry,* 23, no. 2:42–43.

Jones, Jacqueline. 1992. *The Dispossessed: America's Underclass from the Civil War to the Present.* New York: Basic Books.

Katz, Michael B. 1996. *In the Shadow of the Poorhouse: A Social History of Welfare in America.* Rev. ed. New York: Basic Books.

Kenfield, I. 2007. Brazil's Ethanol Plan Breeds Rural Poverty, Environmental Degradation. *Global Research,* March 8. Online at their Web site: http://www.globalresearch.ca/PrintArticle.php?articleId=5012.

Kennan, George. 1902. The Conflict of Sugar Interests. *Outlook,* February 8, 367–70.

King, S. 1977. Sugar Game: Consumer Loses. *New York Times,* November 13, D1.

———. 1978. Sugar States Ask Revival of Quotas. *New York Times,* May 29, F1.

———. 1979a. 17¢ Support for Sugar Is Sought. *New York Times,* February 7, D3.

———. 1979b. Sugar Rise Backed by Carter. *New York Times,* February 16, D1.

Klos, George. 1995. Blacks and the Seminole Removal Debate. In Colburn and Landers 1995, 128–56. Gainesville: University Press of Florida.

Kramer, Peter. 1966. *The Offshores: A Study of Foreign Farm Labor in Florida.* St. Petersburg, FL: Community Action Fund.

Krueger, Anne O. 1993. *Economic Policies at Cross Purposes: The United States and Developing Countries.* Washington, DC: Brookings Institution.

Kuethe, Allan J. 1988. Charles III, the Cuban Military and the Destiny of Florida. In *Charles III: Florida and the Gulf,* ed. Patricia R. Wickman, 64–77. Miami: Count of Gálvez Historical Society.

Kunz, D. 1997. *Butter and Guns: America's Cold War Economic Diplomacy.* New York: Free Press.

Kyriakoudes, L. 2003. "Lookin' for better all the time": Rural Migration and Urbanization in the South, 1900–1950. In *African American Life in the Rural South, 1900–1950,* ed. R. Hurt, 10–26. Columbia: University of Missouri Press.

Landers, Jane L. 1995. Traditions of African American Freedom and Community in Spanish Colonial Florida. In Colburn and Landers 1995, 17–41.

Leonhardy, T. 1987. Memorandum of a Telephone Conversation between Terrance G. Leonhardy of the Office of Middle American Affairs and Joaquin Meyer of the Cuban Sugar Stabilization Institute, Washington, April 11, 1957. In Glennon 1987, 6:781–82.

Levenstein, Harvey. 1993. *Paradox of Plenty: A Social History of Eating in America.* New York: Oxford University Press.

Light, Stephen S., Lance H. Gunderson, and C. S. Holling. 1995. The Everglades: Evolution of Management in a Turbulent Ecosystem. In *Barriers and Bridges to the Renewal of Ecosystems and Institutions,* ed. Lance H. Gunderson, C. S. Holling, and Stephen S. Light. New York: Columbia University Press.

Linder, Marc. 1987. Farm Workers and the Fair Labor Standards Act: Racial Discrimination in the New Deal. *Texas Law Review* 65, no. 7:1335–93.

Low, Thomas E. 1998. The Chautauquans and Progressives in Florida. *Journal of Decorative and Propaganda Arts* 23 (Florida Theme Issue): 306–21.

Lupfer, S. L. 1905. The Florida Everglades: Their Legal Status, Their Drainage, Their Future Value. *Engineering News* 54, no. 11:278–80.

MacDonald, Scott B., and Georges A. Fauriol. 1991. *The Politics of the Caribbean Basin Sugar Trade.* New York: Praeger.

Mahler, Vincent A. 1986. Controlling International Commodity Prices and Supplies: The Evolution of United States Sugar Policy. In *Food, the State, and International Political Economy: Dilemmas of Developing Countries,* ed. F. LaMond Tullis and W. Ladd Hollis. Lincoln: University of Nebraska Press.

Maidenberg, H. 1964. Beet and Sugar Men Are Waging a Bitter War Over U.S. Market. *New York Times,* March 29, F1.

———. 1965. Price War in Sugar Looms on Horizon. *New York Times,* February 28, F1.

———. 1974. A 5-Pound Bag of Sugar to Cost $2.20 in Month. *New York Times,* August 22, 45.

———. 1976. Corn Sweetener Industry Is Expanding Its Share of the Market. *New York Times,* March 8, 49.

———. 1977. Remember the Sugar Price Uproar? *New York Times,* January 10, 45.

———. 1978. Sugar Prices: Sweeter Trends. *New York Times,* August 7, D5.

Mann, Susan Archer. 1990. *Agrarian Capitalism in Theory and Practice.* Chapel Hill: University of North Carolina Press.

Manuel, Fritzie P. 1942a. Land Development in the Everglades. National Defense Migration Hearings, 77th Congress, 2nd sess., 12863–88.

———. 1942b. Sugar Production in Florida. National Defense Migration Hearings. 77th Cong., 2nd sess., 12955–976.

Markel, R. 1975. *The Politics of Sugar in U.S. Domestic and Foreign Affairs.* PhD. diss., Political Science, University of Notre Dame.

Marquis, C. 2000. Americans Need Better Diet, New Health Study Reports. *New York Times,* May 28, 24.

Mayer, Jane, and José de Cordoba. 1991. Sweet Life: First Family of Sugar Is Tough on Workers, Generous to Politicians. *Wall Street Journal,* July 29, 1, 5.

Maynard, R. 2007. Biofuels Report: Against the Grain. *Ecologist Online,* January 3. Online at http://www.theecologist.org/archive_detail.asp?content_id=834.

Mayo, Nathan. 1928. *Florida, An Advancing State: 1907—1917—1927.* St. Petersburg, FL: Lassing Publishing.

McAvoy, Muriel. 2003. *Sugar Baron: Manuel Rionda and the Fortunes of Pre-Castro Cuba.* Gainesville: University Press of Florida.

McCally, David Philip. 1991. *Cane Cutters in the Everglades.* Master's thesis, Department of History, University of Florida.

———. 1999. *The Everglades: An Environmental History.* Gainesville: University Press of Florida.

McClure, R. 1994. Official Shies From Proposal to End Florida Sugar Farming. *Fort Lauderdale Sun-Sentinel,* January 15, D1.

McConnell, J. 1987. Memorandum of a Conversation, Department of State, Washington, February 15, 1955. In Glennon 1987, 6:791–92.

McCormick, P. V., S. Newman, S. Miao, D. E. Gawlik, D. Marley, K. R. Reddy, T. Fontaine. 2002. Effects of Anthropogenic Phosphorous Inputs on the Everglades. In *The Everglades, Florida Bay, and Coral Reefs of the Florida Keys: An Ecosystem Handbook,* ed., J. W. Porter and K. G. Porter, 83–126. Boca Raton, FL: CRC Press.

McCorvie, Mary R., and Christopher L. Lant. 1993. Drainage District Formation and the Loss of Midwestern Wetlands, 1850–1930. *Agricultural History* 67, no. 4:13–39.

McCoy, Terry L. 1990. *U.S. Policy and the Caribbean Basin Sugar Industry: Implications for Migration.* Working Papers, Commission for the Study of International Migration and Cooperative Development, no. 36. Washington, DC: Commission for the Study of International Migration and Cooperative Development.

McGovern, Joseph J. 1981. *The First Fifty Years.* Clewiston, FL: United States Sugar Corporation.

McNair, J. 2000. Big Sugar: A Mound of Trouble. *Miami Herald,* September 24, 1E.

McWilliams, C. 1945. *Ill Fares the Land: Migrants and Migratory Labour in the United States.* London: Faber and Faber.

Meyer, Hermann, comp. 1910. *Select List of References on Sugar, Chiefly in Its Economic Aspects.* Washington, DC: U.S. Government Printing Office.

Miami Herald. 1939. Editorial. December 3, All Florida sec., 5.

———. 1965a. "I'll get rugged," Holland Warns If Wirtz Holds Line on Migrants. April 3, 15A.

———. 1965b. Who's Responsible for Migrants? May 5, 5K.

———. 1972. Sugar Firms Win UFW Court Battle. November 4, 1B.

———. 1991. Coming Clean on the Everglades. Editorial. May 22, 12A.

———. 1996. AFL-CIO Opposes Penny-Per-Pound Tax on Sugar. July 28, 5B.

Minneapolis Journal. 1930. Receiver Asked for Celotex. September 25, n.p. Preserved in the Fred L. Williamson Manuscript Collection, Special Collections, University of Florida Library.

Mintz, Sidney W. 1985. *Sweetness and Power: The Place of Sugar in Modern History.* New York: Viking Penguin.

Miró Cardona, José. 1962. *In Defense of the Position of Cuba as a Supplier of Sugar to the United States Market.* Miami: Revolutionary Council of Cuba.

Mitchell, P. 1994. Babbitt Warns Everglades Sugar Growers. *Orlando Sentinel,* March 11, A1.

Morgan, Curtis 1990. Earth Day Fest Heartening, Douglas Says. *Miami Herald,* April 15, 4B.

———. 2007a. Judge Slaps Mining. *Miami Herald,* July 30. 1B.

———. 2007b. Rock Fight. *Miami Herald,* April 3, 1B.

Mott, Charles Stewart. 1970. Interview with Studs Terkel. In *Hard Times: An Oral History of the Depression,* by Studs Terkel, 134–36. New York: Pantheon Books.

Mott, Frank. 1957. *A History of American Magazines, 1885–1905.* Vol. 4. Cambridge, MA: Harvard University Press.

———. 1968. *A History of American Magazines.* Vol. 5. *Sketches of 21 Magazines, 1905–1930.* Cambridge, MA: Harvard University Press.

Mullendore, William Clinton. 1941. *History of the United States Food Administration.* Stanford, CA: Stanford University Press.

Myrick, Herbert. 1897. *Sugar: A New and Profitable Industry in the United States for Capital, Agriculture and Labor to Supply the Home Market Yearly with $100,000,000 of Its Product.* New York: Orange Judd.

———. 1907. *The American Sugar Industry.* New York: Orange Judd.

Nestle, M. 2002. *Food Politics: How the Food Industry Influences Nutrition and Health.* Berkeley and Los Angeles: University of California Press.

New York Times. 1894. Editorial. January 8, 4.

———. 1896. Sugar from Florida. September 6, 15.

———. 1905. Make Our Possessions Producers of Coffee. April 1, 8.

———. 1942. Florida Seeks Sugar Data. January 31, 30.

———. 1943. Florida Sugar Crop Lacks Help. February 8, 27.

———. 1946a. C. R. Bitting Resigns and F. P. Tralles Succeeds Him. May 22, 36.

———. 1946b. Sugar Company Employees Protest Change in Company Head. May 23, 34.

———. 1960. Sugar Curbs Scored. October 5, 65.

———. 1964. U.S. Accuses Cuba of Sugar Scheme. April 28, 18.

———. 1965a. Sugar Industry Leaders Agree on Bid for Revisions in Quotas. March 29, 53.

———. 1965b. Sugar Bill Wins First House Test. October 13, 62.

———. 1965c. U.S. Sugar Costs Called Excessive. October 15, 63.

———. 1965d. Senate Votes Sugar Quota Bill; Battle is Looming with Cooley. October 21, 43.

———. 1967. Editorial: The Everglades Mess. May 14, E10.

———. 1969. Cane Allotments Increased. April 29, 21.

———. 1972a. Judge to Rule on Legality of Importing Cane Cutters. September 24, 52.

———. 1972b. Chavez Seeks Investigation of Florida's Sugar Industry. November 20, 43.

———. 1974a. Butz Sees Changes in Sugar Program. February 22, 43.

———. 1974b. "Freer Market" for Sugar Urged by Industrial Users. February 20, 46.

———. 1974c. '75 Sugar Quota Set by President. November 19, 61.

———. 1975. "New Direction" to Cuba. March 4, 32.

———. 1997. Editorial: A Lifeline for the Everglades. November 17, A18.

———. 2007. Castro Again Chides U.S. on Ethanol Plan, April 5, A6.

Newbegin, R. 1987. Memorandum from the Director of the Office of Middle American Affairs (Newbegin) to the Deputy Assistant Secretary of State for Inter-American Affairs (Sparks), March 15, 1955. In Glennon 1987, 6:796–98.

Nielsen, Waldemar A. 1972. *The Big Foundations.* New York: Columbia University Press.

Ong, A., and S. Collier. 2005. *Global Assemblages: Technology, Politics, and Ethics as Anthropological Problems.* Malden, MA: Blackwell.

O'Tuathail, G. 2002. Post–Cold War Geopolitics: Contrasting Superpowers in a World of Global Dangers. In *Geographies of Global Change: Remapping the World,* ed. R. Johnston, P. Taylor, and M. Watts, 174–89. 2nd ed. Oxford: Blackwell.

Overfield, Richard A. 1990. Science Follows the Flag: The Office of Experiment Stations and American Expansion. *Agricultural History,* 64, no. 2:31–40.

Paasi, A. 1991. Deconstructing Regions: Notes on the Scales of Spatial Life. *Environment and Planning A* 23:239–56.

———. 2001. Europe as a Social Process and Discourse: Considerations of Place, Boundaries and Identity. *European Urban and Regional Studies* 8, no. 1:7–28.

———. 2002. Place and Region: Regional Worlds and Words. *Progress in Human Geography* 26, no. 6:802–11.

Parker, B. 1993. 'Glades to mold U.S. policy. *Fort Myers News-Press,* February 23, 1.

Pascal, Joan, and Harold G. Tipton. 1942. Vegetable Production in South Florida. National Defense Migration Hearings. 77th Cong., 2nd sess., 12888–955.

Patterson, Gordon. 1997. Raising Cane and Refining Sugar: Florida Crystals and the Fame of Fellsmere. *Florida Historical Quarterly* 75, no. 4:408–28.

Paulson, S., L. Gezon, and M. Watts. 2003. Locating the Political in Political Ecology: An Introduction. *Human Organization* 62, no. 3: 205–17.

Peet, R. 1996. A Sign Taken for History: Daniel Shay's Memorial in Petersham, Massachusetts. *Annals of the Association of American Geographers* 86, no. 1:21–43.

Peet, R., and M. Watts. 1996. Liberation Ecology: Development, Sustainability, and Environment in an Age of Market Triumphalism. In *Liberation Ecologies: Environment, Development, Social Movements,* ed. R. Peet and M. Watts, 1–45. London: Routledge.

Pfeiffer, Jack. c. 1970. *Official History of the Bay of Pigs Operation.* Vol. 3. Online at the Web site of David M. Barrett, Ph. D., Department of Political Science, Villanova University: http://www14.homepage.villanova.edu/david/barrett/bop.html.

Phillips, R. Hart. 1955. Soviet Purchases Appeal to Cuba. *New York Times,* June 19, 30.

———. 1962. Cuban Exiles Growing Sugar Cane in Florida. *New York Times,* January 19, 57.

Porter, John. 1905. Address to the Inter-State Sugar Cane Growers. In *Proceedings of the Third Annual Convention of the Inter-State Sugar Cane Growers Association,* Inter-State Sugar Cane Growers, 43–46. Montgomery, AL, January 25–26.

Porter, Kenneth W. 1971. *The Negro on the American Frontier.* New York: Arno Press and the *New York Times.*

Pred, A. 1984. Place as Historically Contingent Process: Structuration and the Time-Geography of Becoming Places. *Annals of the Association of American Geographers* 74:279–97.

Prinsen Geerligs, H. C. 1912. *The World's Cane Sugar Industry, Past and Present.* Manchester: Norman Rodger Altrincham.

Proctor, Samuel. 1993. *Napoleon Bonaparte Broward: Florida's Fighting Democrat.* Gainesville: University Press of Florida. (Orig. pub. 1950.)

———. 1977. General editor's preface. In *The Humble Petition of Denys Rolle, Esq.,* by Denys Rolle, v–ix. A facsimile reproduction of the 1765 edition. Gainesville: University Presses of Florida.

Purse, D. G. 1906. Report of the President. *Proceedings of the Fourth Annual Convention of the Inter-State Sugar Cane Growers Association,* Inter-State Sugar Cane Growers, 28–33. Mobile, AL, February 7–9.

Putnam, J., and J. Allshouse. 1998. U.S. Per Capita Food Supply Trends. *Food Review* 21, no. 3:2–11.

Quaife, Milo. 1948. Editorial Introduction. In Hanna and Hanna 1948, i–ii.

Reese, J., comp. 1929. *Opening of the Nation's Sugar Bowl.* Clewiston, FL: Press of the Clewiston News.

Revkin, A. 2002. Stockpiling Water for a River of Grass. *New York Times,* March 26, F1.

Richardson, Bonham C. 1992. *The Caribbean in the Wider World, 1492–1992.* Cambridge: Cambridge University Press.

Robbins, P. 2003. Political Ecology in Political Geography. *Political Geography* 22, no. 6: 641–45.

Robbins, W. 1979. Lobbyists Worked Offstage to Shape Sugar Laws. *New York Times,* January 15, A1.

Rohrbough, Malcolm J. 1990. *The Trans-Appalachian Frontier.* Belmont, CA: Wadsworth. (Orig. pub. 1978.)

Rohter, Larry. 1993. Florida Growers and U.S. Reach Everglades Pact. *New York Times,* July 14, A1.

———. 2006. With Big Boost From Sugar Cane, Brazil Is Satisfying Its Fuel Needs. *New York Times,* April 10, A1.

Roll, Eric. 1956. *The Combined Food Board: A Study in Wartime International Planning.* Stanford, CA: Stanford University Press.

Rose, Rufus. 1903. Sugar Growing and Manufacturing in Florida. In *Proceedings of the Interstate Sugar Cane Growers First Annual Convention,* Interstate Sugar Cane Growers, 40–52. Macon, GA, May 6–9. Macon, GA: Smith and Watson.

———. 1906. The Production of Sugar in the South—Practically Considered. In *Proceedings of the Fourth Annual Convention of the Inter-State Sugar Cane Growers Association,* Inter-State Sugar Cane Growers, 125–33. Mobile, AL, February 7–9.

Rosenberg, Carol. 2005. Bay of Pigs Plotters Predicted Failure. Online at the Web site of the *Miami Herald,* posted August 11: http://www.miami.com/mld/miamiherald/news.

Rothenberg, Daniel. 1998. *With These Hands: The Hidden World of Migrant Farmworkers Today.* New York: Harcourt Brace.

Ruane, Laura. 1991. Big Sugar: Industry Struggles With Bittersweet Image. *Fort Myers News Press,* December 2, Business, 1.

Rugaber, W. 1974. Sugar Act Is Killed by House; Revival in Senate Possible. *New York Times,* June 6, 1.

Rutenberg, J., and L. Rohter. 2007. Bush and Chavez Spar at Distance Over Latin Visit. *New York Times,* March 10, A1.

Rutter, Frank R. 1902. The Sugar Question in the United States. *Quarterly Journal of Economics* 17 (November): 44–81.

St. Petersburg Times. 1993. Editorial: The Disturbing Everglades Deal. September 28.

Salisbury, S. 2005. Ethanol Subsidy Can't Woo Sugar to Accept CAFTA. *Palm Beach Post,* June 17, 1D.

———. 2007. Giant Sugar Mill Looks to Sweet Success. *Palm Beach Post,* January 25, 1C.

Salley, George H. n.d. *A History of the Florida Sugar Industry.* Booklet, n.p.

Santaniello, N. 2005. Report Labels Sugar Plant as No. 1 Polluter. *South Florida Sun-Sentinel,* January 6, 1B.

Sassen, S. 2006. *Territory, Authority, Rights: From Medieval to Global Assemblages.* Princeton, NJ: Princeton University Press.

Sayer, Andrew, and Richard Walker. 1992. *The New Social Economy: Reworking the Division of Labor.* Cambridge, MA: Blackwell.

Sayler, Charles. 1905. *A Report on the Progress of the Beet-Sugar Industry in the United States in 1904.* U.S. Department of Agriculture, Report No. 80. Washington, DC: U.S. Government Printing Office.

Schafer, David L. 1995. "Yellow silk ferret tied round their wrists": African Americans in British East Florida, 1763–1784. In Colburn and Landers 1995, 71–103.

Schene, M. 1981. Sugar along the Manatee: Major Robert Gamble, Jr. and the Development of the Gamble Plantation. *Tequesta* 41:69–81.

Schlesinger, Arthur M. 1957. *The Age of Roosevelt: The Crisis of the Old Order, 1919–1933.* Boston: Houghton Mifflin.

———. 1959. *The Age of Roosevelt: The Coming of the New Deal.* Boston: Houghton Mifflin.

Schmitt, E. 2000. Everglades Restoration Plan Passes House, with Final Approval Seen. *New York Times,* October 20, A19.

Schulten, Susan. 2001. *The Geographical Imagination in America, 1880–1950.* Chicago: University of Chicago Press.

Scientific American Supplement. 1900. Sugar as Food. 40, no. 1272:20389.

———. 1901. Sugar as Food. 52, no. 1339:21461.

Scott, J. 1998. *Seeing Like a State: How Certain Schemes to Improve the Human Condition Have Failed.* New Haven, CT: Yale University Press.

Sears, W. J. 1928. Letter from Congressman Sears to Louis Morgan, Publisher, Clewiston News, December 17 (unpaged). Reprinted in Reese 1929.

Shabecoff, P. 1973. Florida Cane Cutters: Alien, Poor, Afraid. *New York Times,* March 12, 24.

———. 1976. What Ford Hopes Carter Will Keep. *New York Times,* August 13, F17.

Shofner, Jerrell H. 1981. The Legacy of Racial Slavery: Free Enterprise and Forced Labor in Florida in the 1940s. *Journal of Southern History* 47, no. 3:411–26.

Sierra Club. 2004. Sierra Club: Bush Administrations in Tallahassee and Washington Have Abandoned Everglades Restoration. Press Release, January 24. Online at the Sierra Club Web site: http://www.sierraclub.org/pressroom/releases/pr2004-01-27.asp.

Silva, M., and C. Zaneski. 1999. Farmland Bought for Glades. *Miami Herald,* January 9, 1A.

Simpson, C. T. 1920. *In Lower Florida Wilds: A Naturalist's Observations on the Life, Physical Geography, and Geology of the More Tropical Part of the State.* New York: J. P. Putnam's Sons.

Sitterson, J. Carlyle. 1953. *Sugar Country: The Cane Sugar Industry in the South, 1753–1950.* Lexington: University of Kentucky Press.

Sklar, F., C. McVoy, R. VanZee, D. E. Gawlik, K. Tarboton, D. Rudnick, S. Miao, T. Armentano. 2002. The Effects of Altered Hydrology on the Ecology of the Everglades. In *The Everglades, Florida Bay, and Coral Reefs of the Florida Keys: An Ecosystem Handbook,* ed. J. W. Porter and K. G. Porter, 39–82. Boca Raton, FL: CRC Press.

Smith, M., and W. Ames. 1925 (June). *Report on Proposed Cane Sugar Development on Lands of the Sugarland Development Company Situated Southwest of Lake Okeechobee Florida.* Consulting Engineers, Havana and New York. Special Collections, University of Florida.

Smith, Neil. 1990. *Uneven Development: Nature, Capital and the Production of Space.* Oxford: Basil Blackwell.

Snyder, G. H. 1994. Soils of the EAA. In *Everglades Agricultural Area (EAA),* ed. A. B. Bottcher and F. T. Izuno, 27–41. Gainesville: University Press of Florida.

Sorrentrue, J. 2005. Sugar Firms Deny Trying to Influence Vote on Annexation. *Palm Beach Post,* April 21, 1A.

Southern Sugar Company. 1928. *Operations of the Southern Sugar Company.* Chicago: Southern Sugar Company.

Spencer, C. Lyman. 1918. *The Sugar Situation.* Jacksonville, FL: Drew Press.

———. 1919. The Low Cost of Foodstuffs of the South East. *Florida Planter* 1, no. 2:1, 12.

State of Florida. 1994. *Everglades Forever Act,* Section 373.4592 of the Florida Statutes.

Stoll, S. 1998. *The Fruits of Natural Advantage: Making the Industrial Countryside in California.* Berkeley and Los Angeles: University of California Press.

Stubbs, W. C. 1907. The Cane Sugar Industry. In Myrick 1907, 18–28.

Sturgill, Claude C. 1977. Introduction. In *The Humble Petition of Denys Rolle, Esq.,* by Denys Rolle, xi–x. A facsimile reproduction of the 1765 edition. Gainesville: University Presses of Florida.

Sugar Association, The. 2002. Online at Web site of The Sugar Association, News Room Overview: http://www.sugar.org. (Accessed May 18, 2002.)

———. 2007. Online at the Web site of The Sugar Association, About Us, Member Companies: http://www.sugar.org. (Accessed April 20, 2007.)

Sugar Journal 1961. Florida Sugar. Vol. 24, no. 7:16.

Sugar y Azúcar. 1994. Mechanization in the Sugarcane Fields. January, 20–23, 26.

Tampa Tribune. 1993. Editorial: Everglades Agreement Too Sweet. September 11, 8.

Tanner, Helen Hornbeck. 1989. *Zespedes in East Florida, 1784–1790.* Jacksonville: University of North Florida Press. (Orig. pub. 1963)

Taussig, F. W. 1931. *Some Aspects of the Tariff Question.* Cambridge, MA: Harvard University Press.

Taylor, P. J. 1997. "Appearances Notwithstanding, We Are All Doing Something Like Political Ecology." *Social Epistemology* 11, no. 1:111–27.

Tebeau, Charlton W. 1971. *A History of Florida.* Coral Gables, FL: University of Miami Press.

Tejada y Sainz, Juan de Dios. 1941. *Azúcar en la Florida.* Havana: Ucar, García & Cía.

Thomas, H. 1971. *Cuba: The Pursuit of Freedom.* New York: Harper & Row.

Thomas, J. 1981. Florida's Refugees Challenging Plan to Use West Indians in Cane Harvest. *New York Times,* October 11, 22.

Time. 1961. Sugar Fever. January 27, n.p. Online at www.time.com/time/magazine/article/0,9171,826823,00.html.

Trigger, D. 1997. Mining, Landscape and the Culture of Development Ideology in Australia. *Ecumene* 4, no. 2:161–80.

Tsing, A. 2005. *Friction: An Ethnography of Global Connection.* Princeton, NJ: Princeton University Press.

Tucker, R. 2000. *Insatiable Appetites.* Berkeley and Los Angeles: University of California Press.

USSC. 1944. *The Everglades, Agro-Industrial Empire of the South.* Clewiston, FL: United States Sugar Corporation.

———. 1992/1993. *The Company.* Clewiston, FL: United States Sugar Corporation.

———. 1993. *Background Briefing on Everglades Negotiations.* December 8. Clewiston, FL: United States Sugar Corporation.

———. 1994. President's Corner. *New Horizons* 12, no. 2:2. (USSC newsletter)

U.S. Commission on Agricultural Workers. 1991. Hearing, Belle Glade Florida, February 14, 1991. Washington, DC: Government Printing Office.

U.S. Cuban Sugar Council. 1947. *Bulletin.* February.

———. 1948. *Sugar Facts and Figures.* New York: U.S. Cuban Sugar Council

U.S. Department of Agriculture (USDA). 1978. *A History of Sugar Marketing Through 1974.* U.S. Department of Agriculture, Economics, Statistics, and Cooperatives Service, Agricultural Economic Report No. 382. Washington, DC: Government Printing Office.

———. 2005. *Crop Production, 2004 Summary.* U.S. Department of Agriculture, National Agricultural Statistics Service. Washington, DC: Government Printing Office.

U.S. Department of Commerce. 1917. *The Cane Sugar Industry.* U.S. Department of Commerce, Bureau of Foreign and Domestic Commerce, Misc. Series No. 53. Washington, DC: Government Printing Office.

U.S. GAO. 1992. *Foreign Farm Workers in the U.S.: Department of Labor Action Needed to Protect Florida Sugar Cane Workers.* HRD-92-95, June.

———. 1993. *Sugar Program: Changing Domestic and International Conditions Require Changes.* RCED-93-94, April.

———. 1995. *Restoring the Everglades: Public Participation in Federal Efforts.* RCED-96-5, October.

———. 2000. *Comprehensive Everglades Restoration Plan: Additional Water Quality Projects May Be Needed and Could Increase Costs.* RCED-00-235, September.

———. 2007. *South Florida Ecosystem, Restoration Is Moving Forward but Is Facing Significant*

Delays, Implementation Challenges, and Rising Costs. Report to the Committee on Transportation and Infrastructure, House of Representatives. GEO-07-520, May.

U.S. House. 1955. Committee on Agriculture. *Review and Extension of Sugar Act of 1948.* Report No. 1348. 84th Cong., 1st sess.

———. 1961. Committee on Agriculture. *Special Study on Sugar.* A Report of the Special Study Group on Sugar of the U.S. Department of Agriculture. 87th Cong., 1st sess. Committee Print, February 14.

———. 1983. Subcommittee on Labor Standards. Job Rights of Domestic Workers: the Florida Sugar Cane Industry. 98th Congress, 1st sess. Committee Print.

———. 1991. Committee Print. *Report on the Use of Temporary Foreign Workers in the Florida Sugar Cane Industry.* Prepared for the Committee on Education and Labor. 102nd Congress, 1st sess., July 1991.

U.S. Newswire. 1994. U.S. Government Reaches Settlement Agreement with Flo-Sun in Everglades Lawsuit. January 14.

U.S. Senate. 1858. *Memorial of the Legislature of the State of Iowa in Favor of the Repeal of the Duty on Sugar.* 35th Cong., 1st sess., Misc. Doc. 239.

———. 1906. *Production and Commercial Movement of Sugar.* 59th Cong., 1st sess., Doc. 250.

———. 1911. *Everglades of Florida: Acts, Reports, and Other Papers State and National, Relating to the Everglades of the State of Florida and Their Reclamation.* 62nd Cong., 1st sess., Doc. 89.

———. 1914. *Florida Everglades: Report of the Florida Everglades Engineering Commission to the Board of Commissioners of the Everglades Drainage District and The Trustees of the Internal Improvement Fund.* 63rd Cong., 2nd sess., Doc. 379.

———. 1929. Tariff Act of 1929: Hearings before a Subcommittee of the Committee on Finance United States Senate. Vol. 5. Schedule 5, June 26–28. 71st Cong., 1st sess.

———. 1930. *Agricultural Possibilities of the Florida Everglades.* 71st Congress, 2nd sess., Doc. 85.

———. 1931. Lobby Investigation: Hearings before a Subcommittee of the Committee on the Judiciary United States Senate, Part 11, February 25, November 23–24. 71st Cong., 3rd sess.

———. 1956. Committee on Finance. Sugar Act Extension Hearings. 84th Cong., 1st sess., H.R. 7030.

———. 1978. Subcommittee on Immigration of the Committee on the Judiciary. *The West Indies (BWI) Temporary Alien Labor Program: 1943–1977.* By Joyce C. Vialet. 95th Cong., 2nd sess., Senate Committee Print, Study Paper.

Vileisis, Ann. 1997. *Discovering the Unknown Landscape: A History of America's Wetlands.* Washington, DC: Island Press.

Wade, Michael G. 1995. *Sugar Dynasty: M.A. Patout & Son, Ltd., 1791–1993.* Lafayette: Center for Louisiana Studies, University of Southwestern Louisiana.

Wiles, Robert. 1916. *Cuban Cane Sugar—A Sketch of the Industry, from Soil to Sack, Together with a Survey of Circumstances Which Combine to Make Cuba the Sugar Bowl of the World.* Indianapolis: Bobbs Merrill.

Wiley, Harvey W. 1891. The Muck Lands of the Florida Peninsula. In the *Report of the Secretary of Agriculture,* 163–171. Washington, DC: U.S. Department of Agriculture.

———. 1898. The True Meaning of the New Sugar Tariff. *Forum* 24 (February): 689–97.

———. 1901. Testimony of Dr. Harvey W. Wiley. In Report of the Industrial Commission on Agriculture and Agricultural Labor, 638–49. U.S. House. 57th Congress, 1st sess., vol. 10, Doc. 179. Washington, DC: U.S. Government Printing Office.

Wilkinson, Joseph Biddle. 1902. Prospects of Domestic Sugar Production. *Gunton's Magazine* 22 (February): 131–36.

Wilkinson, Alec. 1989. *Big Sugar: Seasons in the Cane Fields of Florida.* New York: Knopf.

Willcox, O. W. 1936. *Can Industry Govern Itself? An Account of Ten Directed Economies.* New York: W. W. Norton.

Williams, Eric. 1984. *From Columbus to Castro: The History of the Caribbean.* New York: Vintage Books. (Orig. pub. 1970.)

Wilson, James. 1898. Should the United States Produce its Sugar? *Forum* 25 (March): 1–10.

Woloson, Wendy A. 2002. *Refined Tastes: Sugar, Confectionery, and Consumers in Nineteenth-Century America.* Baltimore, MD: Johns Hopkins University Press.

Woodward, C. Vann. 1971. *Origins of the New South, 1877–1913.* 2nd ed. Baton Rouge: Louisiana State University Press.

Wright, A., and W. Wolford. 2003. *To Inherit the Earth: The Landless Movement and the Struggle for a New Brazil.* Oakland, CA: Food First.

Wright, G. 1987. *Old South, New South: Revolutions in the Southern Economy Since the Civil War.* New York: Basic Books.

Young, Clarence H., and William A. Quinn. 1963. *Foundation for Living: the Story of Charles Stewart Mott and Flint.* New York: McGraw-Hill.

Zaneski, C. 1996a. Sugar Tax Foes Recruit Hastings, Jesse Jackson. *Miami Herald,* October 24, 1A.

———. 1996b. Hope for the Everglades: Gore Offers Plan for $1.5 Billion Rebirth. *Miami Herald,* February 26, 1A.

———. 1999a. Big Ecological Guns Fault Plan. *Miami Herald,* January 30, 1A–11A.

———. 1999b. Park Attacks Plan to Restore Glades. *Miami Herald,* January 16, 1A–23A.

———. 2001. Anatomy of a Deal. *Audubon* 103, no. 4:48–53.

Zanetti, Oscar, and García, Alejandro. 1998. *Sugar and Railroads: A Cuban History, 1837–1959.* Trans. Franklin W. Knight and Mary Todd. Chapel Hill: University of North Carolina Press.

Ziewitz, Kathryn, and June Wiaz. 2004. *Green Empire: The St. Joe Company and the Remaking of Florida's Panhandle.* Gainesville: University Press of Florida.

Index

Page numbers in italics indicate illustrative material.